Mathematical Analysis of Groundwater Resources

Mathematical Analysis of Groundwater Resources

Bruce Hunt
Department of Civil Engineering,
The University of Canterbury, Christchurch, New Zealand.

Butterworths
London Boston Durban Singapore Sydney Toronto Wellington

First published 1983

British Library Cataloguing in Publication Data

Hunt, Bruce
Mathematical analysis of groundwater resources.
1. Groundwater flow—Mathematical models
I. Title
551.49 GB1197.7

ISBN 0–408–01399–0

Library of Congress Cataloging in Publication Data

Hunt, Bruce
Mathematical analysis of groundwater resources.

Bibliography: p.
Includes index.
1. Groundwater flow—Mathematics. I. Title.
GB1197.7.H86 1983 551.49'01'51 83–11900
ISBN 0–408–01399–0

Printed in Great Britain at the
University Press, Cambridge

Preface

This book is a by-product of ten years of teaching, research and consulting in groundwater flow at the University of Canterbury in Christchurch, New Zealand. Christchurch is in the heart of a region that draws virtually all of its water for agricultural, domestic and industrial purposes from groundwater aquifers, and this makes it natural that the book be written from a water resources viewpoint. Most of the book has been used, in one form or another, to teach civil engineering students both in the final undergraduate year and at the masters degree level. Since engineers must be equipped to carry out quantitative calculations in various areas of applied physics, and since the writer is, by nature and training, a mathematically inclined engineer, the book concentrates upon the use of mathematics to describe and solve problems in groundwater flow. The writer makes no apology for this, since engineers who do quantitative work in groundwater flow must be able to understand and use mathematics. Nevertheless, many engineers tend to regard mathematics with a sense of distaste and fear, and the writer has found one of his greatest challenges in teaching to be an attempt to make mathematics appear clear, interesting and useful to engineers. It is hoped that this goal has been achieved herein.

Chapter I introduces definitions and basic mathematical concepts that are used throughout the text. Chapter II joins the physics and mathematics by deriving the partial differential equations of movement and a number of different boundary conditions. Some of the various combinations of these governing equations and boundary conditions that are necessary for well-posed problems are explored in chapter III by using specific examples and corresponding uniqueness proofs. Chapter IV is almost entirely concerned with the use of finite differences and computers to obtain approximate solutions of problems similar to those posed in the previous chapter, with a brief section at the end of the chapter upon the use of Hele-Shaw experiments for problems with a free surface. The inverse problem, the problem of determining aquifer parameters for use in solving boundary-value problems, is discussed in chapter V. Chapter VI contains an introduction to the theory of scattering (dispersion) of neutrally-buoyant pollutants in aquifers. Complex variable theory is used in chapter VII to obtain exact solutions for two-dimensional, steady-flow problems. This chapter also provides some of the mathematical background that is necessary for the use of the Laplace transform to obtain exact solutions for unsteady flow in the final chapter. Problem sets are given at the end of every chapter, and four appendices introduce mathematical topics, define mathematical notation and provide worked examples that should prove helpful to students. The writer normally covers chapters I through V and part of chapter VI in a masters course of 30 lectures, and parts of chapters VII and VIII have been used in a theoretical mechanics course for undergraduate students in their final year.

The book has been written to serve as a teaching text rather than a research monograph. As a result, only those references which, in the writer's opinion, might be needed by a student for an introduction to any particular subject have been included. The coverage of numerical techniques has been largely restricted to finite-difference methods because, in the writer's opinion, they are easier for a beginning student to understand and use than finite-element methods and are more generally applicable than integral equation methods. The coverage of inverse methods, which are still in a state of flux, has been restricted to basic methods that are easily grasped by a beginning student and that have been shown to work. The chapter on pollution contains material that has attracted major research interest only in more recent years, and it, too, reflects a number of the writer's personal views and prejudices. The last two chapters, which cover the exact solution of groundwater flow problems, introduce the use of only several out of a whole range of mathematical solution techniques that can sometimes be used for these problems. Here again, emphasis is placed upon giving students a basic introduction to the use of these techniques rather than attempting to provide a catalogue of solutions for future applications.

The writer would like to acknowledge the help of Dr. Lew Isaacs, who made a number of helpful suggestions and proof read much of the manuscript. The exceptionally good quality of the typing by Mrs. Jan Stewart and of the drafting by Mrs. Val Grey is shown by the fact that the book has been reproduced directly from photographs of their work. Finally, the writer would like to thank his friends, colleagues and students at Canterbury, whose friendship, constructive criticism and help in many different ways have made it possible to write this book.

Bruce Hunt,
Christchurch, New Zealand
1983

Contents

CHAPTER PAGE

1 Introduction

1. The Scope of Study

Groundwater flow, the movement of water beneath the earth's surface, is a subject that finds practical applications in a number of different contexts. For example, agricultural engineers study water movement through soil because plant life requires water for its growth. In this context, water movement usually occurs as a mixture of air and water through a porous earth matrix, a phenomenon known as partially-saturated flow. Drainage engineers are interested in groundwater movement because they must locate drains and ditches to lower groundwater levels. Soils engineers analyse groundwater movement because groundwater levels and velocities are major causes of embankment failures. Finally, water resource engineers are concerned with groundwater because it is one of the most convenient, economical and widely distributed sources of fresh water for agricultural, domestic and industrial uses. Most of the writer's applied work has been in the area of water resource engineering, which is the application area that serves as a focal point for this text.

The quantitative analysis of groundwater problems usually follows a certain number of well-defined steps. First, a particular problem is identified. For example, a local water authority may want to know how much water can be abstracted by a particular consumer without lowering groundwater levels by an unacceptable amount — a state of affairs that could lead to larger abstraction costs for other consumers, pollution of the fresh water with sea water or a general subsidence of ground levels. As a second example, someone might want to know whether a potential pollution source, such as a proposed refuse disposal site, is likely to endanger the water supply of near-by users. In general, this first step motivates the investigation and determines questions that must be answered in the analysis.

The second and third steps, the formulation of a boundary-value problem and the collection of field data, respectively, are usually interlinked. This is because a correctly-posed boundary-value problem cannot be constructed without first knowing such basic facts as the location and type of aquifer boundaries, sources of recharge, etc. On the other hand, collecting large amounts of field data before the problem is defined mathematically can lead to large expenditures of time and money in collecting data that is of little or no use. The translation of a physical problem into a set of mathematical statements, a procedure that is addressed in chapters II and III of this text, is a great simplification of a very complex physical phenomenon. Thus, a very important consequence of the problem formulation stage is that attention is focused upon only those items that are absolutely essential for an understanding of the physical problem. In other words, the problem formulation helps an investigator to organise his thinking about a particular problem and to decide where money and effort are best spent for gathering field data.

The third step, collecting field data, is usually the most expensive and time-consuming part of an investigation. This is also a part of the investigation that uses the skills and experiences of a large number of people in different fields. For example, geologists investigate and interpret geological evidence to obtain an insight into the location and nature of aquifer boundaries. Surveyors may be needed to determine elevations and locations of well heads, and well drillers may be needed to drill observation wells. Hydrologic technicians are often needed to run pump tests and measure water levels in wells, flow rates in streams and rivers and recharge rates from rainfall. Chemists may be needed to analyse and interpret groundwater samples, and people with experience in geophysical methods may carry out electrical resistivity or seismic surveys to help locate aquifer boundaries or interfaces between fresh water and sea water. Chapter V deals with one important aspect of this step, the calculation of aquifer parameters from pump-test and water-level data. Geological, geophysical and chemical methods are not discussed herein, though, and a student wanting general information about these methods is advised to consult other texts, such as Bouwer (1978), Freeze and Cherry (1979), Todd (1980) and Walton (1970).

The fourth, and final, step consists of solving the boundary-value problem that was formulated in the second step. This step calculates answers to the questions that were posed in the first step, and data gathered in the third step is either incorporated directly into the formulation and solution of the problem or is used as evidence to show that the mathematical equations model the aquifer behaviour correctly. This question of model verification is usually answered by obtaining a solution of the governing equations for a historical event that has been measured in the field and then comparing the calculated and experimental solutions. Chapters IV, VII and VIII deal with some different methods for solving the equations in this fourth step. The finite-difference method is discussed first in Chapter IV because it is an easily understood method that can provide approximate solutions under very general circumstances. Mathematical, or analytical, solution techniques are discussed in chapters VII and VIII. These methods are less generally applicable and are usually more difficult for beginning students to obtain. On the other hand, analytical solutions are usually easier and more economical to use and interpret than numerical solutions, and they are often useful as standards to test the accuracy of numerical models.

The solution of groundwater pollution problems, which is introduced in Chapter VI, usually follows the same four steps that have just been outlined. One chief difference between pollution and hydraulics problems in groundwater, however, is that the governing equation contains a complicated dispersion tensor and convective, or transport, terms. This makes the solution of the equations in step four relatively difficult, and the writer's preference under these conditions is to make use of analytical approximations rather than more sophisticated, and more complicated, numerical techniques. Not all workers feel this way, though, and students who are interested in making extensive use of numerical methods in pollution problems are advised to consult Fried (1975).

2. Definition of Terms

The sketch in Fig. 1.1 allows us to define some terms that are commonly used in groundwater problems. Groundwater, which is almost always moving and is seldom, if ever, stagnant, moves through a porous matrix that is called an *aquifer*. Most commercially exploitable aquifers consist of sand and gravel, although considerable quantities of water can sometimes be obtained from other geological formations such as limestones containing fractures and solution channels, porous sandstone and volcanic rocks, such as basalt, that can contain interconnected pores that form as the result of gas bubbles in the cooling lava. Aquifers in most water resource problems have vertical thicknesses that are very small compared with their horizontal dimensions. For example, one of the aquifers in the Canterbury Plains region of New Zealand has a thickness of about 30 metres and a horizontal width of about 30,000 metres, which gives a width-to-thickness ratio of 1,000. This very large slenderness ratio is typical for many water resource problems and justifies the use of a mathematical approximation that is introduced in section 8 as the *Dupuit approximation*.

The top aquifer in Fig. 1.1 is an *unconfined aquifer* because its upper boundary is a free surface. A free surface is usually idealized in water resource problems as a surface of atmospheric pressure that has a zero thickness. In actual fact, however, a free surface consists of a finite-width zone of partially-saturated flow with a thickness that is usually small compared with the total aquifer thickness. This fact is one of the principal reasons why sandbox models are of limited use for modeling free-surface flows in the laboratory. Capillarity, or surface tension, often creates a free surface in these models with a thickness that is no longer small when compared with the model aquifer thickness. Aquifers that have no free surface, such as the bottom aquifer in Fig. 1.1, are called *confined aquifers*.

An aquifer *permeability* is defined by Darcy's law in section 6 to be the ratio of a flux (flow rate) velocity to a piezometric head gradient. Thus, permeability has units of velocity and is a measure of the ability of an aquifer to transmit water under a given piezometric head gradient. Only degrees of permeability are met in application, and the term *impermeable* means that the permeability of the material under consideration is at least several orders of magnitude smaller than the material with which it is being compared. (For the uninitiated, one order of magnitude is a multiplicative factor of 10, two orders of magnitude is a factor of 100, etc.) For example, clays

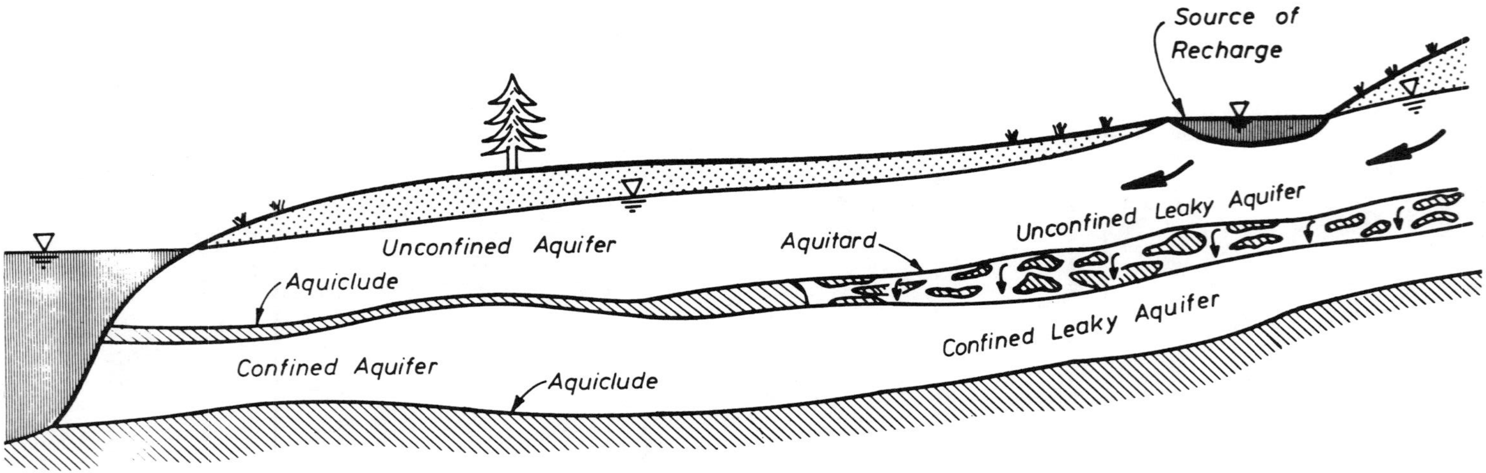

Fig. 1.1 - A definition sketch for some terms that occur in groundwater analysis.

and silts are usually considered impermeable when compared with sand and gravel aquifers.

The bottom boundary of the lower aquifer in Fig. 1.1 is impermeable and, therefore, is called an *aquiclude*. The right half of the boundary between the top and bottom aquifers is semi-permeable and is called an *aquitard*. Differences in piezometric heads on both sides of an aquitard will drive water in a vertical direction through the aquitard. In this case, the aquifers on both sides of the aquitard are called *leaky aquifers*.

A river serves as a recharge source in Fig. 1.1, although rainfall percolating downward to an unconfined aquifer, water ponded on the surface of the earth or vertical leakage between aquifers can also serve as natural sources of recharge. All aquifers must have a recharge source that replenishes water at an average rate that is equal to the aquifer outflow. (Otherwise, the aquifer will eventually dry up!) Sometimes, when the groundwater consumption is seasonal and excess surface water is available during the off season, water is recharged artificially. When the aquifer is confined, artificial recharge is usually carried out with wells. On the other hand, unconfined aquifers are usually recharged artifically by ponding water on the surface of the earth directly above the aquifer. In either case, special precautions must be taken to make sure that the well screen or pond bottom does not clog with sediment and algae. Todd (1980) gives a fairly extensive discussion of artifical recharge techniques, and the numerical and mathematical solutions of these types of problems are considered in Chapters IV and VIII of this text.

One of the important distinctions between confined and unconfined aquifers is that the water and aquifer structure in a confined aquifer must be treated as compressible, whereas compressibility effects in unconfined aquifers are usually small enough to be neglected. This fact may seem surprising to someone who is being introduced to the subject for the first time, but the experimental evidence is convincing. For example, heavy weights placed on top of confined aquifers cause water levels to rise in nearby wells. This is observed under field conditions in wells tapping confined aquifers adjacent to railroad tracks or tidal estuaries. Also, water levels in wells tapping confined aquifers are observed to decrease with an increase in atmospheric pressure since only a portion of the change in atmospheric pressure is transmitted through the overburden to the underlying aquifer, [Examples of this type of experimental evidence are shown in Wisler and Brater (1959).] Compressibility effects will be included when we derive mass conservation equations in section 8 by setting the difference between the inflow and outflow from an aquifer element equal to the change in storage within the element. This change in storage for an unconfined aquifer occurs when the free surface rises or falls, but a change in storage for a confined aquifer can only occur as the result of fluid and aquifer compressibility.

Problems are sometimes encountered with diminishing groundwater supplies, salt-water intrustion, land subsidence or groundwater pollution when water is pumped from wells. For example, too much pumping in an area can lower groundwater levels and make it either more expensive or impossible to continue pumping the same amount of water. Along a seacoast, fresh water exits from an aquifer over the top of a heavier wedge of salt-water that extends inland. The salt-water wedge moves either seaward or landward as the fresh-water flow rate exiting from the aquifer increases or decreases, respectively. Thus, salt-water intrusion and pollution of the fresh-water supply can be caused by pumping enough fresh-water from coastal wells to cause the salt-water wedge to move inland and intersect abstraction wells. In some parts of the world, excessive pumping from confined aquifers has led to a lowering of the ground surface, something which can have disastrous consequences if the area is bordered by the sea or other large body of water. Finally, groundwater supplies can become polluted as the result of polluted surface water recharging an aquifer. All of these possibilities suggest that the study of groundwater flow is a worth-while exercise for any one who intends to become involved in the investigation or management of water resources.

3. Mathematical Preliminaries

This section will be used to introduce several mathematical tools and concepts. The reader who would like a different or more detailed introduction to these topics can consult any of the better advanced calculus texts, such as Hildebrand (1962) or Kreyszig (1979).

A vector will be denoted by placing a bar above the letter that represents the vector. Thus, a vector joining the coordinate origin to the points (x, y, z) will be written

$$\bar{r} = x\hat{i} + y\hat{j} + z\hat{k} \tag{1.1}$$

in which the broken bars above the $\hat{i}$, $\hat{j}$ and $\hat{k}$ denote vectors of unit magnitude or length. The $\hat{i}$, $\hat{j}$ and $\hat{k}$ in Eq. 1.1 are unit base vectors that point in the positive x, y and z directions, respectively. To be consistent with this notation, the vector operator del will be written

$$\bar{\nabla} = \hat{i}\,\frac{\partial}{\partial x} + \hat{j}\,\frac{\partial}{\partial y} + \hat{k}\,\frac{\partial}{\partial z} \tag{1.2}$$

The operator del is not a true vector since it fails to obey all of the laws of vector algebra. For example, the vectors $\bar{A}$ and $\bar{B}$ have the commutative property $\bar{A}.\bar{B} = \bar{B}.\bar{A}$ that is not shared by del ($\bar{\nabla}.\bar{A} \neq \bar{A}.\bar{\nabla}$), and a separate algebra must be defined for del. Thus, if $\bar{A}$ is a vector function and ϕ a scalar function, the divergence of $\bar{A}$ and gradient of ϕ are defined respectively, as

$$\overline{\nabla}.\overline{A} = \frac{\partial A_x}{\partial x} + \frac{\partial A_y}{\partial y} + \frac{\partial A_z}{\partial z} \tag{1.3}$$

$$\overline{\nabla}\phi = \hat{i}\,\frac{\partial \phi}{\partial x} + \hat{j}\,\frac{\partial \phi}{\partial y} + \hat{k}\,\frac{\partial \phi}{\partial z} \tag{1.4}$$

Even though del is not a true vector, the gradient of a scalar function is seen from Eq. 1.4 to be a true vector function.

A three-dimensional, curved line can always be represented in parametric form

$$x = x(s),\ y = y(s),\ z = z(s) \tag{1.5}$$

in which the parameter, s, can be chosen as arc length measured from some fixed point on the line. In this case, Eq. 1.1 and 1.5 show that the displacement vector, $\overline{r}$, from the coordinate origin to any point on the line is a function of s. Since the secant to a short segment of arc points in the same direction as the tangent to the arc when the length of the arc segment approaches zero, Fig. 1.2 shows that

$$\overline{r}(s + \Delta s) - \overline{r}(s) \simeq \Delta s\ \hat{e}_t \text{ as } \Delta s \to 0 \tag{1.6}$$

in which $\hat{e}_t$ is the unit tangent to the arc. Thus, the unit tangent to the arc is given by

$$\hat{e}_t = \underset{\Delta s \to 0}{\text{Limit}}\ \frac{\overline{r}(s + \Delta s) - \overline{r}(s)}{\Delta s} = \frac{d\overline{r}}{ds} \tag{1.7}$$

$$= \frac{dx}{ds}\,\hat{i} + \frac{dy}{ds}\,\hat{j} + \frac{dz}{ds}\,\hat{k}$$

Finally, Eqs. 1.4 and 1.7 can be used to compute

$$\hat{e}_t.\overline{\nabla}\phi = \frac{\partial \phi}{\partial x}\frac{dx}{ds} + \frac{\partial \phi}{\partial y}\frac{dy}{ds} + \frac{\partial \phi}{\partial z}\frac{dz}{ds} = \frac{d\phi}{ds} \tag{1.8}$$

Equation 1.8 gives the "directional derivative" of ϕ. In words, it states that a unit vector dotted with $\overline{\nabla}\phi$ yields the derivative of ϕ in the direction of the unit vector.

A simple application of Eq. 1.8 can be used to show that $\overline{\nabla}\phi$ is normal to the surface $\phi(x, y, z)$ = constant. If Eq. 1.5 is the equation of any line imbedded in this surface, as shown in Fig. 1.3, then

$$\phi[x(s),\ y(s),\ z(s)] = \text{constant} \tag{1.9}$$

Thus, since $d\phi/ds = 0$ along this line, Eq. 1.8 gives

$$\hat{e}_t.\overline{\nabla}\phi = 0 \tag{1.10}$$

But $\hat{e}_t$ has a unit magnitude, and if the magnitude of $\overline{\nabla}\phi$ is not zero, then Eq. 1.10 shows that $\overline{\nabla}\phi$ is normal to the line. Since this same result holds for all lines imbedded in the surface, it can only be concluded that $\overline{\nabla}\phi$ is normal to the surface and that a unit normal is given by

$$\pm\,\hat{e}_n = \overline{\nabla}\phi/|\overline{\nabla}\phi| \tag{1.11}$$

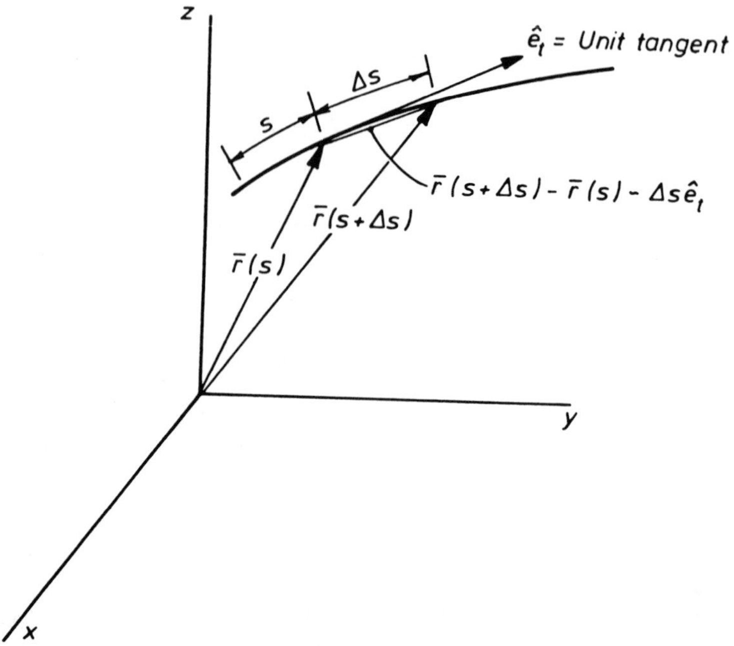

Fig. 1.2 - Calculation of the unit tangent for a curve expressed parametrically by Eq. 1.5.

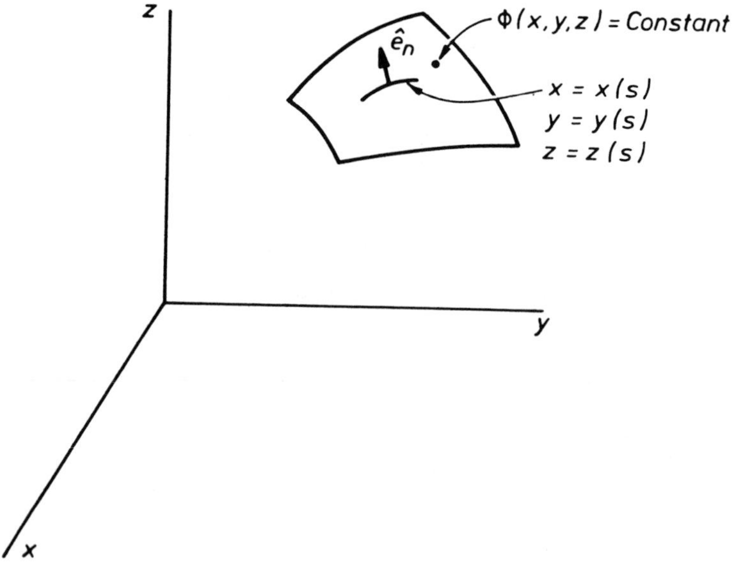

Fig. 1.3 - Calculation of the unit normal for a surface with the equation $\phi(x, y, z)$ = constant.

in which $|\overline{\nabla}\phi|$ denotes the magnitude of $\overline{\nabla}\phi$ and the positive or negative sign gives the direction of $\hat{e}_n$ along a line normal to the surface ϕ = constant. If one chooses n as arc length normal to this surface and positive in the direction of increasing ϕ, then dotting both sides of Eq. 1.11 with $\hat{e}_n$ and using Eq. 1.8 gives

$$\frac{d\phi}{dn} = |\overline{\nabla}\phi| \tag{1.12}$$

In words, Eq. 1.12 shows that the magnitude of $|\overline{\nabla}\phi|$ equals the change in ϕ with respect to distance along a line normal to the surface ϕ = constant, and it is the reason for calling $\overline{\nabla}\phi$ the gradient of ϕ. It is also worth pointing out that the maximum rate of change of ϕ occurs in the direction of $\hat{e}_n$.

Next, some definitions and relationships involving surface and volume integrals will be reviewed. Let F be either a vector or scalar function of x, y and z, and let $\forall$ be a volume enclosed by a surface denoted by S. The outward normal to S will be denoted by $\hat{e}_n$, as shown in Fig. 1.4. If the volume, $\forall$, is subdivided into N tiny elements, each of volume $\Delta\forall_i$, then the volume integral of F is defined as

$$\int_{\forall} F \, d\forall = \underset{\substack{\Delta V_i \to 0 \\ N \to \infty}}{\text{Limit}} \sum_{i=1}^{N} F_i \, \Delta\forall_i \tag{1.13}$$

in which F_i denotes the value of F calculated at the midpoint of the element $\Delta\forall_i$. Likewise, the surface integral of F is defined as

$$\int_{S} F \, dS = \underset{\substack{\Delta S_i \to 0 \\ N \to \infty}}{\text{Limit}} \sum_{i=1}^{N} F_i \, \Delta S_i \tag{1.14}$$

in which F_i is the value of F calculated at the midpoint of the surface element, ΔS_i. An important observation is that

$$\int_{\forall} F \, d\forall = 0 \tag{1.15}$$

implies that

$$F = 0 \text{ for all x, y, z in } \forall \tag{1.16}$$

when one of two sets of circumstances applies. If it is known that $F \geqq 0$ for all points in $\forall$, then Eq. 1.15 implies Eq. 1.16 since each and every term in the sum on the right side of Eq. 1.13 is greater than or equal to zero. Alternatively, if Eq. 1.15 is known to hold for all arbitrary choices of $\forall$, then Eq. 1.16 must also be true since $\forall$ can always be chosen so that F has the same sign at all points within $\forall$. If neither of these two conditions holds, then Eq. 1.15 does not imply Eq. 1.16 and it can only be assumed that the positive and negative terms in the sum on the right side of Eq. 1.13 cancel.

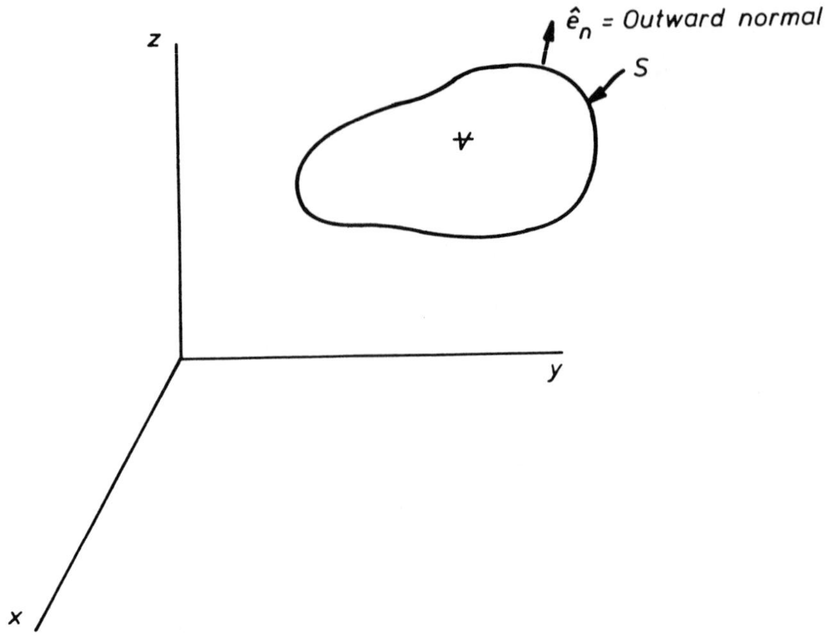

Fig. 1.4 - A volume, ∀, enclosed by a surface, S, with an outward normal, $\hat{e}_n$.

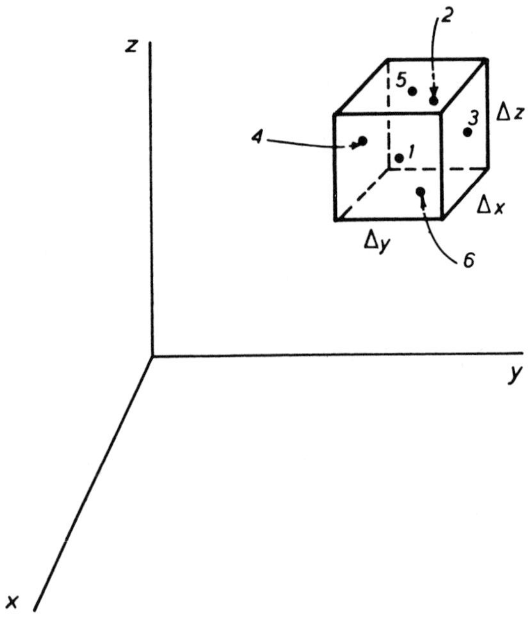

Fig. 1.5 - A sub-element, $\Delta \forall_i$, of the volume, ∀, shown in Fig. 1.4.

A relationship known as the divergence theorem can be used to rewrite a volume integral as a surface integral when the integrand of the volume integral can be expressed as either the divergence of a vector or the gradient of a scalar. For example, if ϕ is a scalar function of x, y and z, and if $\forall$ is subdivided into N tiny elements similar to the element shown in Fig. 1.5, then a straightforward approximation of the following surface integral gives

$$\int_{\Delta S_i} \phi \hat{e}_n \, dS = (\phi_1 - \phi_2)\,\hat{i}\,(\Delta y \Delta z) + (\phi_3 - \phi_4)\,\hat{j}\,(\Delta x \Delta z) + (\phi_5 - \phi_6)\,\hat{k}\,(\Delta x \Delta y) \tag{1.17}$$

If the first, second and third terms on the right side of Eq. 1.17 are multiplied by $\Delta x/\Delta x$, $\Delta y/\Delta y$ and $\Delta z/\Delta z$, respectively, Eq. 1.17 can be re-written in the form

$$\int_{\Delta S_i} \phi \hat{e}_n \, dS = (\overline{\nabla}\phi)_i \, \Delta\forall_i, \quad (\Delta\forall_i = \Delta x \Delta y \Delta z) \tag{1.18}$$

The final form of the divergence theorem can be obtained by writing an equation similar to Eq. 1.18 for each of the N sub-elements of $\forall$ and adding the equations together to obtain

$$\sum_{i=1}^{N} \int_{\Delta S_i} \phi \hat{e}_n \, dS = \sum_{i=1}^{N} (\overline{\nabla}\phi)_i \, \Delta\forall_i \tag{1.19}$$

The right side of Eq. 1.19 becomes a volume integral as $\Delta V_i \to 0$ and $N \to \infty$. On the left side, one is left with contributions only over the exterior surface, S, of $\forall$ since contributions from interior, adjacent sides of the sub-elements cancel (ϕ is assumed to be single-valued in $\forall$, and $\hat{e}_n$ differs by a minus sign on two adjacent sides). Thus, the divergence theorem takes the form

$$\int_S \phi \hat{e}_n \, dS = \int_\forall \overline{\nabla}\phi \, d\forall \tag{1.20}$$

A similar calculation using the integrand $\overline{F}.\hat{e}_n$ in the surface integral yields another variant of the same theorem:

$$\int_S \overline{F}.\hat{e}_n \, dS = \int_\forall \overline{\nabla}.\overline{F} \, d\forall \tag{1.21}$$

Equations 1.20 and 1.21 hold under very general conditions for ϕ, $\overline{F}$ and $\forall$. In particular, ϕ and $\overline{F}$ must be finite and continuous on S and in $\forall$ and must have finite and continuous first derivatives in $\forall$. The volume, $\forall$, may be either simply or multiply connected. Finally, it is worth pointing out that these same results hold in two dimensions as well provided that the volume and surface integrals are replaced with surface and line integrals, respectively.

REFERENCES

Bouwer, H. 1978. Groundwater Hydrology, McGraw-Hill Book Co., New York, pp. 137-155, 370-378.

Freeze, R.A. and Cherry, J.A. 1979. Groundwater, Prentice-Hall, Inc., Englewood Cliffs, pp. 80-165, 238-297, 306-314.

Fried, J.J. 1975. Groundwater Pollution, Elsevier Scientific Publishing Co., New York, pp. 246-261.

Hildebrand, F.B. 1962. Advanced Calculus for Applications, Prentice-Hall, Inc., Englewood Cliffs, Ch.6.

Kreyzsig, E. 1979. Advanced Engineering Mathematics, Fourth Edition, John Wiley & Sons, New York, Chs. 8 and 9.

Todd, D.K., 1980. Groundwater Hydrology, Second Edition, John Wiley & Sons, New York, pp. 267-312, 409-455, 458-488.

Walton, W.C., 1970. Groundwater Resource Evaluation, McGraw-Hill Book Co., New York, pp. 36-41, 55-97, 439-481.

Wisler, C.O. and Brater, E.F., 1959. Hydrology, Second Edition, John Wiley & Sons, Inc., New York, pp. 140-143.

PROBLEMS

1. Darcy's law for an isotropic, heterogeneous aquifer can be written

$$\bar{u} = -k\bar{\nabla}h$$

in which $k(x, y, z)$ = permeability and $h = \frac{P}{\rho g} + z$ = piezometric head. Use Eq. 1.11 to deduce the direction of the velocity vector, $\bar{u}$, in relation to surfaces of constant piezometric head, h.

2. Euler's equations for an inviscid, incompressible fluid flow can be written in the form

$$-g\bar{\nabla}h = \bar{a}$$

in which g = gravitational constant, h = piezometric head and $\bar{a}$ = acceleration vector. What is the direction of $\bar{a}$ in relation to surfaces of constant h?

3. Pressures are calculated in a motionless fluid by integrating

$$-\bar{\nabla}P + \rho\bar{g} = 0$$

in which P = pressure, ρ = mass density and $\bar{g}$ = vector with a magnitude equal to the gravitational constant, g, and which points towards the center of the earth. What is the direction of $\bar{g}$ in relation to surfaces of constant P?

4. A cube is formed by the six plane surfaces $x = 0$, $x = D$, $y = 0$, $y = D$, $z = 0$ and $z = D$. Calculate directly, by integration, the volume integral

$$I_1 = \int_{\forall} \overline{\nabla}.\overline{r}\ d\forall$$

in which $\overline{r}$ is given by Eq. 1.1 and $\forall$ is the interior of the cube. Then use direct integration to calculate the surface integral

$$I_2 = \int_S \overline{r}.\hat{e}_n\ dS$$

in which S = surface of the cube and $\hat{e}_n$ = outward normal. Finally, compare I_1 and I_2 to see if Eq. 1.21 is satisfied for this particular case.

5. Let $\overline{F} = \overline{r}/|\overline{r}|^3$ in Eq. 1.21, in which $\overline{r}$ is given by Eq. 1.1. Calculate separately, by direct integration,

$$I_1 = \int_{\forall} \overline{\nabla}.\overline{F}\ d\forall \text{ and } I_2 = \int_S F.\hat{e}_n\ dS$$

in which $\forall$ is a sphere of unit radius with its center at the origin. Compare the values of I_1 and I_2 to show that $I_1 \neq I_2$, and then explain why Eq. 1.21 is not satisfied for this particular choice of $\forall$ and $\overline{F}$.

6. Suppose that the Laplace equation

$$\nabla^2\phi = 0, \quad \left(\nabla^2 = \overline{\nabla}.\overline{\nabla} = \frac{\partial^2}{\partial x^2} + \frac{\partial^2}{\partial y^2} + \frac{\partial^2}{\partial z^2}\right)$$

is to hold at each and every point of a finite region, $\forall$. Integrate this equation throughout $\forall$, use Eq. 1.21 to convert the volume integral to a surface integral and use Eq. 1.8 to obtain a condition that must be satisfied by ϕ. What is the physical interpretation of this result if ϕ is the velocity potential for an irrotational fluid flow? (Hint: The velocity vector is given by $\overline{u} = \overline{\nabla}\phi$.)

2 The Equations of Groundwater Flow

4. Continuum Concepts

Most analytical work in engineering makes the assumption of a continuum. For example, fluids and solids are composed of discrete molecules separated by measureable distances. In applications, though, it is impossible to use mathematical models for fluids and solids that describe the behaviour of individual molecules. Instead, we use the mathematics of continuous functions for our models, a mathematics that assumes that material characteristics and properties are distributed continuously for the smallest imaginable quantity of the fluid or solid. (Otherwise, it would be impossible to calculate spatial derivatives of mass densities, velocities, stresses, etc.) Thus, we expect to obtain good agreement between calculations and experiment only when the experimental variables are averaged over an area whose characteristic dimension is large when compared with the spacing between molecules.

Groundwater flow requires that the concept of a continuum be carried one step further. Since it is neither possible nor desirable to calculate water movement through the individual pores and passages of any aquifer matrix, we use continuous mathematics and assign continuous values for mass densities, velocities and pressures throughout both void and solid portions of the aquifer. As a result, we can expect reasonable comparisons between theory and experiment only if the experimental variables are measured and averaged over regions whose characteristic dimensions are large compared with characteristic scales of the aquifer heterogeneities.

Limitations placed upon groundwater flow calculations by the continuum concept are relatively common. One example is a limestone aquifer in which flow occurs through fractures and solution channels. If the spacing of these fractures and solution channels is large compared to the diameter of a well, it is entirely unreasonable to expect to use continuous mathematics to describe experimental variables in the neighbourhood of the well. On the other hand, it is reasonable to use this mathematics to calculate average values for the same aquifer if the solution domain has dimensions that are large compared with the spacing of the fractures and solution channels. As another example, consider the problem of carrying out calculations for a heterogeneous aquifer. If the aquifer has horizontal dimensions that are of the order of kilometers and aquifer heterogeneities have a scale that is of the order of a meter or more, it is virtually impossible to obtain aquifer permeabilities by using laboratory permeameters,which have a characteristic scale of centimeters. Instead, aquifer permeabilities must be measured with large-scale, in-place field tests.

5. Flux and Pore Velocities

Two different velocities are used for two different purposes in groundwater flow calculations. A pore velocity, $d\bar{r}(t)/dt$, is defined as the velocity of an average fluid particle as it moves through an aquifer. Different fluid particles released in the neighbourhood of the same point in an aquifer will travel slightly different paths, and the pore velocity is the result of averaging the velocities of these different particles. Dispersion (scattering) results from the fact that many particles deviate from the path of the average fluid particle, with each fluid particle arriving at the same downstream cross section at a different time and place. Pore velocities are used most extensively in pollution calculations, where it is necessary to calculate the position and travel times of pollutant particles that have been released upstream.

A flux velocity, $\bar{u}$, is defined as a volumetric flow rate, or flux, divided by the cross sectional area normal to the flow. This cross sectional area is the total area and includes the area of both voids and solids. Thus, if the aquifer porosity is denoted by σ, the flow rate, Q, through an area, A can be calculated by using either the pore or flux velocity

$$Q = \sigma A \frac{d\bar{r}}{dt}.\hat{e}_n = A\bar{u}.\hat{e}_n \tag{2.1}$$

in which $\hat{e}_n$ is a unit vector normal to A. Hence, the pore and flux velocities are related by the equation

$$\sigma \frac{d\bar{r}}{dt} = \bar{u} \tag{2.2}$$

The flux velocity, $\bar{u}$, is used in most calculations that are not concerned with pollution studies.

The porosity, σ, that appears in Eqs. 2.1 and 2.2 is an area porosity that is defined as the area of voids or pore divided by the total surface area of an aquifer sample. This porosity, however, is difficult to measure experimentally. On the other hand, a volumetric porosity is defined as the ratio of volume of voids to the total volume of an aquifer sample. If x is distance along the axis of a cylindrical aquifer sample, and A is the total (constant) cross sectional area at right angles to this axis, the average area porosity of the cylinder is

$$\sigma_{av} = \frac{1}{L}\int_0^L \sigma(x)\,dx = \frac{\int_0^L A\sigma(x)\,dx}{AL} \tag{2.3}$$

in which L is the length of the cylindrical sample. But the numerator on the right side of Eq. 2.3 is the volume of voids in the sample, and the denominator is the total sample volume. Thus, the average area porosity equals the volumetric porosity, and no distinction will be made in the future between area and volume porosities.

6. Darcy's Law

Darcy's law is actually an equation that was discovered experimentally by Henry Darcy in 1856. However, some heuristic reasoning with the equations of classical fluid mechanics can be used to suggest that Darcy's law is a simplified form of Newton's momentum equation. This is also of interest because it suggests the form that Darcy's law should take when the flow is compressible or when the aquifer is anisotropic.

In a fluid flow with no porous medium present, an application of Newton's second law to a volume, $\forall$, of fluid moving with the flow gives

$$-\int_S P\hat{e}_n\,dS + \int_\forall \rho\bar{g}\,d\forall + \int_\forall \bar{f}\,d\forall = \int_\forall \rho\bar{a}\,d\forall \tag{2.4}$$

The symbols in Eq. 2.4 are P = fluid pressure (positive for compression), ρ = fluid mass density, $\bar{f}$ = viscous force per unit volume and $\bar{a}$ = acceleration vector. The gravitational vector, $\bar{g}$, has a magnitude equal to the gravitational constant (9.81 meters/sec.^{2} at mean sea level) and points towards the center of the earth. A simple application of the divergence theorem (Eq. 1.20) to the surface integral gives

$$\int_\forall (-\bar{\nabla}P + \rho\bar{g} + \bar{f} - \rho\bar{a})\,d\forall = 0 \tag{2.5}$$

But this must hold for an arbitrary choice of $\forall$. Thus, as explained after Eq. 1.15, the integrand must vanish at each and every point within the fluid.

$$-\bar{\nabla}P + \rho\bar{g} + \bar{f} - \rho\bar{a} = 0 \tag{2.6}$$

Equation 2.6 is an equation that describes fluid motion when no porous matrix is present. However, the next step is to invoke the hypothesis of a continuum and assume that Eq. 2.6 also describes fluid motion through a porous matrix if the variables that appear in it are interpreted as values that have been averaged throughout a region whose scale is large when compared with the scale of the aquifer heterogeneities.

Velocities in groundwater flow are usually very small, of the order of one to ten meters per day, and accelerations are also negligibly small. Thus, Eq. 2.6 can be approximated with

$$-\overline{\nabla}P + \rho\overline{g} + \overline{f} = 0 \tag{2.7}$$

Finally, the force exerted by a solid sphere upon laminar flow is shown by Stokes law to be directly proportional to the product of the fluid velocity and viscosity. Since the force exerted by the sphere upon the fluid is in the direction opposite to $\overline{u}$, this suggests that $\overline{f}$ can be approximated with

$$\overline{f} = -\frac{\mu}{k_0}\overline{u} \tag{2.8}$$

in which μ = dynamic or absolute fluid viscosity and k_0 is a proportionality factor that has units of length squared and is called the physical, or intrinsic, permeability. The introduction of Eq. 2.8 into Eq. 2.7 gives a general form of Darcy's law that holds for compressible and incompressible flow.

$$\overline{u} = -\frac{k_0}{\mu}(\overline{\nabla}P - \rho\overline{g}) \tag{2.9}$$

The intrinsic permeability, k_0, is proportional to the square of the diameter of the aquifer matrix particles and depends only upon the aquifer matrix geometry.

When the flow can be considered incompressible, the gravitational term in Eq. 2.9 possesses a potential:

$$\rho\overline{g} = \overline{\nabla}(\rho\overline{g}.\overline{r}) \tag{2.10}$$

In this case, since ρ and g are constants, Eq. 2.10 can be inserted in Eq. 2.9 and the resulting equation can be manipulated into the form

$$\overline{u} = -K\overline{\nabla}h, \quad \left(K = \frac{\rho g k_0}{\mu}, \; h = \frac{P}{\rho g} - \hat{g}.\overline{r}\right) \tag{2.11}$$

The variable K has units of a velocity and is called the coefficient of permeability. Some typical values for K, as given, for example, by Harr (1962) and Polubarinova-Kochina (1962), are shown in Table 1. The variable h has units of length and is called the piezometric head, and the unit vector $\hat{g}$ points in the direction of $\overline{g}$. (For example, when the z axis is positive in the upward direction, $\hat{g} = -\hat{k}$ and $\hat{g}.\overline{r} = -z$). Equation 2.11 states that $\overline{u}$ is orthogonal to surfaces of constant h, as shown by Eq. 1.11. Experiments indicate that Darcy's law, like Stoke's law, becomes invalid for Reynolds numbers greater than about 1 to 12. Muskat (1946) gives a discussion of this experimental evidence.

- Table 1 -

Typical Permeabilities for Different Aquifer Materials

Aquifer Material	K(mm/sec)
Clean gravel	10 and greater
Clean, coarse sand	10 - 0.1
Well-graded, sand mixture	0.1 - 0.01
Fine sand, silt	0.01 - 0.0001
Clay	0.0001 and smaller

7. Darcy's Law for Anisotropic Aquifers

Equation 2.11 was obtained by assuming that the coefficient of $\bar{u}$ in Eq. 2.8 does not depend upon the direction of $\bar{u}$. As a result, the permeability, K, in Eq. 2.11 is a scalar quantity that does not depend upon the direction of $\bar{u}$ at any point in an aquifer. An aquifer which has this property is called isotropic.

There are some aquifers, known as anisotropic aquifers, for which the permeability changes its magnitude at a fixed point as the velocity changes its direction. This condition of anisotropy can occur as the result of flattened or oblong particles in an aquifer orientating themselves so that their longitudinal axes are all in the same direction or by a layered effect in which layers of less permeable material are interbedded with more permeable layers (see problem 3). Causes of these conditions include the artificial compaction of granular material in layers, as in embankment construction, or the settlement of granular material through a fluid, such as water or air. On a regional scale, it is common to find confined and leaky aquifers separated by horizontal layers of less permeable material. When the characteristic scale of the region for which calculations are being carried out is of the same order of magnitude as the aquifer thicknesses, then the aquifer is most correctly thought of as heterogeneous and isotropic (i.e. - K varies with x, y and z but not with the direction of $\bar{u}$). On the other hand, most water resource calculations are carried out for regions in which these aquifer thicknesses are very small when compared with a characteristic dimension of the solution domain. In these cases, one must use a "bulk" permeability for the layered aquifer system, with a larger permeability occurring in a direction parallel to the layering and a smaller permeability normal to the layering.

The considerations mentioned in the last paragraph suggest that Darcy's law can be written in the following form when the coordinate axes are normal and parallel to the principal directions of anisotropy:

$$\bar{u} = -K_x \frac{\partial h}{\partial x}\hat{i} - K_y \frac{\partial h}{\partial y}\hat{j} - K_z \frac{\partial h}{\partial z}\hat{k} \tag{2.12}$$

Physical reasoning suggests that two of the permeabilities in Eq. 2.12 will be equal. A more general way to write Eq. 2.12 is to introduce the use of a permeability tensor, $\underline{K}$:

$$\bar{u} = -\underline{K}.\bar{\nabla}h \tag{2.13}$$

The tensor, $\underline{K}$, takes the following form in the principal coordinate system:

$$\underline{K} = K_x\hat{i}\hat{i} + K_y\hat{j}\hat{j} + K_z\hat{k}\hat{k} \tag{2.14}$$

Care must be taken when computing the dot product of a tensor to not invert the order of the unit base vectors in the tensor and to take the dot product from either the right or left side, as indicated. An introduction to the use of this direct notation for tensors is given by Malvern (1969).

If calculations are carried out by using an (x_1, x_2, x_3) Cartesian coordinate system with the unit base vectors $\hat{e}_1$, $\hat{e}_2$ and $\hat{e}_3$ that do not point in the principal directions of the permeability tensor, then $\bar{\nabla}h$, $\underline{K}$ and $\bar{u}$ are given by

$$\bar{\nabla}h = \hat{e}_1 \frac{\partial h}{\partial x_1} + \hat{e}_2 \frac{\partial h}{\partial x_2} + \hat{e}_3 \frac{\partial h}{\partial x_3} \tag{2.15}$$

$$\begin{aligned}\underline{K} = {} & K_{11}\hat{e}_1\hat{e}_1 + K_{12}\hat{e}_1\hat{e}_2 + K_{13}\hat{e}_1\hat{e}_3 \\ & + K_{21}\hat{e}_2\hat{e}_1 + K_{22}\hat{e}_2\hat{e}_2 + K_{23}\hat{e}_2\hat{e}_3 \\ & + K_{31}\hat{e}_3\hat{e}_1 + K_{32}\hat{e}_3\hat{e}_2 + K_{33}\hat{e}_3\hat{e}_3\end{aligned} \tag{2.16}$$

$$\begin{aligned}\bar{u} = -\underline{K}.\bar{\nabla}h = {} & -\hat{e}_1\left(K_{11}\frac{\partial h}{\partial x_1} + K_{12}\frac{\partial h}{\partial x_2} + K_{13}\frac{\partial h}{\partial x_3}\right) \\ & -\hat{e}_2\left(K_{21}\frac{\partial h}{\partial x_1} + K_{22}\frac{\partial h}{\partial x_2} + K_{23}\frac{\partial h}{\partial x_3}\right) \\ & -\hat{e}_3\left(K_{31}\frac{\partial h}{\partial x_1} + K_{32}\frac{\partial h}{\partial x_2} + K_{33}\frac{\partial h}{\partial x_3}\right)\end{aligned} \tag{2.17}$$

The values of K_{ij} can be computed from Eqs. 2.14 and 2.16 by taking dot products. For example,

$$\begin{aligned}K_{12} = \hat{e}_1.\underline{K}.\hat{e}_2 = {} & K_x\,(\hat{e}_1.\hat{i})\,(\hat{i}.\hat{e}_2) + K_y\,(\hat{e}_1.\hat{j})\,(\hat{j}.\hat{e}_2) \\ & + K_z\,(\hat{e}_1.\hat{k})\,(\hat{k}.\hat{e}_2)\end{aligned} \tag{2.18}$$

The dot products that appear in Eq. 2.18 give the cosines of the angles between the coordinate axes of the (x, y, z) and (x_1, x_2, x_3) systems. Thus, it can be shown, in general, that

$$K_{\alpha\beta} = K_x \cos(x, x_\alpha) \cos(x, x_\beta) + K_y \cos(y, x_\alpha) \cos(y, x_\beta) + K_z \cos(z, x_\alpha) \cos(z, x_\beta) \quad (2.19)$$

Equation 2.19 shows that $K_{\alpha\beta} = K_{\beta\alpha}$, which means that the permeability tensor is symmetric. Also, note that $\bar{u}$ in Eq. 2.17 is no longer orthogonal to contours of constant h.

In closing this section, it should be pointed out that very few analyses of water resource systems have been carried out in which the effects of anisotropy have been included. In fact, the only study of an actual aquifer system that the author is aware of is the work on a Long Island, New York aquifer by Collins, Gelhar and Wilson (1972). The reason for the scarcity of applications in this area is simply that it is very difficult to obtain enough field data to permit calculation of the principal values and directions of the permeability tensor.

8. Conservation of Mass Equations

Darcy's law, when written in scalar component form, consists of three scalar equations with four unknowns: three unknown velocity components and the piezometric head, h. A fourth scalar equation with these same four unknowns is obtained by writing a conservation of mass statement. This conservation of mass statement is usually called a "continuity" equation, and the form which it takes depends upon the set of assumptions that is made in its derivation.

Free Surface Flow

For general, free-surface flow problems in which the compressibility of the fluid and aquifer can be neglected, the conservation of mass statement reduces to a conservation of volume statement:

$$\int_S \bar{u} \cdot \hat{e}_n \, dS = 0 \quad (2.20)$$

The dot product in the integrand of Eq. 2.20 is positive for outflow and negative for inflow, and Eq. 2.20 simply states that the difference between outflow and inflow through S must vanish. An application of the divergence theorem, Eq. 1.21, gives

$$\int_{\forall} \bar{\nabla} \cdot \bar{u} \, d\forall = 0 \quad (2.21)$$

But Eq. 2.21 must hold for all arbitrary choices of $\forall$, which, as explained following Eq. 1.16, means that

$$\bar{\nabla} \cdot \bar{u} = 0 \text{ for all } x, y, z \text{ in } \forall \quad (2.22)$$

Equation 2.22 is the desired continuity equation, and the use of Eq. 2.11 to eliminate $\bar{u}$ gives one equation for the scalar unknown, h.

$$\bar{\nabla} \cdot (K \bar{\nabla} h) = 0 \quad (2.23)$$

When the permeability, K, is constant, Eq. 2.23 reduces to the Laplace equation.

Confined Flow

In confined flow problems, both the aquifer and fluid must be considered compressible. In this case, the conservation of mass statement becomes

$$\int_S \rho\bar{u}.\hat{e}_n \, dS + \frac{\partial}{\partial t}\int_{\forall} \rho\sigma \, d\forall = 0 \qquad (2.24)$$

The surface integral gives the mass flux of fluid out through S, and the derivative of the volume integral gives the rate of increase of fluid mass within $\forall$. The divergence theorem allows Eq. 2.24 to be rewritten in the form

$$\int_{\forall} \left[\bar{\nabla}.(\rho\bar{u}) + \frac{\partial(\rho\sigma)}{\partial t}\right] d\forall = 0 \qquad (2.25)$$

Since Eq. 2.25 must hold for an arbitrary choice of $\forall$, the continuity equation is

$$\bar{\nabla}.(\rho\bar{u}) + \frac{\partial(\rho\sigma)}{\partial t} = 0 \qquad (2.26)$$

In groundwater flow problems, the fluid and aquifer are normally considered only slightly compressible. This means that we can linearize Eq. 2.26 by setting

$$\rho = \rho_0 + \rho', \quad (\rho'/\rho_0 << 1) \qquad (2.27)$$

$$\sigma = \sigma_0 + \sigma', \quad (\sigma'/\sigma_0 << 1) \qquad (2.28)$$

in which the primed variables are relatively small changes created by a change in pressure. Substituting Eqs. 2.27 - 2.28 into Eq. 2.26 and retaining only the first-order terms gives

$$\rho_0 \, \bar{\nabla}.\bar{u} + \frac{\partial(\sigma_0\rho' + \rho_0\sigma')}{\partial t} = 0 \qquad (2.29)$$

A similar substitution into Darcy's law for compressible flow, Eq. 2.9 gives

$$\bar{u} = - K\bar{\nabla}h \qquad (2.30)$$

in which K and h are calculated by using ρ_0 for ρ. Since ρ' and σ' are functions of the fluid pressure, Eqs. 2.29 - 2.30 can be combined and rewritten in the form

$$\rho_0\bar{\nabla}.(K\bar{\nabla}h) = \frac{d(\sigma_0\rho' + \rho_0\sigma')}{dP}\frac{\partial P}{\partial t} \qquad (2.31)$$

Finally, P in the definition for h, as given by Eq. 2.11, can be substituted into the right side of Eq. 2.31 and the result divided by ρ_0 to obtain

$$\overline{\nabla}.(K\overline{\nabla}h) = S_s \frac{\partial h}{\partial t} \tag{2.32}$$

in which the specific storage, S_s, is given by

$$S_s = g \frac{d(\sigma_0\rho' + \rho_0\sigma')}{dP} \tag{2.33}$$

The specific storage has units of length^{-1}, and it is worth pointing out the Eqs. 2.32 and 2.23 become identical for steady flow.

The specific storage, S_s, defined by Eq. 2.33 is really an elasticity constant that must be determined by in-place, field tests. A number of authors, for example, Bear (1979), carry out a rather detailed set of calculations to show that

$$S_s = \rho_0 g\ [\alpha(1 - \sigma_0) + \sigma_0\beta] \tag{2.34}$$

The constants α and β are the bulk coefficients of compressibility for the aquifer and water, respectively, (α and β are, therefore, reciprocals of the bulk moduli of elasticity for the aquifer and water). Values of β, or its reciprocal, are well known and are tabulated as functions of temperature in fluid mechanics texts. Values for α, however, depend primarily upon the change created in σ by a shifting of the granular material in the aquifer. Thus, values for α can not be expected to be constant from one aquifer to the next, and S_s must be calculated from field data by comparing solutions of Eq. 2.32 with experimentally measured solutions for h. On the other hand, since $\alpha(1-\sigma_0)$ in Eq. 2.34 is a positive constant, Eq. 2.34 can be used to calculate a lower bound for S_s. For example, for $\rho_0 = 1000\ kg/m^3$, $g = 9.81\ m/s^2$, $\sigma_0 = 0.1$ and $\beta = 4.74 \times 10^{-10}\ m^2/N$

$$\rho_0 g\sigma_0\beta = 4.65 \times 10^{-7}\ m^{-1} \tag{2.35}$$

This means that an experimentally measured value of S_s should satisfy the inequality

$$S_s \geq 4.65 \times 10^{-7}\ m^{-1} \tag{2.36}$$

In practice, values of S_s for a fully-confined aquifer are sometimes three orders of magnitude larger than the lower bound indicated by Eq. 2.36, which suggests that aquifer compressibility often plays a dominant role in determining the magnitude of S_s.

<u>The Dupuit Approximation for Unconfined Flow</u>

Equations 2.23 and 2.32 permit, in principle the calculation of a solution for any fully-saturated groundwater flow problem. Unfortunately, however, many groundwater resource problems require a solution for three-dimensional flows in which horizontal dimensions of the aquifer are orders of magnitude larger than vertical dimensions. Under these conditions it is usual to make a simplification known as the Dupuit approximation.

The Dupuit approximation assumes (a) that velocities are horizontal and (b) that piezometric heads do not change within the aquifer along any vertical line. In actual fact, since Darcy's law shows that the vertical velocity component is proportional to the derivative of h with respect to the vertical coordinate, either one of these two assumptions leads to the other as a result. Thus, the Dupuit approximation is sometimes spoken of as the assumption of a hydrostatic pressure distribution along vertical lines. Some authors also refer to this approximation as the hydraulic theory or the assumption of sheet seepage.

If the vertical coordinate is taken as z, then $h = h(x, y, t)$ has the physical meaning of being the elevation of the free surface above some arbitrarily chosen, horizontal plane. The velocity, $\bar{u}$, has components only in the horizontal directions, but the permeability, K, may be a function of x, y and z. Integrating Darcy's law, Eq. 2.11, with respect to z from the bottom aquifer boundary at $z = z_0(x, y)$ to the free surface at $z = z_0(x, y) + B(x, y, t)$ gives

$$\bar{q} = -T\bar{\nabla}h \qquad (2.37)$$

in which $B(x, y, t)$ = saturated aquifer thickness, which is a function of h, and $\bar{q}$ = vector with the direction of $\bar{u}$ and a magnitude equal to the flow rate (flux) per unit arc length in the horizontal (x, y) plane. The transmissivity or transmissibility, T, is given by

$$T(x, y, t) = \int_{z_0}^{z_0 + B} K(x, y, z)\, dz \qquad (2.38)$$

Equation 2.37 is an integrated form of Darcy's law that must be combined with a conservation of mass statement to obtain as many equations as unknowns. Note that T has dimensions of length^2/time.

The conservation of mass statement for a control volume with vertical sides takes the form

$$\int_{\Gamma} \bar{q}.\hat{e}_n ds + \frac{\partial}{\partial t}\int_A Sh\, dA + \int_A u_z\, dA = 0 \qquad (2.39)$$

in which A and Γ are the projections of the control volume and control volume boundary, respectively, upon the horizontal (x, y) plane. The vertical flux velocity, u_z, is taken as positive for outflow through either the free surface or the bottom aquifer boundary, and the storage coefficient, S, is an effective porosity that is slightly less than the actual porosity, σ, since not all of the pore volume in an aquifer is vacated or filled by water as the free surface falls or rises, respectively. The storage coefficient S, is dimensionless and may have slightly different magnitudes for the same aquifer, depending upon whether the free surface is falling or rising. The first term in Eq. 2.39 is the net volumetric outflow through the vertical sides of the control volume, the second term is the increase in storage within the control volume that results from a raising of the free surface and the third term is the volumetric outflow in the vertical direction through either the free surface or bottom aquifer boundary.

An application of the divergence theorem to Eq. 2.39 gives

$$\int_A \left(\bar{\nabla}.\bar{q} + S\frac{\partial h}{\partial t} + u_z\right) dA = 0 \tag{2.40}$$

Since Eq. 2.40 must hold for arbitrary choices of A, the continuity equation takes the form

$$\bar{\nabla}.\bar{q} + S\frac{\partial h}{\partial t} + u_z = 0 \tag{2.41}$$

Finally, Eq. 2.37 can be used to eliminate $\bar{q}$.

$$\bar{\nabla}.(T\bar{\nabla}h) = S\frac{\partial h}{\partial t} + u_z \tag{2.42}$$

Various approximations are often inserted for u_z in Eq. 2.42. For example, u_z is taken as a specified, positive number for evaporation or as a negative number for recharge from either rainfall or irrigation water. Leakage through the bottom aquifer boundary is usually modelled by setting

$$u_z = \left(\frac{K'}{B'}\right)(h - h') \tag{2.43}$$

in which K' and B' are the permeability and thickness, respectively, of the bottom aquitard and h' is the piezometric head along the bottom boundary of the aquitard. Outflow from a well at the point (x_0, y_0) can be modelled by writing

$$u_z = Q\delta(x - x_0)\,\delta(y - y_0) \tag{2.44}$$

in which Q is the well flow rate and the Dirac delta function, δ, is defined as a function that vanishes everywhere except in the neighbourhood of (x_0, y_0). Near (x_0, y_0) the product $\delta(x - x_0)\,\delta(y - y_0)$ approaches infinity in such a way as to give

$$\int_A \delta(x - x_0)\,\delta(y - y_0)\, dA = 1 \tag{2.45}$$

when the region A contains the point (x_0, y_0). In numerical solution methods, the singularity in Eq. 2.44 is usually "smeared out" by writing

$$\begin{aligned} u_z &= 0, \quad (x, y \text{ outside of } A) \\ &= \frac{Q}{A}, \quad (x, y \text{ within } A) \end{aligned} \tag{2.46}$$

in which A is a small, finite area that surrounds the well.

The Dupuit Approximation for Confined Flow

Equations 2.37 and 2.38 remain unchanged for a confined aquifer except that B is the confined aquifer thickness. The continuity equation has the form

$$\int_{\Gamma} \rho\bar{q}.\hat{e}_n ds + \frac{\partial}{\partial t}\int_{A} \rho\sigma B dA + \int_{A} \rho u_z dA = 0 \qquad (2.46)$$

The first integral is the net mass flux out through the vertical sides of the control volume, the second term is the increase in fluid mass stored within the aquifer as a result of the compressibility of both the fluid and aquifer and the third term is the net mass flux out through the top or bottom aquifer boundary. An application of the divergence theorem to this arbitrarily chosen control volume yields

$$\bar{\nabla}.(\rho\bar{q}) + \frac{\partial(\rho\sigma B)}{\partial t} + \rho u_z = 0 \qquad (2.47)$$

Since the aquifer and fluid are normally considered only slightly compressible, an argument very similar to the one following Eq. 2.26 leads to the continuity equation

$$\bar{\nabla}.\bar{q} + S\frac{\partial h}{\partial t} + u_z = 0 \qquad (2.48)$$

in which the storage coefficient, S, is a measure of the elasticity of the fluid and aquifer and is defined as

$$S = g\frac{d(\rho_0\sigma_0 B' + \rho_0\sigma' B_0 + \rho'\sigma_0 B_0)}{dP} \qquad (2.49)$$

The variables B', σ' and ρ' are relatively small changes in B, σ and ρ that are caused by changes in pressure, P. A comparison of Eqs. 2.33 and 2.49 shows that S is dimensionless and that S has the lower bound

$$S \geqq B_0 S_s \qquad (2.50)$$

The equality in Eq. 2.50 applies only if B' = 0.

The use of Eq. 2.37 to eliminate $\bar{q}$ from Eq. 2.48 gives the final result

$$\bar{\nabla}.(T\bar{\nabla}h) = S\frac{\partial h}{\partial t} + u_z \qquad (2.51)$$

Equation 2.51 appears, at first glance, to be identical with Eq. 2.42. Two basic differences, however, exist between Eqs. 2.42 and 2.51. First, the transmissivity, T, depends upon h in Eq. 2.42 but is independent of h in Eq. 2.51. Second, the storage coefficient, S, is an effective porosity, with an order of magnitude of 0.1, in Eq. 2.42 but an elasticity constant, with an order of magnitude between 10^{-7} and 10^{-4}, in Eq. 2.51. Thus, Eq. 2.42 is nonlinear but Eq. 2.51 is linear, and

$\partial h/\partial t$ will be orders of magnitude larger in solutions of Eq. 2.51 than in solutions of Eq. 2.42. The expressions for u_z in Eq. 2.51 remain essentially the same as those in Eqs. 2.43 - 2.46 except that Eq. 2.43 can be given the additional interpretation of representing outflow from a spring if h' is taken as the elevation of the free surface in the spring.

9. Boundary Conditions

The solution of a groundwater resources problem usually consists of obtaining an exact or approximate solution to either Eq. 2.23, 2.32, 2.42 or 2.51 subject to certain boundary conditions and an initial condition if the problem involves unsteady flow. Boundaries for any given problem are, if possible, always chosen to coincide with some natural geological or physical boundary where the applied boundary conditions can be assumed to remain invariant with time. Examples include impermeable boundaries, which permit no flow across the boundary, and the edges of reservoirs, lakes, rivers and seas, which are often connected to an aquifer in such a way as to force piezometric levels in the aquifer to approach water levels at the edge of the body of water. It cannot always be assumed that this latter condition applies at every reservoir, lake, river or sea boundary, and, in the final analysis, this is a question that can only be settled with experimental field data of piezometric head levels. In some instances it is not possible or practical to choose a location for part of the boundary that meets these requirements. Then one usually attempts to locate the boundary at a point that is relatively distant from the region of most interest, and calculations are carried out in the hope that an incorrect boundary condition along this boundary will exert only a minor influence upon the behaviour of the solution in the region of interest.

The use of either Eq. 2.11 or 2.37 and Eq. 1.8 shows that the requirement $\bar{u}.\hat{e}_n = 0$ along an impermeable boundary is equivalent to requiring

$$\frac{dh}{dn} = 0 \tag{2.52}$$

in which n = arc length normal to the impermeable boundary. Along the edge of a water body that is connected to an aquifer, the boundary condition is

$$h = f \tag{2.53}$$

in which f is a prescribed function equal to the elevation of the water surface above the coordinate origin.

Sometimes problems are met in which discontinuities in either K or T occur along internal boundaries. One would expect, from physical arguments, that pressures and normal fluxes should be continuous along these boundaries. Since the gravitational potential, $-\hat{g}.\bar{r}$, is continuous across these internal boundaries, these two conditions can be written

$$h_+ = h_- \tag{2.54}$$

$$T_+ \frac{dh_+}{dn} = T_- \frac{dh_-}{dn} \text{ or } K_+ \frac{dh_+}{dn} = K_- \frac{dh_-}{dn} \tag{2.55}$$

in which the plus and minus signs denote values obtained by approaching the boundary of discontinuity from one side or the other. Eq. 2.55 requires normal flux velocity components to be identical on both sides of the discontinuity in K or T, and Eq. 2.54 will be used in problem 6 to show that tangential velocity components are discontinuous. Thus, the velocity vector magnitude and direction will be discontinuous along these internal boundaries.

The boundary condition for a surface of seepage must sometimes be applied in problems for which the Dupuit assumption is not made. An example of a seepage surface is shown in Fig. 2.1 along CD, where seepage through the embankment exits along the downstream embankment surface. This is actually a surface of specified h since pressures are atmospheric along CD. Thus, since the y axis is positive in the upward direction, $h = P/\rho g + y$ and the correct boundary condition is

$$h(x, y, z, t) = y, \quad (x, y, z \text{ on CD}) \tag{2.56}$$

Equation 2.56 has been obtained by assuming that the atmospheric pressure is zero (gauge pressures). It should also be emphasised that normal velocity components along CD are not zero, so that CD is not a streamline.

Boundary conditions along a free surface are not needed when the Dupuit approximation is used since these boundary conditions have been incorporated into the partial differential equation, Eq. 2.42. However, problems which require the solution of Eq. 2.23 will require boundary conditions along a free surface. Two boundary conditions are always imposed along any free surface because the problem has an additional unknown: the unknown geometry of the free surface must be calculated as part of the solution of the problem. The first boundary condition requires that pressures be atmospheric along the free surface and is identical with Eq. 2.56. The second boundary condition, in its most general form, requires that a conservation of volume statement be satisfied for a control volume chosen along the free surface. As an example, Fig. 2.2 shows a thin control volume along a free surface that is being recharged by vertical seepage with a flux velocity R. The dashed boundaries of the control volume are fixed, and the free surface is moving with a pore velocity of $d\bar{r}/dt$. Thus, a conservation of volume statement which becomes exact as the control volume thickness normal to

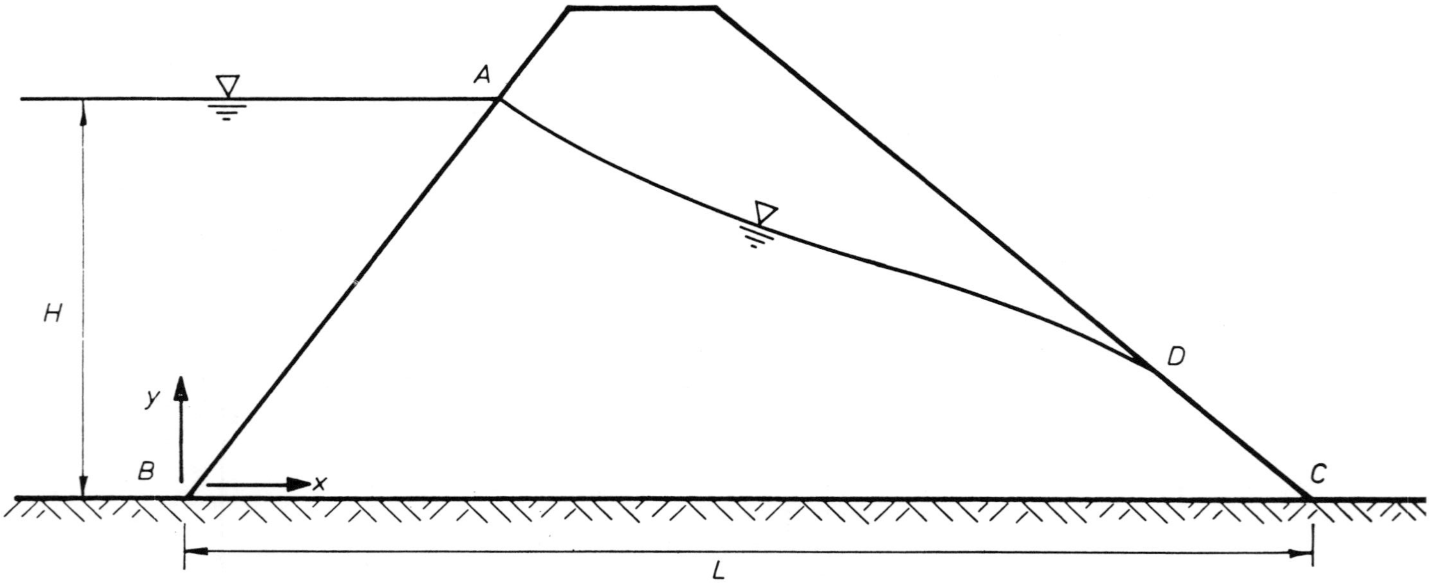

Fig. 2.1 - Seepage through an embankment.

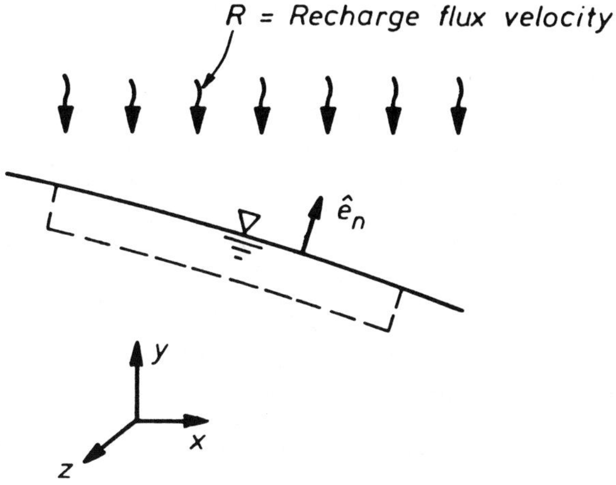

Fig. 2.2 - Recharge along a free surface.

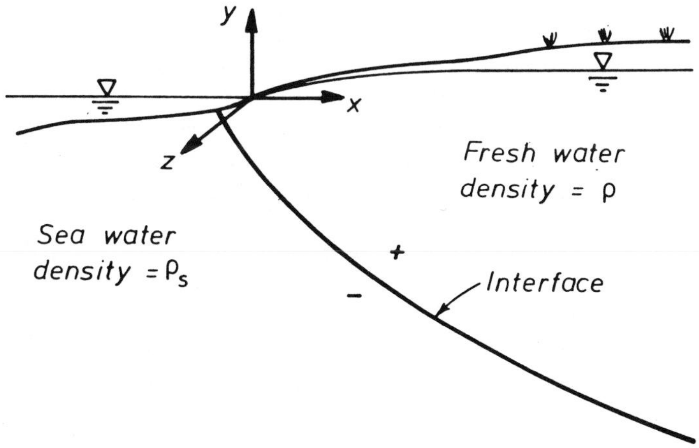

Fig. 2.3 - An interface between fresh water and sea water along a sea coast.

the free surface approaches zero is

$$\bar{u}.\hat{e}_n + R\hat{j}.\hat{e}_n = \sigma \frac{d\bar{r}}{dt} . \hat{e}_n \tag{2.57}$$

Since $\hat{e}_n$ is the outward normal to the free surface, multiplying both sides by a small time increment allows Eq. 2.57 to be interpreted as the result of setting the sum of inflows through the bottom and top control volume boundaries respectively, equal to the change in storage within the control volume that results from a rise in position of the free surface.

Equations 2.56 and 2.57 are sufficient boundary conditions for a free surface problem, but sometimes it is more convenient to replace Eq. 2.57 with an equivalent equation obtained from Eqs. 2.56 and 2.57. Since Eq. 2.56 gives the equation of the free surface, Eq. 1.11 shows that Eq. 2.57 can be rewritten in the form

$$(\bar{u} + R\hat{j}) . \bar{\nabla}(h - y) = \sigma \frac{d\bar{r}}{dt} . \bar{\nabla}(h - y) \tag{2.58}$$

Since x(t), y(t) and z(t) are the coordinates of a point on the free surface, Eq. 2.56 can be differentiated with respect to time to obtain

$$0 = \frac{d(h - y)}{dt} = \frac{\partial(h - y)}{\partial x}\frac{dx(t)}{dt} + \frac{\partial(h - y)}{\partial y}\frac{dy(t)}{dt} + \frac{\partial(h - y)}{\partial z}\frac{dz(t)}{dt} + \frac{\partial(h - y)}{\partial t} \tag{2.59}$$

But, since $\partial y/\partial t = 0$, Eq. 2.59 can be written as

$$0 = \frac{d\bar{r}}{dt} . \bar{\nabla}(h - y) + \frac{\partial h}{\partial t} \tag{2.60}$$

Thus, using Eq. 2.60 to eliminate $d\bar{r}/dt$ on the right side of Eq. 2.58 and using Eq. 2.11 to eliminate $\bar{u}$ gives the final result

$$(-K\bar{\nabla}h + R\hat{j}) . \bar{\nabla}(h - y) = -\sigma \frac{\partial h}{\partial t} \tag{2.61}$$

Equation 2.61 contains h and the coordinates of the free surface as its only unknown. However, it is a nonlinear equation, a fact which is emphasized by computing the gradient and the dot product and rewriting Eq. 2.61 in its more usual form

$$K\bar{\nabla}h.\bar{\nabla}h + R = (K + R)\frac{\partial h}{\partial y} + \sigma \frac{\partial h}{\partial t} \tag{2.62}$$

Thus, Eq. 2.56 and either Eq. 2.57 or 2.62 are the two boundary conditions that must be applied along a free surface.

Boundary conditions along an interface between fresh water and sea water, as shown in Fig. 2.3, are analogous to the boundary conditions for a free surface. The interface is usually modelled as a discontinuity in density, and since its geometry must be found as part of the problem solution, it needs two boundary conditions that require continuity of pressure and normal flux velocity. If the y axis points upwards, then $h = P/\rho g + y$ and the continuity of pressure requirment becomes

$$\rho(h - y)_+ = \rho_s(h - y)_- \tag{2.63}$$

in which ρ and ρ_s are the densities of fresh and sea water, respectively, and the plus and minus signs denote the result of approaching the interface through fresh and sea water, respectively. Since K in Eq. 2.11 depends upon ρ and μ, the requirement for continuity of normal flux velocity is

$$K_+ \frac{dh_+}{dn} = K_- \frac{dh_-}{dn} \tag{2.64}$$

For problems in which the interface is stationary, pressures in the sea water are everywhere hydrostatic. This means that $h_- = 0$, and Eq. 2.63 reduces to

$$h(x, y, z) = - \varepsilon y, \quad \left(\varepsilon = \frac{\rho_s}{\rho} - 1\right) \tag{2.65}$$

Equation 2.64 reduces to the requirement that the normal derivative of h vanish along the interface, and the method used in problem 8 can be used to rewrite this boundary condition in the form

$$\bar{\nabla} h . \bar{\nabla} h + \varepsilon \frac{\partial h}{\partial y} = 0 \tag{2.66}$$

The reader is asked to derive Eq. 2.66 in problem 9.

A very useful result, which is known as the Ghyben-Herzberg approximation, follows from Eq. 2.65 by making the Dupuit approximation. Then, since vertical velocity components are assumed negligible and piezometric heads are constant along any vertical line, the left side of Eq. 2.65 is identical with the elevation of the free surface. Since the dimensionless constant, ε, has a magnitude of about 1/40, Eq. 2.65 states that the interface lies below mean sea level a distance of 40 times the elevation of the free surface above mean sea level. A surface of seepage is always ignored in the Dupuit approximation and this, together with the Ghyben-Herzberg approximation, means that the free surface, mean sea level and the interface all meet at the same point along a coast.

REFERENCES

Bear, J. 1979. Hydraulics of Groundwater, McGraw-Hill, New York, pp. 84-92.

Collins, M.A., Gelhar, L.W. and Wilson, J.L. 1972. "Hele-Shaw Model of Long Island Aquifer System," ASCE Journal of the Hydraulics Division, Vol. 98, No. HY9, September, pp. 1701-1714.

Harr, M.E. 1962. Groundwater and Seepage, McGraw-Hill, New York, p. 8.

Malvern, L.E. 1969. Introduction to the Mechanics of a Continuous Medium, Prentice-Hall, Englewood Cliffs, Ch. 2.

Muskat, M. 1946. The Flow of Homogeneous Fluids through Porous Media, J.W. Edwards, Inc., Ann Arbor, pp. 55-69.

Polubarinova-Kochina, P.Ya. 1962. Theory of Ground Water Movement, translated by J.M.R. de Wiest, Princeton University Press, Princeton, pp. 16, 341-343.

PROBLEMS

1. Write an expression for the piezometric head, h, when(a) the positive y axis points upward, (b) the positive y axis points downward and (c) the x axis dips at an angle θ below the horizontal, as shown in the sketch. Hint: Write $\hat{g}$ in each coordinate system and then insert $\hat{g}.\bar{r}$ in the definition given in Eq. 2.11. The vector $\bar{r}$ is defined by Eq. 1.1.

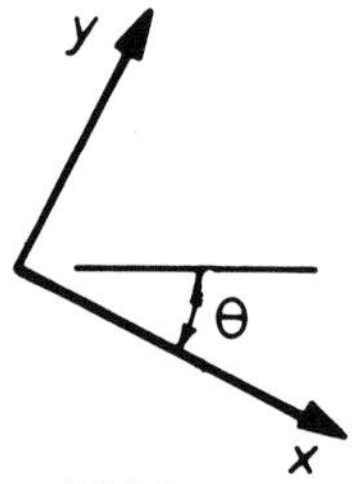

PROB. 1

2. An average piezometric gradient in the northern portion of the Canterbury Plains of New Zealand is about 10 meters in 3 Kilometers, and a typical porosity is about 0.1. Calculate, in units of meters/day, ranges for the flux and pore velocities corresponding to ranges of permeability given in Table 1.

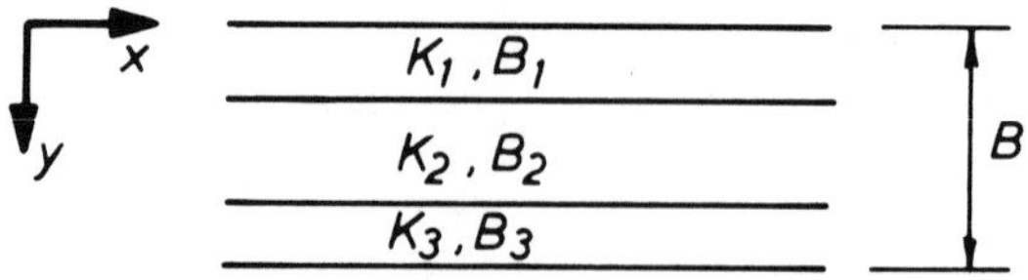

PROB. 3

3. Following Polubarinova-Kochina (1962), consider N aquifer layers with permeabilities and thicknesses of K_i and B_i for each layer.

(a) Assume that the flow is parallel to the layers, and note that dh/dx will be the same for each layer. Equate the sum of the flow rates through the individual layers to

$$K_x B \frac{dh}{dx}$$

Hence, show that the bulk permeability, K_x, is

$$K_x = \frac{1}{B} \sum_{i=1}^{N} K_i B_i$$

(b) Assume that the flow is normal to the layers, and note that the vertical flux velocity, u_y, is the same in each layer. Equate the sum of the changes in h across the individual layers to

$$\frac{u_y B}{K_y}$$

In this way, show that the bulk permeability, K_y, is given by

$$K_y = \frac{B}{\sum_{i=1}^{N} \frac{B_i}{K_i}}$$

(c) The following proof shows that $K_x \geqq K_y$. The method of proof is to assume that $K_x \geqq K_y$ and to show that the result of this assumption reduces to a statement that is true under all conditions. The statement $K_x \geqq K_y$ can be written

$$\sum_{i=1}^{N} K_i \ell_i \geqq \frac{1}{\sum_{j=1}^{N} \frac{\ell_j}{K_j}}, \quad \left(\ell_i = \frac{B_i}{B}\right)$$

Therefore,

$$\sum_{i=1}^{N} K_i \ell_i \sum_{j=1}^{N} \frac{\ell_j}{K_j} - 1 \geqq 0$$

or, when the product is computed and terms regrouped,

$$(\ell_1^2 + \ell_2^2 + \ldots\ldots + \ell_N^2 - 1) + \sum_{\substack{i,j=1 \\ i \neq j}}^{N} \ell_i \ell_j \left(\frac{K_i}{K_j} + \frac{K_j}{K_i}\right) \geqq 0$$

But the definition of $\ell_i \equiv B_i/B$ leads to the result

$$(\ell_1 + \ell_2 + \ldots\ldots + \ell_N)^2 = (1)^2 = 1$$

An expansion of this equation gives

$$(\ell_1^2 + \ell_2^2 + \ldots\ldots + \ell_N^2 - 1) = -2 \sum_{\substack{i,j=1 \\ i \neq j}}^{N} \ell_i \ell_j$$

Thus, inserting this last result into the inequality gives

$$\sum_{\substack{i,j=1 \\ i \neq j}}^{N} \ell_i \ell_j \left(\frac{K_i}{K_j} + \frac{K_j}{K_i} - 2 \right) \geqq 0$$

Finally, factoring out the ratio K_j/K_i leads to the end result

$$\sum_{\substack{i,j=1 \\ i \neq j}}^{N} \ell_i \ell_j \left(\frac{K_j}{K_i} \right) \left(\frac{K_i}{K_j} - 1 \right)^2 \geqq 0$$

This is a statement that is always true. It also shows that the inequality ($K_x > K_y$) holds if at least one of the values of K_i differs from the other values of K_i, and that the equality ($K_x = K_y$) holds only if all values of K_i are the same.

4. The permeability tensor, $\underline{K}$, is shown by Eq. 2.19 to be symmetric ($K_{\alpha\beta} = K_{\beta\alpha}$).

(a) Show by direct calculation, that $\underline{K}.\overline{\nabla}h = \overline{\nabla}h.\underline{K}$ if $\underline{K}$ is given by

$$\underline{K} = \hat{i}\hat{i} + 2(\hat{i}\hat{j} + \hat{j}\hat{i}) + 3\hat{j}\hat{j}$$

(b) Show, by direct calculation, that $\underline{A}.\overline{\nabla}h \neq \overline{\nabla}h.\underline{A}$ for a non-symmetric tensor, $\underline{A}$, given by

$$\underline{A} = \hat{i}\hat{i} + 2\hat{i}\hat{j} - \hat{j}\hat{i} + 3\hat{j}\hat{j}$$

(c) The symmetric tensor $\underline{1} = \hat{i}\hat{i} + \hat{j}\hat{j} + \hat{k}\hat{k}$ is called the identity tensor because its dot product with any vector, from either side, always gives back the same vector. Show, by direct calculation, that

$$\underline{1}.\overline{\nabla}h = \overline{\nabla}h.\underline{1} = \overline{\nabla}h$$

The identity tensor will be met in groundwater pollution studies when it becomes necessary to calculate the dispersion tensor.

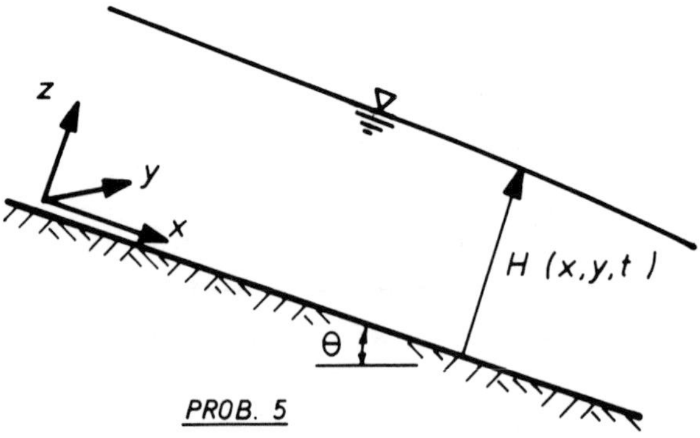

PROB. 5

5. Consider a free surface flow down a sloping aquifer in which the z axis is normal to the bottom aquifer boundary, the y axis is horizontal and the x axis points down the slope. The free surface is nearly parallel to the bottom boundary, which means that the Dupuit approximation is equivalent to requiring that $u_z = 0$ and

h = h(x, y, t) only.

(a) Use the definition for h in Eq. 2.11 to calculation an expression for h in terms of H. Then substitute this result into Eq. 2.37 to show that

$$\bar{q} = -T\cos\theta\bar{\nabla}H + T\sin\theta\hat{i}$$

in which $\bar{\nabla}H$ only contains derivatives in the x and y directions. Assume that K = constant, so that Eq. 2.38 gives T = KH. Thus, if H is constant, this equation shows that the uniform flux velocity is $K\sin\theta$ down the slope.

(b) Note that Eq. 2.41 remains unchanged provided that h is replaced with H(x, y, t). Thus, substitute the calculated expression for $\bar{q}$ and divide the result by S to obtain

$$\bar{\nabla}.(D\bar{\nabla}H) = U_x \frac{\partial H}{\partial x} + \frac{\partial H}{\partial t} + \frac{u_z}{S}$$

in which $D = KH\cos\theta/S$ and $U_x = K\sin\theta/S$. Thus, although h in Eqs. 2.42 and 2.51 satisfies a heat diffusion equation, H for this sloping aquifer satisfies a heat transport equation in which the transport velocity is the pore velocity U_x and D is the diffusion coefficient.

6. Tangential and normal velocity components along an internal boundary of discontinuity in K are given by

$$\bar{u}.\hat{e}_t = -K\frac{dh}{ds}, \quad \bar{u}.\hat{e}_n = -K\frac{dh}{dn}$$

in which s and n are arc length along and normal to the boundary, respectively. Use these equations together with Eqs. 2.54 - 2.55 to show that

$$(\bar{u}.\hat{e}_n)_+ = (\bar{u}.\hat{e}_n)_-$$

$$\left(\frac{u.\hat{e}_t}{K}\right)_+ = \left(\frac{u.\hat{e}_t}{K}\right)_-$$

Note that the first equation shows that normal components of $\bar{u}$ are continuous and that the second equation shows that tangential velocity components are not continuous. Thus, the direction and magnitude of the velocity vector will be discontinuous along the discontinuity in K.

7. Show, for either a confined or unconfined aquifer that is anisotropic, that h satisfies

$$\bar{\nabla}.(\underline{T}.\bar{\nabla}h) = S\frac{\partial h}{\partial t} + u_z$$

in which the transmissivity tensor is

$$\underline{T} = \int_{z_0}^{z_0 + B} \underline{K}(x, y, z)\, dz$$

and $\underline{K}$ is given by either Eq. 2.14 or 2.16. Probably most applications of this equation result from a condition of horizontal stratification, in which case $K_x = K_y > K_z$. In this case, show that the above equation reduces to either Eq. 2.42 or 2.51.

8. Boundary conditions along a free surface in steady flow with no recharge are given by Eq. 2.56 and the requirement that

$$\bar{u}.\hat{e}_n = 0$$

Compute $\bar{u}$ from Eq. 2.11 and $\hat{e}_n$ from Eqs. 1.11 and 2.56 and show that the above boundary condition is identical with the result obtained by setting $R = \partial h/\partial t = 0$ in Eq. 2.62.

9. Since Eq. 1.8 shows that $dh/dn = \bar{\nabla}h.\hat{e}_n$, use Eq. 2.65 and the method explained in problem 8 to obtain Eq. 2.66.

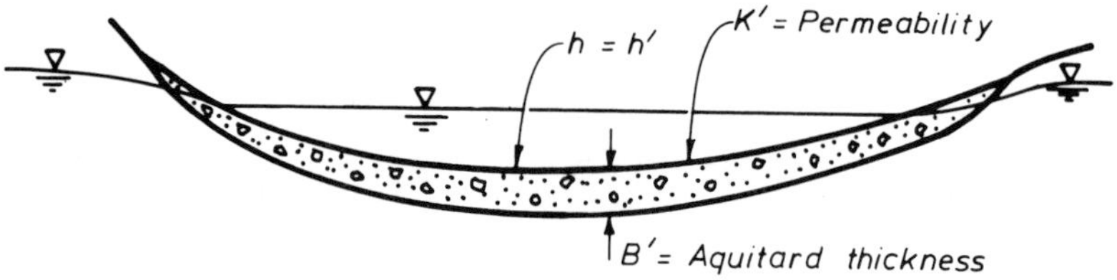

PROB. 10

10. A boundary condition that is not described in the text and which is used very seldom in groundwater problems occurs when groundwater seeps into, or out of, a river or reservoir which has a layer of less permeable sediment upon its bottom. Let K' and B' be the permeability and thickness of this sediment layer, let K be the permeability of the aquifer beneath and let h and h' be piezometric heads along the bottom and top boundaries of the sediment layer, respectively. Show that the boundary condition along the bottom boundary of the sediment layer (an aquitard) is

$$K \frac{dh}{dn} + \left(\frac{K'}{B'}\right) h = \left(\frac{K'}{B'}\right) h'$$

in which n = arc length in the direction of the outward normal to the aquifer. In heat conduction problems, this is often called a "radiation" boundary condition.

3 The Formulation of Boundary-Value Problems

10. Regional Problems

Regional problems are nearly always problems which are solved by making the Dupuit approximation. This is because aquifers for large regions usually have vertical thicknesses that are very small when compared with lateral aquifer dimensions, and this, in turn, means that changes in h in the vertical direction are very small compared with the maximum change in h in the horizontal direction.

A typical regional problem is shown in Fig. 3.1. An unconfined, sand and gravel aquifer which lies beneath the flood plain of a river is bounded on the right by a river and on the left by a relatively impermeable embankment. Sediment in the river bed is permeable, and field data shows that piezometric head levels in the aquifer vary in a continuous manner near the river edge and meet water levels in the river. A well is to withdraw a known flow rate at (x_0, y_0), and the objective is to predict the changes in aquifer piezometric levels that this abstraction will create.

The following equations give a mathematical statement of the problem in Fig. 3.1:

$$\overline{\nabla}.(T\overline{\nabla}h) = S\frac{\partial h}{\partial t} + \sum_i Q_i\ \delta(x - x_i)\ \delta(y - y_i) \tag{3.1}$$

$$h(x, y, 0) = I(x, y) \tag{3.2}$$

$$h = f \text{ for } x, y \text{ on ABC} \tag{3.3}$$

$$\frac{dh}{dn} = 0 \text{ for } x, y \text{ on CDA} \tag{3.4}$$

Equation 3.1 is the partial differential equation (Eq. 2.42) that must be satisfied at all points in the region, and the summation must include the flow rates, Q_i, of every well in the aquifer. Equation 3.2 is an initial condition that requires h to agree with the measured distribution of h before pumping starts in the well at (x_0, y_0). Equation 3.3 requires piezometric heads in the aquifer to agree with river levels along the river edge, and Eq. 3.4 requires that the boundary CDA be impermeable. The boundary of the solution domain is shown as a dashed line in Fig. 3.1.

There are difficulties that are inherent in trying to formulate and solve Eqs. 3.1 - 3.4. First, Eq. 3.1 is non-linear, a fact which becomes most apparent if the horizontal aquifer bottom coincides with the (x, y) plane and if the permeability is considered constant in the vertical direction. Then Eq. 2.38 gives

$$T(x, y, t) = K(x, y)\ h(x, y, t) \tag{3.5}$$

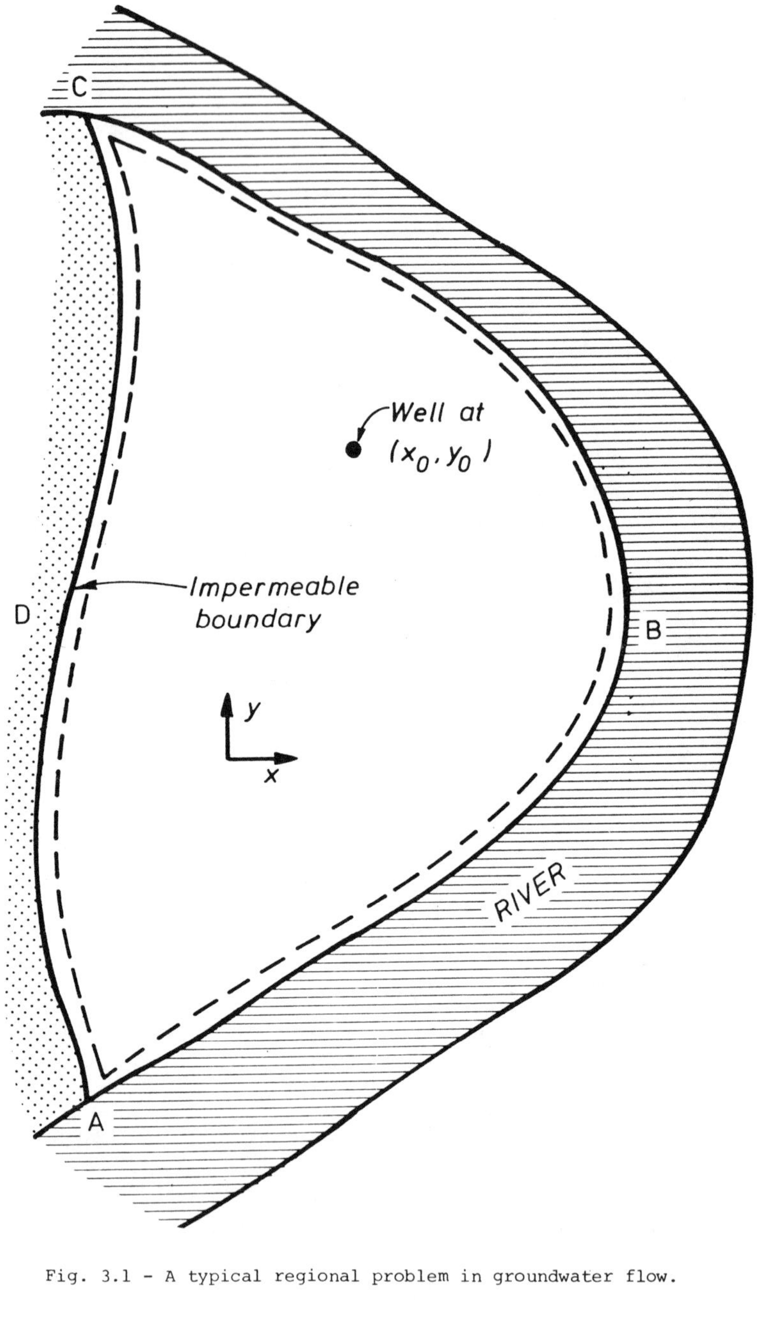

Fig. 3.1 - A typical regional problem in groundwater flow.

Second, it is expensive and time consuming to measure all of the values of Q_i, I(x, y) and f that are needed for this formulation. Third, the solution of Eqs. 3.1 - 3.4 will give h rather than the change in h which is required for the problem solution.

In some instances it may be impossible to do anything else other than try to solve Eqs. 3.1 - 3.4. In a great many applications, though, computed changes in h will be small when compared with saturated aquifer thicknesses throughout most of the region of interest. Under these circumstances, it is possible and desirable to linearize Eq. 3.1 by using throughout all of the calculations a distribution for T that has been computed at some earlier time. Then T(x, y) will not change with h, Eqs. 3.1 - 3.4 are linear and the superposition principle for any linear system shows that Eqs. 3.1 - 3.4 can be replaced with the following much simpler set of equations:

$$\overline{\nabla}.(T\overline{\nabla}h) = S \frac{\partial h}{\partial t} + Q_0 \delta(x - x_0) \delta(y - y_0) \quad (3.6)$$

$$h(x, y, 0) = 0 \quad (3.7)$$

$$h = 0 \text{ for } x, y \text{ on ABC} \quad (3.8)$$

$$\frac{dh}{dn} = 0 \text{ for } x, y \text{ on CDA} \quad (3.9)$$

The advantages of this system of equations are (1) the system of equations is linear, which means that their analytical or numerical solution is much easier to obtain, (2) only the changes in Q_i, I and f are needed, which makes a tremendous savings in the expense and time required to obtain field data and (3) Eqs. 3.6 - 3.9 compute directly the change in h created by withdrawing a flow Q_0 from the well at (x_0, y_0). If h is needed for the problem solution, then the computed change in h can be added algebraically to values of h measured in the field before pumping commences. Usually, however, the change in h is the variable of most interest. Also, it should be pointed out that the calculated perturbation in h may, in other applications, be the result of perturbations in f rather than in the values of Q_i.

Before closing this discussion, we will consider the question of uniqueness. It's important to know that the solution of a boundary-value problem is unique for two reasons: first, it assures us that solutions of the same set of equations obtained by using different methods will yield the same numerical values of h within the solution domain, and, second, it suggests that the number and type of boundary conditions are appropriate for the solution of the differential equation. For example, it settles the question of whether we should prescribe one or two initial conditions or whether the tangential velocity component should be prescribed in addition to the normal component along impermeable boundaries.

A formal proof of uniqueness will be carried out by first assuming that the solution is not unique. This means that two different solutions, h_1 and h_2, will each satisfy the following equations:

$$\overline{\nabla}.(T\overline{\nabla}h_i) = S\frac{\partial h_i}{\partial t} + \left(\frac{K'}{B'}\right)(h_i - h') + Q\,\delta(x - x_0)\,\delta(y - y_0) - R \quad (3.10)$$

$$h_i(x, y, 0) = I(x, y) \quad (3.11)$$

$$h_i = f \text{ for } x, y \text{ on } \Gamma_1 \quad (3.12)$$

$$\frac{dh_i}{dn} = g \text{ for } x, y \text{ on } \Gamma_2 \quad (3.13)$$

The functions f, g, I, h', Q and R are identical for i = 1 and 2, and the entire boundary, Γ, has been split into the sum of Γ_1 and Γ_2. It will be assumed that Eq. 3.10 has been linearized, so that subtraction of the two sets of equations for h_1 and h_2 shows that their difference, $h = h_1 - h_2$, is a solution of a homogeneous set of equations.

$$\overline{\nabla}.(T\overline{\nabla}h) = S\frac{\partial h}{\partial t} + \left(\frac{K'}{B'}\right)h \quad (3.14)$$

$$h(x, y, 0) = 0 \quad (3.15)$$

$$h = 0 \text{ for } x, y \text{ on } \Gamma_1 \quad (3.16)$$

$$\frac{dh}{dn} = 0 \text{ for } x, y \text{ on } \Gamma_2 \quad (3.17)$$

Equations 3.14 - 3.17 have the solution h = 0, and if it can be shown that this is the only solution, then $h_1 - h_2 = 0$ at all points within $\Gamma = \Gamma_1 + \Gamma_2$ and the solution is unique.

Multiplication of Eq. 3.14 by h gives

$$h\overline{\nabla}.(T\overline{\nabla}h) = Sh\frac{\partial h}{\partial t} + \left(\frac{K'}{B'}\right)h^2 \quad (3.18)$$

Manipulation of the derivatives allows Eq. 3.18 to be rewritten in the form

$$\overline{\nabla}.(hT\overline{\nabla}h) - T\overline{\nabla}h.\overline{\nabla}h = \frac{\partial}{\partial t}\left(\frac{1}{2}Sh^2\right) + \left(\frac{K'}{B'}\right)h^2 \quad (3.19)$$

Integration of Eq. 3.19 over the region, A, bounded by $\Gamma = \Gamma_1 + \Gamma_2$ and use of the divergence theorem gives

$$\int_\Gamma Th\frac{dh}{dn}\,ds - \int_A T\overline{\nabla}h.\overline{\nabla}h\,dA = \frac{d}{dt}\int_A \frac{1}{2}Sh^2\,dA + \int_A \left(\frac{K'}{B'}\right)h^2\,dA \quad (3.20)$$

Finally, integration of Eq. 3.20 with respect to t from t = 0 to t and use of Eq. 3.15 gives

$$\int_0^t \int_\Gamma Th \frac{dh}{dn} \, ds \, dt = \int_A \frac{1}{2} Sh^2 \, dA + \int_0^t \int_A [T\overline{\nabla}h.\overline{\nabla}h + \left(\frac{K'}{B'}\right) h^2] \, dA \, dt \qquad (3.21)$$

But the left hand side of Eq. 3.21 vanishes by virtue of Eqs. 3.16 - 3.17, and the integrands of the integrals on the right side are all greater than or equal to zero since S, T and K'/B' are all greater than or equal to zero. Thus, for the first reason explained after Eq. 1.16, the integrand of each integral must vanish at all points in A for all values of $t > 0$, which means that $h = h_1 - h_2 = 0$ is the only solution to Eqs. 3.14 - 3.17.

It should be pointed out that the proof just given also holds for steady flow, since the equation for steady flow can be obtained simply by setting $S = 0$. However, if $(K'/B') = 0$, too, then either Eq. 3.20 or 3.21 shows that

$$|\overline{\nabla}h|^2 = 0 \text{ in } A \qquad (3.22)$$

which means that $h = h_1 - h_2 =$ constant in A. This constant can be calculated as zero if $\Gamma_1 \neq 0$ since h vanishes on Γ_1. However, if $\Gamma_1 = 0$, then the constant cannot be calculated and two solutions may differ by an unknown, additive constant.

In closing, it should also be pointed out that requiring $K'/B' > 0$ for the proof just given is not a restriction that might be removed by using some more general method of proof. For example, the following eigenvalue problem has an infinite number of different solutions:

$$\nabla^2 h = -\lambda^2\pi^2 h \qquad (3.23)$$

$$h = 0 \text{ for } x, y \text{ on } \Gamma \qquad (3.24)$$

If Γ is chosen as the square bounded by the straight lines $x = \pm 1$, $y = \pm 1$, then solutions to Eqs. 3.23 - 3.24 are given by

$$h = A_{m,n} \sin(m\pi x) \sin(n\pi y) \qquad (3.25)$$

in which $A_{m,n}$ is an arbitrary constant and m and n are any integers which satisfy

$$m^2 + n^2 = \lambda^2 \qquad (3.26)$$

Equation 3.23 is, of course, the result of setting $T = 1$, $S = 0$ and $K'/B' = -\lambda^2\pi^2$ in Eq. 3.14.

11. A Confined Flow Problem

For a second example, we will consider the unsteady, confined flow toward a well in a two-layered aquifer. A sketch of the problem is shown in Fig. 3.2. Flow is withdrawn through a well screen that is placed in the lower portion of the bottom aquifer, and the Dupuit approximation, which assumes horizontal flow, would only be a valid model at relatively

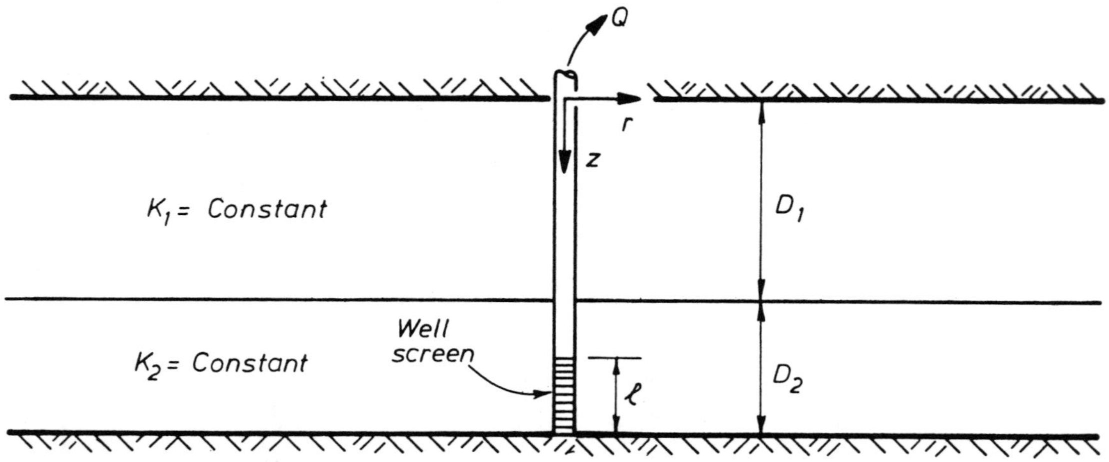

Fig. 3.2 - Unsteady, confined flow toward a well in a two-layered aquifer.

large radial distances from the well. Since the flow is axisymmetric, and since the permeability within each of the two layers is constant, the problem in Fig. 3.2 is described by the following equations:

$$K_i\left[\frac{1}{r}\frac{\partial}{\partial r}\left(r\frac{\partial h_i}{\partial r}\right)+\frac{\partial^2 h_i}{\partial z^2}\right]=S_{s_i}\frac{\partial h_i}{\partial t}\text{ for } i = 1 \text{ and } 2 \tag{3.27}$$

$$h_i(r, z, 0) = 0 \text{ for } i = 1 \text{ and } 2 \tag{3.28}$$

$$h_i \to 0 \text{ as } r \to \infty \text{ for } i = 1 \text{ and } 2 \tag{3.29}$$

$$\frac{\partial h_i}{\partial z} = 0 \text{ on } z = 0 \text{ for } i = 1 \text{ and on } z = D_1 + D_2 \text{ for } i = 2 \tag{3.30}$$

$$\lim_{r\to 0}\left(2\pi r \ell K_2 \frac{\partial h_i}{\partial r}\right) = Q \text{ for } D_1 + D_2 - \ell < z < D_1 + D_2$$
$$= 0 \text{ for } 0 < z < D_1 + D_2 - \ell \tag{3.31}$$

$$h_1 = h_2 \text{ on } z = D_1 \tag{3.32}$$

$$K_1\frac{\partial h_1}{\partial z} = K_2\frac{\partial h_2}{\partial z} \text{ on } z = D_1 \tag{3.33}$$

Equation 3.27 is the axisymmetric form of Eq. 2.32. Equations 3.28 and 3.29 are corresponding initial and boundary conditions, respectively, that require h to vanish everywhere at t = 0 and at infinity for all finite values of t. Thus, since the governing equations are linear, the superposition principle applies and Eqs. 3.28 - 3.29 indicate that the calculated values of h_i will be perturbations of an existing piezometric head distribution. Equation 3.30 requires that the top and bottom boundaries be impermeable, and Eq. 3.31 specifies the normal derivative of h along the cylindrical surface r = 0 by requiring that the flow rate abstracted by the well be uniformly distributed along the well screen. Finally, Eqs. 3.32 - 3.33 require that the pressure and normal flux velocity component be continuous across the discontinuity in K.

A proof of uniqueness for this type of problem can be carried out by using the same approach as the one used in showing uniqueness for Eqs. 3.10 - 3.13. There are, however, some differences in the details. First, an equation similar to Eq. 3.20 is written for each of the two layers. Second, these two equations are added and the two surface integrals of Kh dh/dn along the common boundary cancel as a result of Eqs. 3.32 - 3.33 and the fact that the outward normals of regions 1 and 2 are in opposite directions along this common surface, $z = D_1$. This shows why both Eq. 2.54 and Eq. 2.55 are needed along a boundary of discontinuity in K. Third, control volume surfaces for the top and bottom layers must be closed by two right circular cylinders, one surrounding the z axis near r = 0 and the other having a very large radius. Then h must be assumed to vanish fast enough at large values of r to ensure the existence of all surface and volume integrals as the radii of the large and small circular cylinders approach infinity and zero, respectively. The reader is asked to complete the details of this proof in a problem at the end of the chapter. It should also be pointed out that a numerical solution of this problem would probably use finite values for the radii of the circular cylinders that are used in the uniqueness proof.

12. A Sea Water Intrusion Problem

Assume that the river edge, ABC, in Fig. 3.1 represents the sea-coast along an aquifer that is recharged by a rainfall flux velocity of R. If the flow is steady and the Dupuit approximation is made, the problem is modelled with the following equations:

$$\overline{\nabla}.(T\overline{\nabla}h) = Q\ \delta(x - x_0)\ \delta(y - y_0) - R \tag{3.34}$$

$$h = 0 \text{ for } x,\ y \text{ on ABC} \tag{3.35}$$

$$\frac{dh}{dn} = 0 \text{ for } x,\ y \text{ on CDA} \tag{3.36}$$

It will be assumed that the well at (x_0, y_0) is the only well in the aquifer, and it will also be assumed, initially, that the problem has been linearized so that T does not change with h. Once h has been calculated, the Ghyben-Herzberg approximation, Eq. 2.65, can be used to calculate the elevation of the salt-water interface below mean sea level. In particular, the interface will intersect the bottom aquifer boundary at the point where

$$h(x,\ y) = \varepsilon D, \quad \left(\varepsilon = \frac{\rho_s}{\rho} - 1 \simeq \frac{1}{40}\right) \tag{3.37}$$

in which D is the constant vertical distance between mean sea level and the bottom aquifer boundary. Equation 3.37 is the equation of a curved line that roughly parallels the coast, ABC.

Strack (1976) points out that a critical condition exists when a point on the line given by Eq. 3.37 lies beneath a saddle point on the free surface. This is because any further increase in Q or decrease in R will decrease h, and the interface will fail to intersect the bottom aquifer boundary in the neighbourhood of the saddle point. When this happens, the steady-state solution will be one in which the salt-water interface extends all the way to the well, thus polluting the well water. This critical condition can be analyzed by solving Eq. 3.37 simultaneously with

$$\overline{\nabla}h = 0 \tag{3.38}$$

Since $z = h(x, y)$ is the equation of the free surface, Eq. 3.38 requires that $\hat{e}_n = \hat{k}$ at the location of the saddle point. Furthermore, Eqs. 3.37 - 3.38 give 3 scalar equations that can be solved simultaneously for the values of Q and the (x, y) coordinates of the toe of the salt-water interface at this critical condition.

Strack (1976) also shows how the non-linear forms of Eqs. 3.34 - 3.38 can be reformulated without a linearizing approximation. Strack assumes a constant permeability in his calculations, but we will show that this is an unnecessary restriction. If it is assumed that K = K(x, y), then Eq. 2.38 and the Ghyben-Herzberg approximation show that T in Eq. 3.34 is given by

$$\begin{aligned} T &= K(h + D) \text{ for } x, y \text{ in region I} \\ &= K\left(1 + \frac{1}{\varepsilon}\right) h \text{ for } x, y \text{ in region II} \end{aligned} \qquad (3.39)$$

in which ε is given in Eq. 3.37. Region II contains the salt-water interface, and region I is the remainder of the aquifer. The dependent variable, h, can now be replaced with a potential function, ϕ, defined by

$$\begin{aligned} \phi &= \frac{1}{2}\left[(h + D)^2 - (1 + \varepsilon)D^2\right] \text{ for } x, y \text{ in region I} \\ &= \frac{1}{2}\left(1 + \frac{1}{\varepsilon}\right) h^2 \text{ for } x, y \text{ in region II} \end{aligned} \qquad (3.40)$$

Thus, rewriting Eqs. 3.34 - 3.38 in terms of ϕ gives

$$\overline{\nabla}.(K\overline{\nabla}\phi) = Q\ \delta(x - x_0)\ \delta(y - y_0) - R \qquad (3.41)$$

$$\phi = 0 \text{ for } x, y \text{ on ABS} \qquad (3.42)$$

$$\frac{d\phi}{dn} = 0 \text{ for } x, y \text{ on CDA} \qquad (3.43)$$

$$\phi(x, y) = \frac{\varepsilon}{2}(1 + \varepsilon)\ D^2 \qquad (3.44)$$

$$\overline{\nabla}\phi = 0 \qquad (3.45)$$

Equations 3.41 - 3.45 are linear in ϕ. Since Eq. 3.44 shows that ϕ is continuous along the common boundary between regions I and II (the toe of the salt-water interface), and since continuity of normal flux velocity components requires that the normal derivative of ϕ be continuous across this common boundary, it is seen that $\phi(x, y)$ can be chosen as the same function throughout the entire aquifer (i.e. - ϕ can be chosen as the solution of Eqs. 3.41 - 3.43 without requiring that ϕ satisfy two additional boundary conditions along the common boundary of regions I II). Furthermore, a uniqueness proof along the lines of the one outlined for the problem in Fig. 3.2 can be used to show that this choice for ϕ in regions I and II is unique.

13. Free Surface Flows

When the Dupuit approximation can not be made, free surface problems require the use of two boundary conditions along the free surface. As an example, the problem shown in Fig. 2.1 is described by the following equations:

$$\overline{\nabla}.(K\overline{\nabla}h) = 0 \qquad (3.46)$$

$$h(x, y, 0) = I(x, y) \tag{3.47}$$

$$h = H \text{ for } x, y \text{ on } AB \tag{3.48}$$

$$\frac{dh}{dn} = 0 \text{ for } x, y \text{ on } BC \tag{3.49}$$

$$h = y \text{ for } x, y \text{ on } CD \tag{3.50}$$

$$h = y \text{ for } x, y \text{ on } DA \tag{3.51}$$

$$K\bar{\nabla}h.\bar{\nabla}h + R = (K + R)\frac{\partial h}{\partial y} + \sigma\frac{\partial h}{\partial t} \text{ for } x, y \text{ on } DA \tag{3.52}$$

Equation 3.46 is the partial differential equation that must be satisfied at all points in the solution domain, Eq. 3.47 is an initial condition, Eq. 3.48 is the result of a known pressure along the reservoir boundary, Eq. 3.49 requires a zero flux through an impermeable boundary and Eqs. 3.50 and 3.51 require pressures to be atmospheric along the surface of seepage and free surface, respectively. Equation 3.52 is a second boundary condition that must be satisfied along the free surface and is identical with Eq. 2.62. Special cases of this condition occur when the rate of recharge, R, along the free surface is zero or when the flow is steady. The latter case would require that Eq. 3.47 and $\partial h/\partial t$ in Eq. 3.52 be omitted.

In numerical solutions it is impossible to satisfy Eqs. 3.51 and 3.52 simultaneously without using some type of iterative approach. For example, in a steady-flow problem the location of the free surface would be guessed. Then, if $R = 0$, only Eq. 3.52 or its equivalent (for $R = \partial h/\partial t = 0$), Eq. 2.52, would be imposed as a boundary condition along the free surface. Finally, the calculated values of h along the free surface would be inspected to see if they agreed with Eq. 3.51. If not, then the free surface position would be adjusted and the process repeated until both Eqs. 3.51 and 3.52 are satisfied.

Numerical solutions for an unsteady flow are probably obtained most easily by replacing Eq. 3.52 with Eq. 2.57. If the position of the free surface is known at the beginning of a small, but finite, time step, then h can be calculated throughout the solution domain by solving Eq. 3.46 and 3.48 - 3.51. The computed values of h can be used to calculate $\bar{u}.\hat{e}_n$ along the free surface, which could be inserted in Eq. 2.57 to calculate the normal velocity of the free surface. An approximate integration with respect to time would then give the location of the free surface at the end of the time step, and the entire cycle could be repeated to obtain the location of the free surface at the end of the following time steps.

A simplification can be made when a free surface is displaced a relatively small distance from a horizontal position. An example is shown in Fig. 3.3, in which irrigation water is recharging an aquifer with a free surface that was initially horizontal. As long as the vertical displacement of the free surface is small compared to the radius of curvature of the free surface, the motion can be considered to be a perturbation from a state of rest and the non-linear terms in Eq. 3.52 can be neglected. Thus, the problem can be described approximately with the following equations:

$$\overline{\nabla}\cdot(K\overline{\nabla}h) = 0 \tag{3.53}$$

$$h(x, 0, 0) = 0 \tag{3.54}$$

$$\frac{\partial h}{\partial y} = 0 \text{ on } y = -D \tag{3.55}$$

$$K\frac{\partial h}{\partial y} + \sigma\frac{\partial h}{\partial t} = 0 \text{ on } y = 0 \text{ for } \frac{L}{2} < |x| < \infty \tag{3.56}$$

$$(K + R)\frac{\partial h}{\partial y} + \sigma\frac{\partial h}{\partial t} = R \text{ on } y = 0 \text{ for } 0 \leqq |x| < \frac{L}{2} \tag{3.57}$$

The remarkable feature of this formulation is that the linearised free surface boundary conditions, Eqs. 3.56 - 3.57, are applied on the initial position of the free surface at t = 0, an approximation which is very similar to the one used to study the movement of small-amplitude waves on the free surface of a reservoir of fluid. It is not difficult to show that Eqs. 3.53 - 3.57 are sufficient to determine a unique solution for h. After h is calculated, then the second boundary condition on the free surface, Eq. 3.51, can be used to calculate the approximate location of the free surface:

$$y = h(x, 0, t) \tag{3.58}$$

Approximate solutions to these type of problems were first obtained by Polubarinova-Kochina (1962).

REFERENCES

Polubarinova-Kochina, P.Ya. 1962. Theory of Ground-Water Movement, Translated by J.M.R. de Wiest, Princeton University Press, Princeton, pp. 559-565.

Strack, O.D.L. 1976. "A Single-Potential Solution for Regional Interface Problems in Coastal Aquifers," Water Resources Research, Vol. 12, No. 6, pp. 1165-1174.

PROBLEMS

1. The Dupuit approximation causes the boundary condition introduced in problem 10 of Chapter II to take the form

$$T\frac{dh}{dn} + \left(\frac{T'}{B'}\right)h = \left(\frac{T'}{B'}\right)h'$$

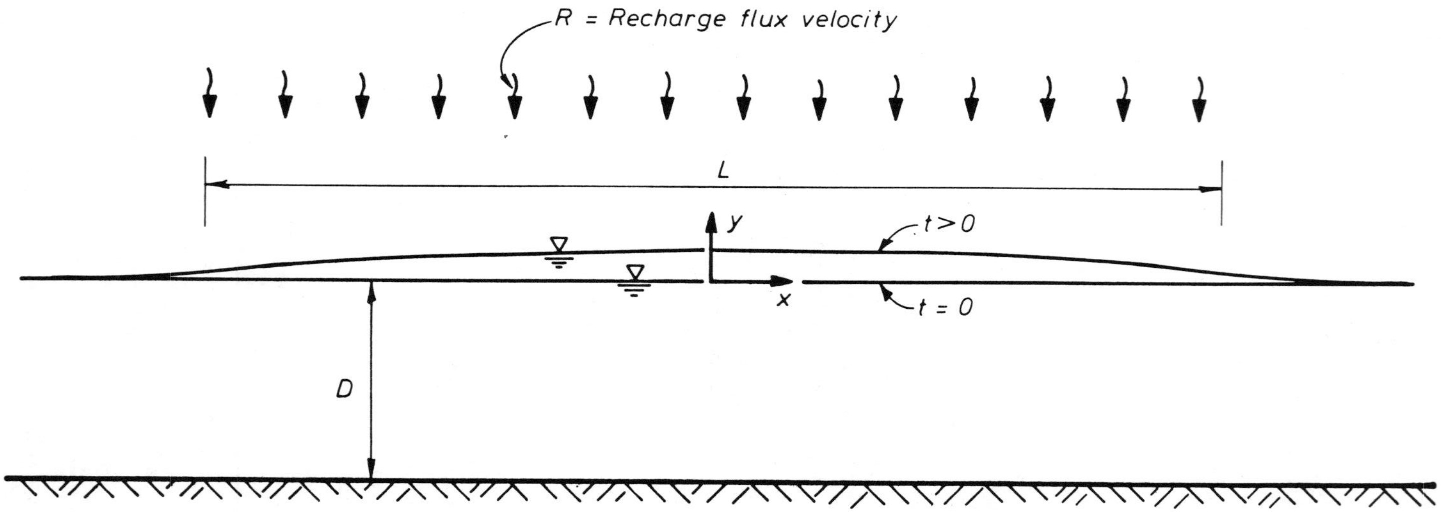

Fig. 3.3 - Recharge to an initially horizontal water table.

Assume that the boundary, Γ, in Eqs. 3.10 - 3.13 has been divided into three parts and that the "radiation" condition just given applies along Γ_3. Prove that the solution is unique if S, T, K'/B' and T'/B' are all greater than or equal to zero.

2. Carry out a uniqueness proof for the problem given by Eqs. 3.27 - 3.33 and shown in Fig. 3.2. If $K_1 = K_2$ and $\ell = D_1 + D_2$, then h is known to have the following asymptotic behaviour for large r:

$$h \simeq c_1(t) \frac{e^{-rc_2(t)}}{r} \text{ as } r \to \infty \text{ for finite t}$$

Assume this same asymptotic behaviour in your uniqueness proof to ensure the existence of all surface and volume integrals.

3. Formulate the problem in Fig. 3.3 by using the Dupuit approximation.

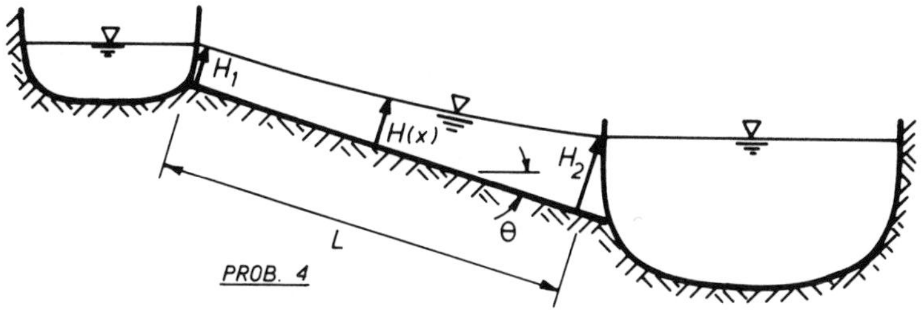

PROB. 4

4. Two-dimensional, steady flow takes place down an inclined aquifer. Use the results of problem 5 in chapter II to formulate this as a linearized problem by using H_1 to calculate D. Solve the ordinary differential equations for H(x) and use this result to calculate the two-dimensional flow rate, q. Note that the correct linearized form for the expression for q has $T = KH_1$ in the first, non-linear term and leaves T = KH unaltered in the second term.

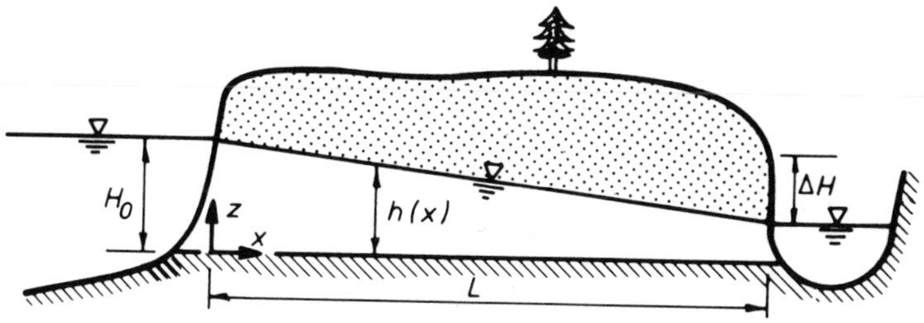

PROB. 5

5. A ditch parallels a river. Use the Dupuit approximation for steady flow to calculate the two-dimensional flow rate that must be removed from the ditch in order that its water level be a distance of ΔH below the river level. Carry out the calculations for the linearized ($T = KH_0$) and non-linear ($T = Kh(x)$) equations and show that the results agree if $\Delta H/H_0 << 1$.

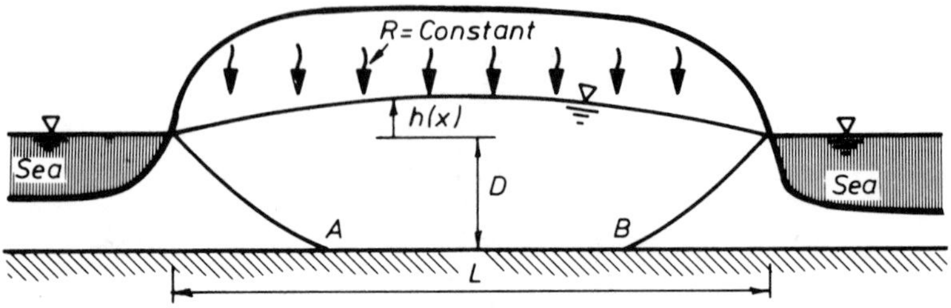

PROB. 6

6. An elongated peninsula has a shallow, unconfined aquifer that is recharged with rainfall at a constant rate, R. Assume a steady flow in which h only varies across the peninsula to calculate the value of R which will make the interface toes, A and B, meet. Carry out the calculations (a) by using the linearized formulation and (b) by using the non-linear formulation. Then compare these results and show that the linearized model overestimates R by a factor of about two.

7. A horizontal, confined aquifer near a seacoast has a constant thickness of B and a bottom boundary that lies a vertical distance of D below mean sea level. (a) Choose a datum at mean sea level and use the Ghyben-Herzberg approximation to show that a point on the interface lies a distance of h/ε below mean sea level. (b) Hence, show that T is given by

$$T = KB \quad \text{for } x,\ y \text{ in region I}$$
$$= K\left(\frac{h}{\varepsilon} - D + B\right) \quad \text{for } x,\ y \text{ in region II}$$

(c) Strack (1976) obtains the following potential for this problem:

$$\phi = Bh - \varepsilon BD + \varepsilon B^2/2 \quad \text{for } x,\ y \text{ in region I}$$
$$= \frac{\varepsilon}{2}\left(\frac{h}{\varepsilon} - D + B\right)^2 \quad \text{for } x,\ y \text{ in region II}$$

in which $\bar{q} = -T\bar{\nabla}h = -K\bar{\nabla}\phi$. Show that ϕ and its normal derivative are continuous at the boundary between regions I and II, which means that ϕ is one and the same function in regions I and II. (d) The correct boundary condition requires that the thickness of fresh water in the aquifer be zero at the point where the flow enters the sea. Show that this corresponds to the requirement $\phi = 0$ at this point.

4 The Approximate Solution of Boundary-Value Problems

14. The Finite-Difference Method

The finite-difference method is one of the oldest, most generally applicable and most easily understood methods of obtaining numerical solutions to steady and unsteady groundwater flow problems. The general method consists of superimposing a finite-difference grid of nodes upon the solution domain, as shown in Fig. 4.1. Each node is given a global identification number, and, in the neighbourhood of each of these nodes, the dependent variable is approximated with a finite-degree polynomial whose coefficients are written in terms of unknown values of the dependent variable at the surrounding nodes. This polynomial is used to obtain an algebraic approximation for the partial differential equation at each interior node and an algebraic approximation for a boundary condition at each node that lies upon or beside the solution domain boundary. In this way one algebraic equation is obtained for each node, and the equations are solved simultaneously to obtain unknown values of the dependent variable at all nodes.

As an example, consider the typical five-node grid shown in Fig. 4.2. The nodes have a constant spacing of Δ in each coordinate direction, and the integers 0 through 4 are used as local node identification numbers. The dependent variable, h, will be represented in the neighbourhood of these nodes with the following polynomial, which is actually a Taylor series expansion about node 0:

$$h(x, y) = h_0 + \frac{\partial h_0}{\partial x} x + \frac{\partial h_0}{\partial y} y + \frac{1}{2!} \frac{\partial^2 h_0}{\partial x^2} x^2 + \frac{\partial^2 h_0}{\partial x \partial y} xy + \frac{1}{2!} \frac{\partial^2 h_0}{\partial y^2} y^2 + \frac{1}{3!} \frac{\partial^2 h_0}{\partial x^3} x^3 + \ldots \tag{4.1}$$

By setting (x, y) equal to the coordinates of nodes 1 through 4 we obtain the four equations

$$h_1 = h_0 + \frac{\partial h_0}{\partial x} \Delta + \frac{1}{2!} \frac{\partial^2 h_0}{\partial x^2} \Delta^2 + \frac{1}{3!} \frac{\partial^3 h_0}{\partial x^3} \Delta^3 + O(\Delta^4) \tag{4.2}$$

$$h_2 = h_0 + \frac{\partial h_0}{\partial y} \Delta + \frac{1}{2!} \frac{\partial^2 h_0}{\partial y^2} \Delta^2 + \frac{1}{3!} \frac{\partial^3 h_0}{\partial y^3} \Delta^3 + O(\Delta^4) \tag{4.3}$$

$$h_3 = h_0 - \frac{\partial h_0}{\partial x} \Delta + \frac{1}{2!} \frac{\partial^2 h_0}{\partial x^2} \Delta^2 - \frac{1}{3!} \frac{\partial^3 h_0}{\partial x^3} \Delta^3 + O(\Delta^4) \tag{4.4}$$

$$h_4 = h_0 - \frac{\partial h_0}{\partial y} \Delta + \frac{1}{2!} \frac{\partial^2 h_0}{\partial y^2} \Delta^2 - \frac{1}{3!} \frac{\partial^3 h_0}{\partial y^3} \Delta^3 + O(\Delta^4) \tag{4.5}$$

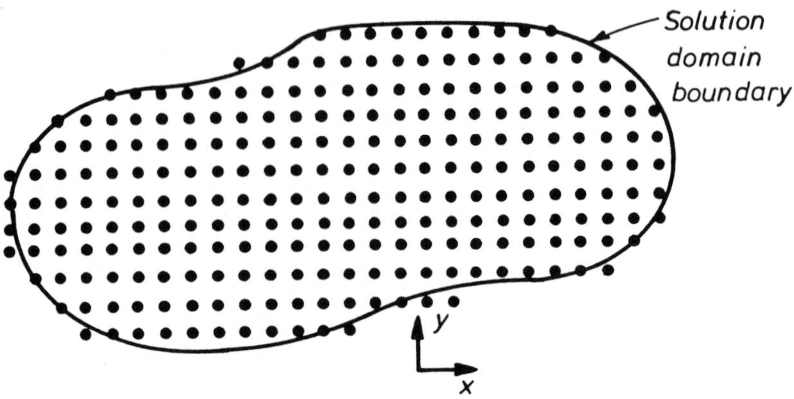

Figure 4.1 – A finite-difference grid superimposed upon a solution domain. Each node is given a different global identification number.

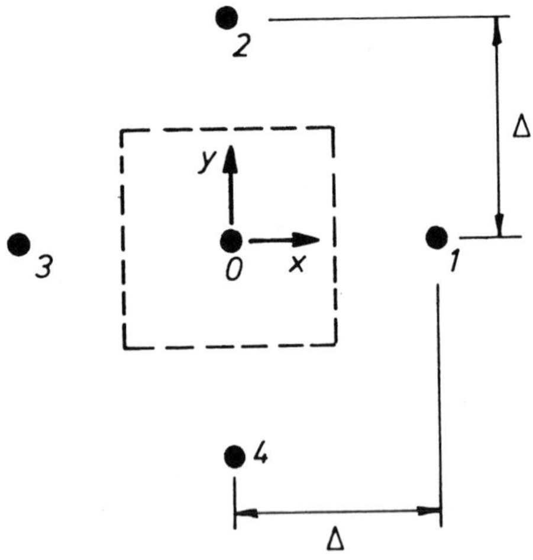

Figure 4.2 - A typical five-node grid with a node spacing of Δ. The integers 0 through 4 are local identification numbers for the nodes.

The order symbol, $0(\Delta^4)$, indicates that the next term in these expansions is a constant multiplied by Δ^4. Adding Eqs. 4.2 and 4.4 and Eqs. 4.3 and 4.5 gives two expressions that can be solved for the second derivatives at node 0:

$$\frac{\partial^2 h_0}{\partial x^2} = \frac{h_1 + h_3 - 2h_0}{\Delta^2} + 0(\Delta^2) \tag{4.6}$$

$$\frac{\partial^2 h_0}{\partial y^2} = \frac{h_2 + h_4 - 2h_0}{\Delta^2} + 0(\Delta^2) \tag{4.7}$$

In a similar way, subtracting Eqs. 4.2 and 4.4 and Eqs. 4.3 and 4.5 gives expressions that can be solved for the first derivatives at node 0:

$$\frac{\partial h_0}{\partial x} = \frac{h_1 - h_3}{2\Delta} + 0(\Delta^2) \tag{4.8}$$

$$\frac{\partial h_0}{\partial y} = \frac{h_2 - h_4}{2\Delta} + 0(\Delta^2) \tag{4.9}$$

Equations 4.6 - 4.9 give numerical approximations to the derivatives of h at the central node, 0, and, therefore, are known as second-order, central-difference approximations. Forward and backward-difference approximations for derivatives can be obtained by substituting Eqs. 4.6 - 4.9 into Eq. 4.1 and then calculating derivatives of Eq. 4.1 at nodes 3 and 1 or nodes 4 and 2, respectively.

One straightforward approach is to simply expand all derivatives that appear in the governing differential equation and replace these derivatives with the algebraic approximations given by Eqs. 4.6 - 4.9. However, a finite-difference approximation to a partial differential equation is often not unique, and , in groundwater flow problems, it is possible to obtain a set of equations with better solution properties by first integrating the partial differential equation throughout the control volume that is shown with a dashed line in Fig. 4.2. Thus, integration of the following equation for steady, two-dimensional flow

$$\overline{\nabla}.(T\overline{\nabla}h) = \left(\frac{K'}{B'}\right)(h - h') + \frac{Q_0}{A} - R \tag{4.10}$$

gives

$$\int_\Gamma T\frac{dh}{dn}\,ds = \int_A \left[\left(\frac{K'}{B'}\right)(h - h') + \frac{Q_0}{A} - R\right] dA \tag{4.11}$$

in which Γ is the control volume boundary, R is the rainfall recharge and Q_0 is the discharge from a well at node 0. An approximate calculation of the integrals in terms of variables at nodes 0 through 4 gives

$$\sum_{i=1}^{4} \frac{1}{2}(T_0 + T_i)\frac{h_i - h_0}{\Delta}\Delta = \left[\left(\frac{K'}{B'}\right)(h_0 - h_0') - R\right]\Delta^2 + Q_0 \tag{4.12}$$

Equation 4.12 can be put in a more convenient form by collecting terms which are coefficients of the unknown values of h. This gives

$$A_1h_1 + A_2h_2 + A_3h_3 + A_4h_4 - A_0h_0 = -B_0 \tag{4.13}$$

in which

$$A_i = \frac{1}{2}(T_0 + T_i) \text{ for } i = 1, 2, 3 \text{ and } 4 \tag{4.14a}$$

$$A_0 = A_1 + A_2 + A_3 + A_4 + \left(\frac{K'}{B'}\right)\Delta^2 \tag{4.14b}$$

$$B_0 = \left[\left(\frac{K'}{B'}\right) h_0' + R\right]\Delta^2 - Q_0 \tag{4.15}$$

In the finite difference calculations, Eq. 4.13 is written, with numerical values for A_i, A_0 and B_0, for each interior node of the finite-difference grid.

Along the boundaries, we will obtain an algebraic approximation to the boundary condition

$$\alpha \frac{dh}{dn} + \beta h = F \tag{4.16}$$

in which α, β and F may, in general, be discontinuous functions. For example, for the problem described by Eqs. 3.1 - 3.4, we would set $\alpha = 0$, $\beta = 1$ and $F = f$ along ABC and $\alpha = 1$, $\beta = 0$, $F = 0$ along CDA. Equation 4.16 will be approximated in the neighbourhood of the three-node grid shown in Fig. 4.3a by using the following first-degree polynomial:

$$h(x, y) = h_0 + \frac{h_0 - h_2}{\Delta} x + \frac{h_0 - h_1}{\Delta} y + 0(\Delta^2) \tag{4.17}$$

The coefficients in Eq. 4.17 have been obtained by evaluating the polynomial at nodes 0, 1 and 2, a result which may be checked by setting (x, y) equal to (0, 0), (0, $-\Delta$) and ($-\Delta$, 0), respectively. Let point P be the nearest point on the boundary to node 0, $\hat{e}_n = (N_x, N_y)$ be the outward normal to the boundary at point P and δ be the distance between points zero and P. The radial coordinate δ is positive or negative, depending upon whether node 0 lies within or outside the solution domain, respectively. Thus, Eq. 4.17 gives

$$\left(\frac{dh}{dn}\right)_P = (\overline{\nabla}h.\hat{e}_n)_P = \frac{h_0 - h_2}{\Delta} N_x + \frac{h_0 - h_1}{\Delta} N_y \tag{4.18}$$

$$(h)_P = h_0 + \frac{h_0 - h_2}{\Delta} N_x\delta + \frac{h_0 - h_1}{\Delta} N_y\delta \tag{4.19}$$

Finally, substituting Eqs. 4.18 - 4.19 into Eq. 4.16 gives the result

$$A_0h_0 - A_1h_1 - A_2h_2 = F_P \tag{4.20}$$

in which F_P denotes the value of F at point P and the coefficients, A_i, are given by

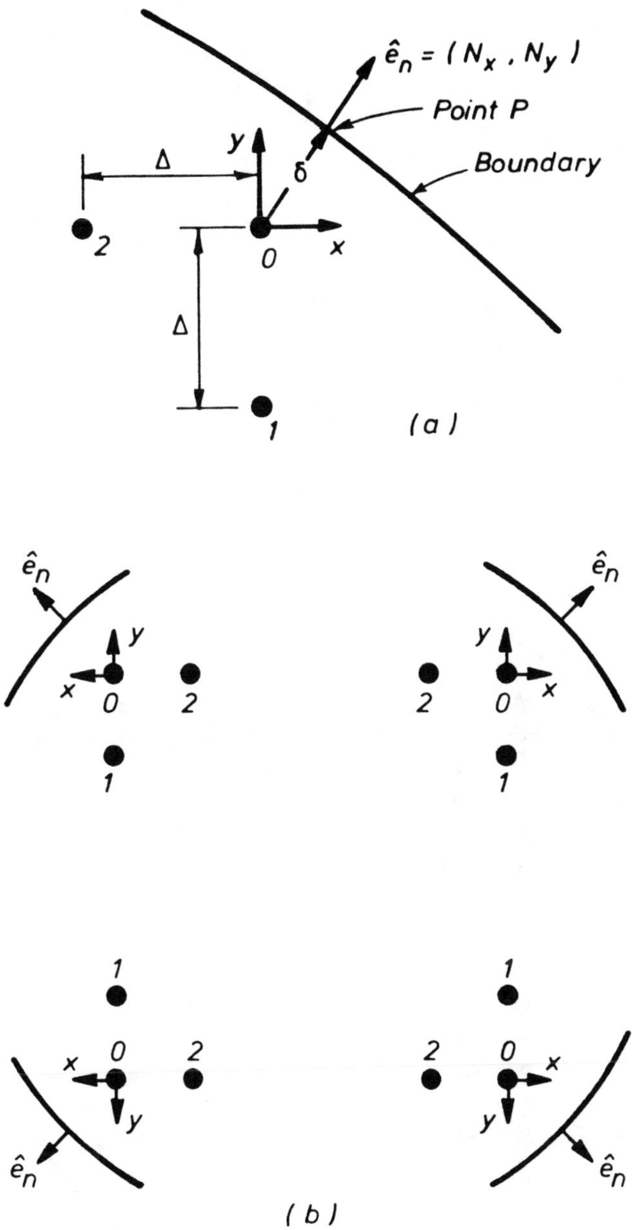

Figure 4.3 - A three-node grid along a boundary

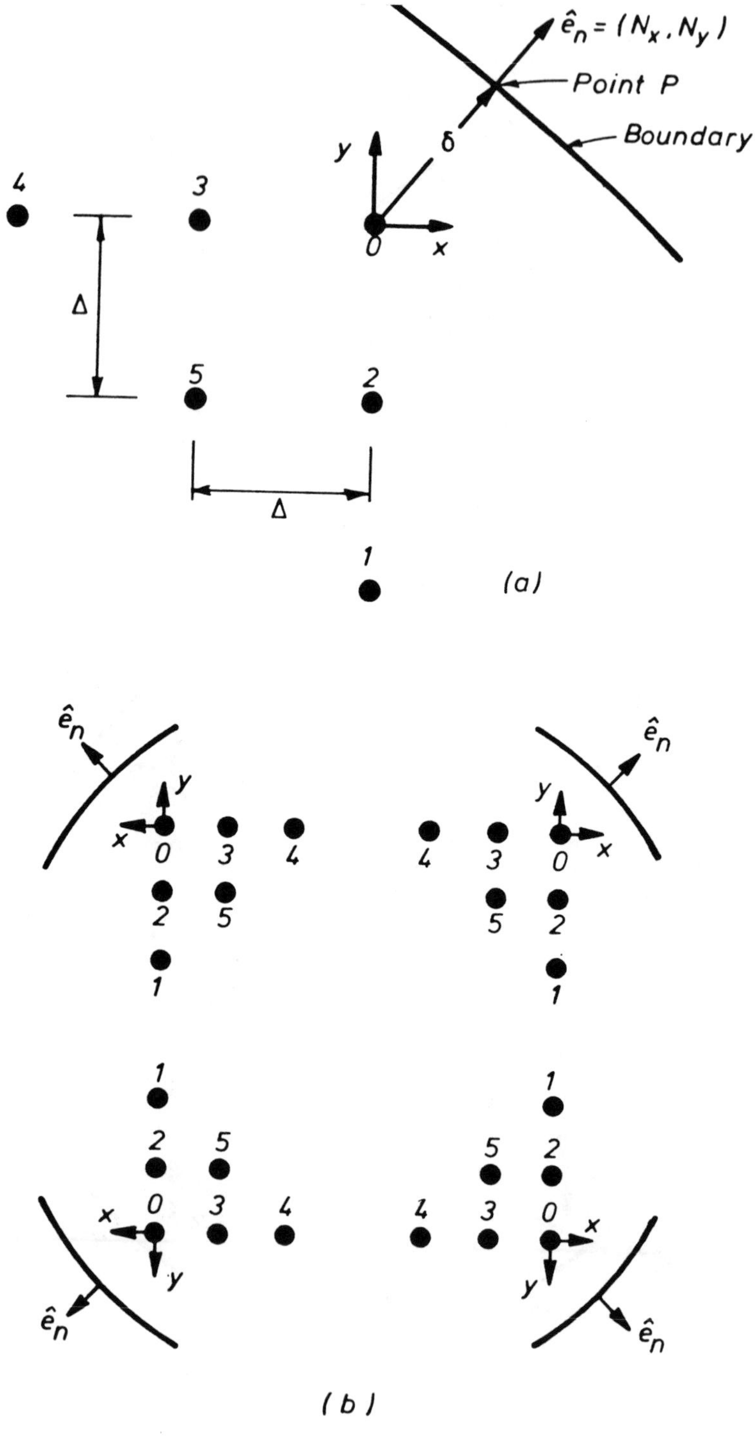

Figure 4.4 - A six-node grid along a boundary

$$A_1 = (\alpha + \beta\delta)\,\frac{N_y}{\Delta} \qquad (4.21)$$

$$A_2 = (\alpha + \beta\delta)\,\frac{N_x}{\Delta} \qquad (4.22)$$

$$A_0 = A_1 + A_2 + \beta \qquad (4.23)$$

Equation 4.20 is the algebraic approximation to Eq. 4.16 that is written for each node that is either on or adjacent to the solution domain boundary. The coordinate systems and local numbering schemes for other boundary orientations are shown in Fig. 4.3b.

Truncation errors in these finite-difference approximations can be obtained by using Taylor series expansions about node 0. In particular, the truncation error for Eq. 4.12 or 4.13 can be calculated by setting

$$T_1 = T_0 + \frac{\partial T_0}{\partial x}\,\Delta + 0(\Delta^2) \qquad (4.24)$$

$$T_2 = T_0 + \frac{\partial T_0}{\partial y}\,\Delta + 0(\Delta^2) \qquad (4.25)$$

$$T_3 = T_0 - \frac{\partial T_0}{\partial x}\,\Delta + 0(\Delta^2) \qquad (4.26)$$

$$T_4 = T_0 - \frac{\partial T_0}{\partial y}\,\Delta + 0(\Delta^2) \qquad (4.27)$$

$$h_1 = h_0 + \frac{\partial h_0}{\partial x}\,\Delta + \frac{1}{2!}\,\frac{\partial^2 h_0}{\partial x^2}\,\Delta^2 + \frac{1}{3!}\,\frac{\partial^3 h_0}{\partial x^3}\,\Delta^3 + 0(\Delta^4) \qquad (4.28)$$

$$h_2 = h_0 + \frac{\partial h_0}{\partial y}\,\Delta + \frac{1}{2!}\,\frac{\partial^2 h_0}{\partial y^2}\,\Delta^2 + \frac{1}{3!}\,\frac{\partial^3 h_0}{\partial y^3}\,\Delta^3 + 0(\Delta^4) \qquad (4.29)$$

$$h_3 = h_0 - \frac{\partial h_0}{\partial x}\,\Delta + \frac{1}{2!}\,\frac{\partial^2 h_0}{\partial x^2}\,\Delta^2 - \frac{1}{3!}\,\frac{\partial^3 h_0}{\partial x^2}\,\Delta^3 + 0(\Delta^4) \qquad (4.30)$$

$$h_4 = h_0 - \frac{\partial h_0}{\partial y}\,\Delta + \frac{1}{2!}\,\frac{\partial^2 h_0}{\partial y^2} - \frac{1}{3!}\,\frac{\partial^3 h_0}{\partial y^3}\,\Delta^3 + 0(\Delta^4) \qquad (4.31)$$

Substitution of Eqs. 4.24 - 4.31 into Eq. 4.12 and division by $\Delta^2 = A$ gives

$$[\bar{\nabla}.(T\bar{\nabla}h) - \left(\frac{K'}{B'}\right)(h - h') - \frac{Q_0}{A} + R]_0 + 0(\Delta^2) = 0 \qquad (4.32)$$

Thus, Eqs. 4.12 and 4.13 are second-order approximations to Eq. 4.10 at node 0.

A similar calculation for Eq. 4.20 can be used to show that it is a first-order approximation to Eq. 4.16. Hunt (1978) has obtained a second-order approximation to Eq. 4.16 by using the six nodes shown in

Fig. 4.4. The final result can be written in the form

$$A_0h_0 + A_1h_1 - A_2h_2 - A_3h_3 + A_4h_4 + A_5h_5 = F_p \qquad (4.33)$$

in which the coefficients, A_i, are given by

$$C = \frac{1}{2}\left(\beta\frac{\delta}{\Delta} + \frac{\alpha}{\Delta}\right) \qquad (4.34)$$

$$D = \frac{\delta}{\Delta}\left(\frac{\beta}{2}\frac{\delta}{\Delta} + \frac{\alpha}{\Delta}\right) \qquad (4.35)$$

$$E = N_x + N_y \qquad (4.36)$$

$$A_0 = \beta + E(3C + DE) \qquad (4.37)$$

$$A_1 = N_y\ (C + DN_y) \qquad (4.38)$$

$$A_2 = 2N_y\ (2C + DE) \qquad (4.39)$$

$$A_3 = 2N_x\ (2C + DE) \qquad (4.40)$$

$$A_4 = N_x\ (C + DN_x) \qquad (4.41)$$

$$A_5 = 2N_xN_yD \qquad (4.42)$$

Despite the fact that Eq. 4.33 has a smaller truncation error than Eq. 4.20, we will prefer Eq. 4.20 for reasons that will be given in the next section.

15. Solution of the Finite-Difference Equations

The system of simultaneous equations that is generated by writing Eq. 4.13 at each interior node and Eq. 4.20 at each boundary node can be solved by either of two methods: a direct elimination method, such as Gaussian elimination, in which the equations are manipulated algebraically into a form that permits the unknowns to be calculated directly, or an iterative method, such as the Gauss-Seidel iteration, in which each estimate for the unknowns is used to calculate a new, improved estimate and calculations are stopped when these estimates cease to change significantly from one cycle to the next. Direct methods can, under certain circumstances, be more efficient. However, some of the applications considered herein may require the simultaneous solution of one or two thousand equations with a sparse matrix and relatively large diagonal terms in the coefficient matrix. Under these conditions, the iterative methods are easier to code for a computer, require considerably less computer storage and use less computational time. Thus, only iterative methods will be considered herein.

As a simple numerical example, we will use the Gauss-Seidel iteration to calculate a solution to the following set of equations:

$$3h_1 \qquad + h_3 = 6$$
$$h_1 + 5h_2 \qquad = 10 \qquad (4.43\ a,b,c)$$
$$h_1 \qquad + 8h_3 = 16$$

The first step consists of solving Eqs. 4.43 for the unknowns that appear along the main diagonal of the coefficient matrix.

$$h_1 = 2 - \frac{1}{3} h_3$$
$$h_2 = 2 - \frac{1}{5} h_1 \qquad (4.44\ a,b,c)$$
$$h_3 = 2 - \frac{1}{8} h_1$$

The second step uses a guess in Eq. 4.44a for the unknown vector, say $(h_1, h_2, h_3) \simeq (0, 0, 0)$, to calculate an improved value for h_1. The result, $h_1 \simeq 2$, is then used in an improved guess for the unknown vector, $(h_1, h_2, h_3) \simeq (2, 0, 0)$, which is substituted into Eq. 4.44b to calculate $h_2 \simeq 2 - \frac{2}{5} = 1.6$ and $(h_1, h_2, h_3) \simeq (2, 1.6, 0)$. Finally, this last estimate for the unknown vector is put into Eq. 4.44c to calculate $h_3 \simeq 2 - \frac{2}{8} = 1.75$ and $(h_1, h_2, h_3) \simeq (2, 1.6, 1.75)$. This completes one cycle in the iteration, and the process is continued until the approximate solution vector ceases to change significantly from one cycle to the next. A total of 5 complete cycles for this particular example gives (1.393, 1.721, 1.826), which can be compared with the exact result $\left(\frac{32}{23}, \frac{198}{115}, \frac{42}{23}\right) \simeq (1.391, 1.722, 1.826)$.

The iterative process just described does not always terminate so quickly with such a happy result. In qualitative terms, the process works well with Eqs. 4.43 because the coefficients of terms off the main diagonal of the coefficient matrix have small magnitudes compared with the main diagonal coefficients. Thus, inserting different approximation vectors in the right side of Eqs. 4.44 creates only small perturbations in the vector of unknowns, and the process converges. On the other hand, if the main diagonal coefficients are relatively small, perturbations in the approximation vector grow into larger perturbations in the next approximation. In this case, continuing the iteration indefinitely leads to approximations for the unknowns that become unbounded.

The qualitative observations just mentioned can be made quantitative. Varga (1962) proves that the Gauss-Seidel iteration is convergent for the system of equations

$$\sum_{j=1}^{N} a_{ij} h_j = b_i \text{ for } i = 1, 2, 3....N \qquad (4.45)$$

if the set of equations is irreducibly diagonally dominant. A set of equations in N unknowns is irreducible if some of the unknowns cannot be found by solving fewer than N equations. A set of equations is irreducibly diagonally dominant if it is irreducible and if

$$|a_{ii}| \geqq \sum_{\substack{j=1 \\ i\neq j}}^{N} |a_{ij}| \quad \text{for } i = 1, 2, 3....N \tag{4.46}$$

provided that the inequality sign holds for at least one equation in the set.

Now it can be seen why Eq. 4.20, rather than Eq. 4.33, is the preferred approximation to the boundary condition given by Eq. 4.16. This is because the set of equations generated by Eq. 4.13 and 4.20 is irreducibly diagonally dominant. The only exception appears, at first glance, to occur if $\alpha = \delta = 0$ at one or more points along the boundary. In this case, the equations are still diagonally dominant, but the set of equations is reducible since one or more of the equations generated from Eq. 4.21 can be solved directly for h_0. This is not a limitation, however, since h_0 is really not an unknown at these boundary points, and the equations generated at these boundary points can be ignored when applying the test for irreducible diagonal dominance.

It should be emphasized, at this point, that the use of Eq. 4.20 rather than Eq. 4.33 results in a loss in accuracy for a gain in the solution characteristics of the approximating set of algebraic equations. This is particularly desirable when boundary locations or other field data are not known with sufficient accuracy to make a more accurate solution worthwhile. On the other hand, the more accurate approximation, Eq. 4.33, may well be worth using for some problems in which more accurate solutions are needed. The theorem on irreducibly diagonally dominant matrices does not state that an iterative solution of the more accurate equations will fail to converge, and, in fact, Hunt (1978) gives several numerical examples of convergence when solving Eq. 4.33 with the Gauss-Seidel iteration.

16. Computer Coding

Probably the biggest difficulty in computer coding is to relate the global and local numbering schemes in a way which has enough flexibility to solve problems in solution domains with curved bounaries. This problem is overcome herein with the following scheme: first, each point that lies on or next to the boundary is assigned a global number between 1 and NB, and each interior point is assigned a global number between NB + 1 and N; second, each of these points is assigned a series of either four or two integers that give the global numbers of surrounding nodes. For a node with a global number I, these integers will be called

ID(I, J), where J goes from 1 through 4 when I is an interior node and from 1 through 2 when node I lies upon or beside the boundary. These integers are the global numbers of the nodes that correspond with the local numbering scheme shown in Figs. 4.2 and 4.3. The following examples for the grid shown in Fig. 4.5 should help to make this clear: for the interior node 154, ID(154, J) = 155, 43, 153 and 198 for J = 1, 2, 3 and 4, respectively; for the boundary node 41, ID(41, J) = 200 and 155 for J = 1 and 2, respectively. It should also be noted that a more precise definition of a boundary node is any node near a boundary for which it is impossible to form a five-node grid like the one shown in Fig. 4.2.

The use of the numbers ID(I, J) makes it relatively simple to introduce a graded mesh in regions where h varies rapidly. Figure 4.6 shows an example of a graded mesh, in which the letters a through j indicate nodes where it is necessary to rotate the coordinate system by 45 degrees in order to obtain a five-node mesh like the one shown in Fig. 4.2. Since the mesh spacing, Δ, will usually appear in the finite-difference approximation to the differential equation, Eq. 4.13, it is necessary to specify the values of ID(I, J) and the value of Δ for each node, but these are the only variables which will change in Eq. 4.13 with a rotation of the coordinate system.

Computer flow-charts for the numerical solution of Eq. 4.10 are shown in Figs. 4.7 and 4.8. Statements in these flow-charts are written in Fortran, and the symbols that appear in the flow-charts are defined as follows:

NB	= number of boundary points;
N	= number of nodes in the entire grid;
ERR	= error criteria;
ID(I, J)	= global grid numbers corresponding to the local numbering scheme;
ALPHA	= α in Eq. 4.20;
BETA	= β " " " ;
F(I)	= F_p " " " ;
XN	= N_x " " " ;
YN	= N_y " " " ;
DP	= δ " " " ;
D(I)	= Δ " " " or Eq. 4.13;
T(I)	= T " Eq. 4.13;
BK(I)	= K'/B' in Eq. 4.13;
HP(I)	= h_0' " " " ;
R(I)	= R " " " ;
Q(I)	= Q_0 " " " ; and
H(I)	= h.

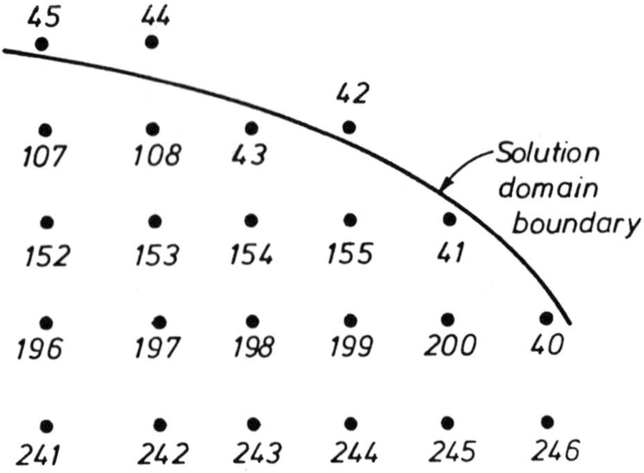

Figure 4.5 - A typical node numbering scheme.

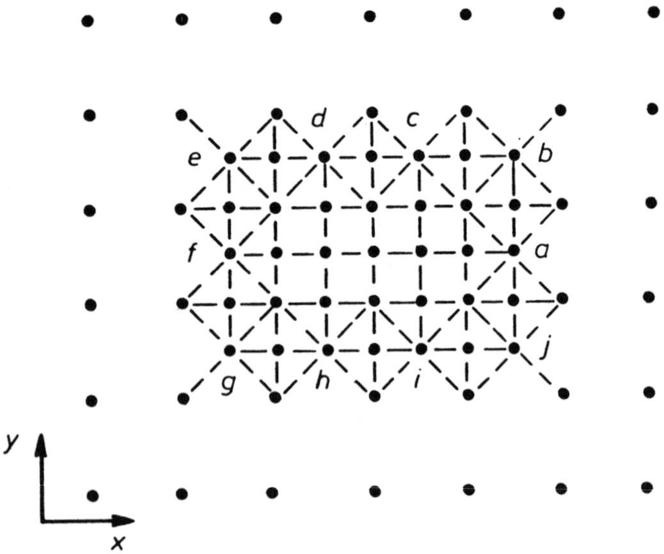

Figure 4.6 - An example of a graded mesh.

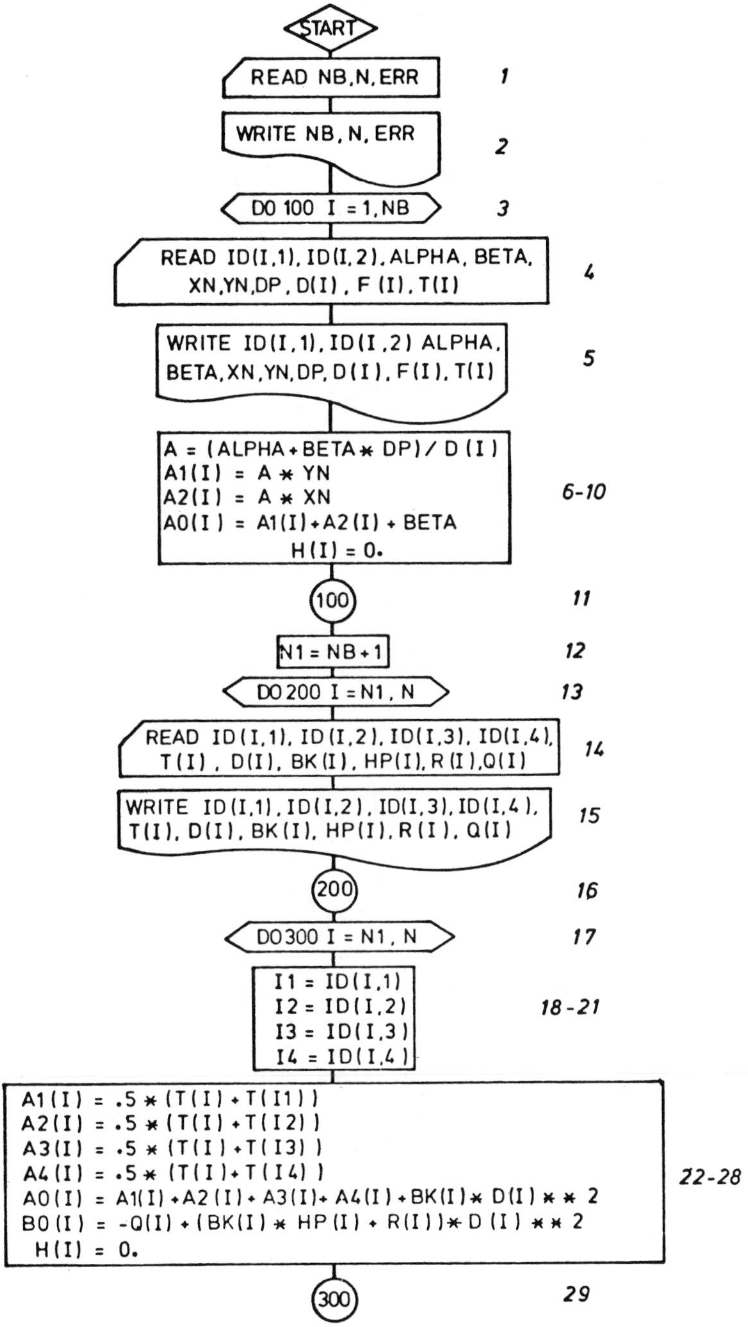

Figure 4.7 - A flowchart for the first half of a computer program for steady flow.

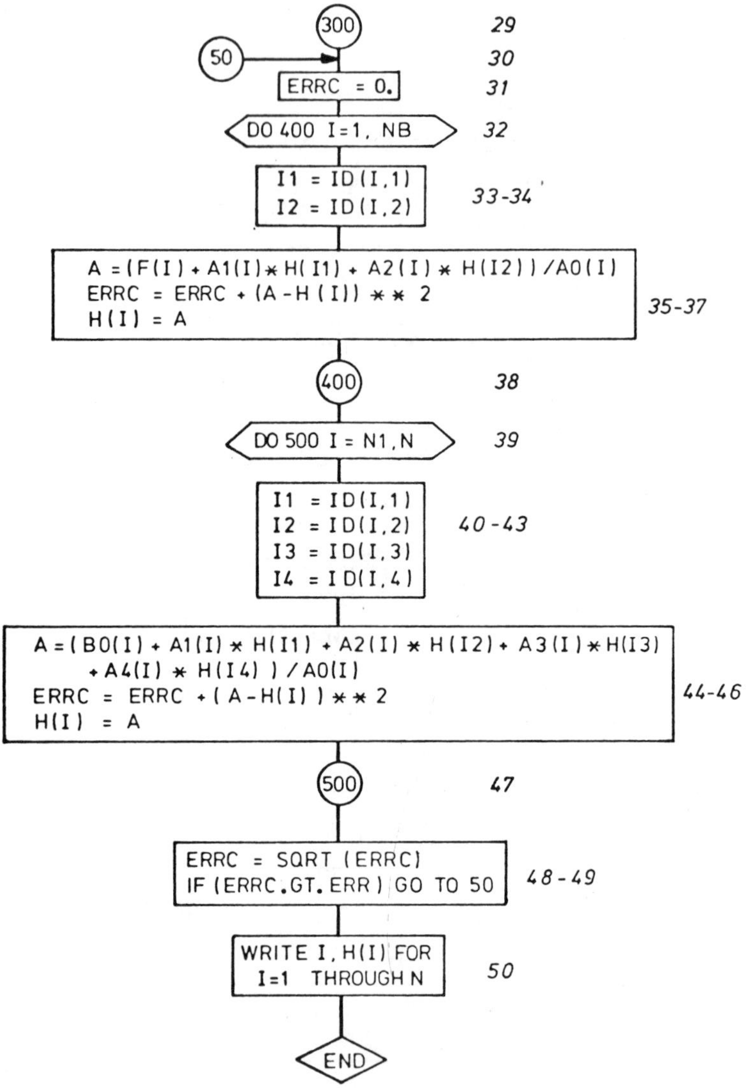

Figure 4.8 - A flowchart for the second half of a computer program for steady flow.

The first half of the flow-chart, shown in Fig. 4.7, reads in data, prints out the same data, computes the coefficients in Eqs. 4.13 and 4.20 at each node and sets starting values of h = 0 at each node. A detailed discussion of this flow-chart follows:

Statements 1-2: The number of boundary nodes, NB, the total number of nodes in the entire grid, N, and an error criteria, ERR, are read in and printed out. The error criteria, ERR, is a small number, say 10^{-4} or 10^{-5}, which determines when the Gauss-Seidel iteration will stop. The safest procedure for choosing ERR is to guess a suitable value and calculate a solution. Then decrease ERR by an order of magnitude (a factor of 10) and see if this creates a significant change in the solution for h.

Statements 4-5 and 14-15: These statements read in and print out data that is needed for each boundary and interior node, respectively. The information for each node is printed as soon as it is read so that an error in the data can be located easily. Data preparation is the most time consuming chore in any of these finite-difference solutions.

Statements 6-10 and 22-28: These statements calculate the coefficients in Eqs. 4.20 and 4.13 for each node and then set starting values of h = 0 at each node.

Statements 18-21: The relationship between the global and local numbering schemes is established.

The second half of the flow-chart, shown in Fig. 4.8, uses the Gauss-Seidel iteration to solve Eqs. 4.13 and 4.20. A detailed discussion of this flow-chart follows:

Statement 31: An estimate of the calculated error, ERRC, is set equal to zero.

Statements 33-34 and 40-43: The relationship between the global and local numbering schemes is established.

Statements 35-37 and 44-46: The terms along the main diagonal of the coefficient matrix are the coefficients of h_0 in Eqs. 4.13 and 4.20. Thus, A in statements 35 and 44 is the newest estimate for h_0 in Eqs. 4.20 and 4.13, respectively. Statements 36 and 45 square the difference between this new estimate for h, which is stored in A, and the previous estimate for h, which is stored in H(I), and add this to the sum ERRC. Statements 37 and 46 replace the old estimate for h with the newest estimate.

Statements 48-49: The square root of ERRC is compared with ERR. If ERRC is greater than ERR, the machine goes back to statement 30 and repeats the calculation. If ERRC is less than or equal to ERR, statement 49 is ignored and the machine is instructed to print out the solution in the next statement, statement 50.

17. Computer Program Modifications:

The computer program described in Figs. 4.7 and 4.8 is relatively simple but general enough to solve most steady-flow problems in groundwater. The reader who has experience in using a computer to solve boundary-value problems, however, may wish to make certain modifications of his own. For example, some people may be bothered by the possibility that statements 30 and 49 might permit a machine to become locked in an infinite number of cycles without ever satisfying the error criteria. This is unlikely, because the Gauss-Seidel iteration for these equations will always converge, but the possibility can be avoided by limiting the number of cycles in the iteration. Other people may feel that a different type of error calculation in statements 31, 36, 45 and 48 is more appropriate. Probably the most worthwhile modification, though, is to use successive over-relaxation rather than the Gauss-Seidel iteration to solve the finite-difference equations.

The reader who is interested in the theory of successive over-relaxation (SOR) methods should consult Smith (1978). The use of SOR can, typically, reduce the number of cycles required for an iterative solution by a factor of two or three, and the required modifications to the flow-chart in Fig. 4.8 are almost trivial. Statements 35 and 44, respectively, are replaced with

```
A = H(I) + W*(F(I) + A1(I)*H(I1) + A2(I)*H(I2) - A0(I)*H(I))/A0(I)   (4.47)

A = H(I) + W*(B0(I) + A1(I)*H(I1) + A2(I)*H(I2) + A3(I)*H(I3)
    + A4(I)*H(I4) - A0(I)*H(I))/A0(I)                                (4.48)
```

The relaxation factor, W, in Eqs. 4.47 - 4.48 has a numerical value between 1 and 2. In fact, setting W = 1 gives the Gauss-Seidel iteration, but an optimum value for W depends upon the distribution and magnitude of terms in the coefficient matrix of the finite-difference equations. Thus, it becomes worthwhile to compute an optimum value for W if a number of solutions are going to be computed for the same aquifer geometry and transmissivity distribution. The optimum value of W can be found by plotting the number of required iteration cycles as a function of W. Smith (1965) also suggests that an approximate value for W can be estimated from

$$W_{optimum} = \frac{2}{1 + \sqrt{1 - \theta}} \qquad (4.49)$$

in which θ is calculated with the Gauss-Seidel iteration from

$$\theta = \underset{n\to\infty}{\text{Limit}} \frac{(\text{ERRC})_{\text{cycle}(n)}}{(\text{ERRC})_{\text{cycle}(n-1)}} \tag{4.50}$$

Equations 4.49 - 4.50 assume that the coefficient matrix has some very special properties that may not always be satisfied in practice. Nevertheless, almost any guess in the range $1 < W < 2$ will usually improve convergence rates, and Eqs. 4.49 - 4.50 frequently give a useful approximation even for coefficient matrices without the required properties.

18. Unsteady Flow Solutions

Unsteady flow is modelled by solutions to the equation

$$\overline{\nabla}.(T\overline{\nabla}h) - \left(\frac{K'}{B'}\right)(h - h') - \frac{Q}{A} + R = S\frac{\partial h}{\partial t} \tag{4.51}$$

The spatial discretization of Eq. 4.51 can be carried out by integrating both sides throughout the area shown with a dashed line in Fig. 4.2 to obtain

$$P\{h_i(t)\} = S\Delta^2 \frac{dh_0(t)}{dt} \tag{4.52}$$

in which the operator, P, is defined as

$$P\{h_i(t)\} = A_1h_1 + A_2h_2 + A_3h_3 + A_4h_4 - A_0h_0 + B_0 \tag{4.53}$$

The coefficients A_i and B_0 are given by Eqs. 4.14 - 4.15. Equation 4.52 can be written for each internal node and Eq. 4.20 for each boundary node to obtain a system of N ordinary differential equations to be solved for values of h_i, as functions of t, at each node.

The usual method of solving the system of ordinary differential equations for h_i is to use finite-difference approximations for the time derivatives. Three commonly used approximations are

$$P\{h_i(t)\} = S\Delta^2 \frac{h_0(t + \Delta t) - h_0(t)}{\Delta t} \tag{4.54}$$

$$P\{h_i(t)\} = S\Delta^2 \frac{h_0(t) - h_0(t - \Delta t)}{\Delta t} \tag{4.55}$$

$$\tfrac{1}{2}\,[P\{h_i(t)\} + P\{h_i(t - \Delta t)\}] = S\Delta^2 \frac{h_0(t) - h_0(t - \Delta t)}{\Delta t} \tag{4.56}$$

Equation 4.54 uses a first-order, forward-difference approximation for the time derivative that allows $h_0(t + \Delta t)$ to be calculated directly from values of $h_i(t)$ without having to solve a set of simultaneous equations. Because of this, Eq. 4.54 is known as an explicit approximation. Equation 4.55 uses a first-order, backward-difference approximation for the time derivative that results in a set of N equations that must be solved simultaneously at the end of every time step. Thus, Eq. 4.55 is known as an implicit approximation. Equation 4.56 is also an implicit approximation, known as the Crank-Nicolson approximation, which is obtained by approximating Eq. 4.51 halfway between the two times, t and $t - \Delta t$. This means that the time derivative is a second-order, central-difference approximation.

At first glance, Eq. 4.54 appears to be the simplest of the three approximations. Unfortunately, however, the growth of numerical instabilities often makes the use of Eq. 4.54 a relatively inefficient process. A simple example of an instability is shown in Fig. 4.9 for a one-dimensional aquifer that is bounded at each end by a reservoir. The reservoir level at one end is maintained at a constant elevation while the water level in the other reservoir is raised instantaneously to a positon that is held throughout the calculations. At $t = t_1 > 0$ a small numerical disturbance is introduced into the solution, perhaps as the result of a round-off error. At values of t greater than t_1, this small disturbance is propagated into many disturbances with amplitudes that grow with time. The correct position of the free surface is shown with a solid line in Fig. 4.9, and the unstable, numerical solution is shown with dots connected by a dashed line.

A theoretical stability analysis can be carried out by assuming that T is a constant. Then, since Eq. 4.51 is linear, the difference between the correct and incorrect solutions (i.e. - the solution error) will satisfy the homogeneous form of Eqs. 4.54 - 4.56, with $h' = Q = R = 0$. Under these circumstances, the operator P reduces to

$$P\{h_i\} = T(h_1 + h_2 + h_3 + h_4) - (4T + \frac{K'}{B'} \Delta^2)\ h_0 \qquad (4.57)$$

Thus, if the solution error is represented with

$$h = \phi(t)\ \exp(i\omega_1 x + i\omega_2 y),\ (i = \sqrt{-1}) \qquad (4.58)$$

then taking the coordinate origin at node 0 gives

$$P\{h_i(t)\} = \phi(t)\ [2T(\cos\ \omega_1\Delta + \cos\ \omega_2\Delta) - (4T + \frac{K'}{B'}\Delta^2)] \qquad (4.59)$$

The amplitude of the solution error is represented by $\phi(t)$ in Eqs. 4.58 - 4.59, and ω_1 and ω_2 are the disturbance frequencies. Since the equations are linear, more general numerical disturbances (errors) can be represented by superimposing functions of the form given by Eq. 4.58. In a rectangular solution domain, the result will be a double Fourier series, and in an infinite solution domain, the result will be a double Fourier integral.

The stability analysis of Eq. 4.54 is carried out by substituting Eq. 4.59 into the left side of Eq. 4.54 and Eq. 4.58 into the right side to obtain

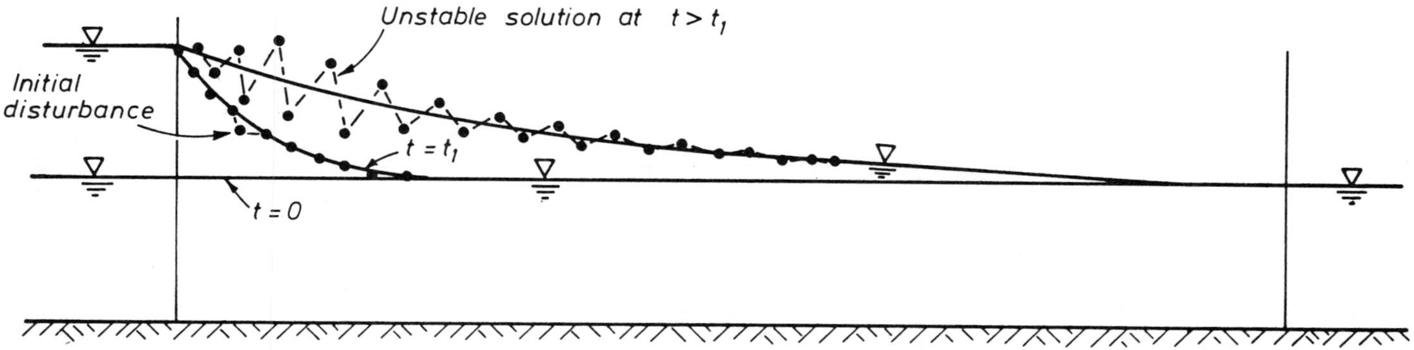

Figure 4.9 - The propagation of an initial disturbance into numerical instability.

$$\phi(t)\ [2T(\cos \omega_1\Delta + \cos \omega_2\Delta) - (4T + \frac{K'}{B'}\Delta^2)]$$

$$= S\Delta^2 \frac{\phi(t + \Delta t) - \phi(t)}{\Delta t} \qquad (4.60)$$

Thus, manipulation of Eq. 4.60 gives

$$\frac{\phi(t + \Delta t)}{\phi(t)} = \frac{2T(\cos \omega_1\Delta + \cos \omega_2\Delta) - (4T + \frac{K'}{B'}\Delta^2) + \frac{S\Delta^2}{\Delta t}}{\frac{S\Delta^2}{\Delta t}} \qquad (4.61)$$

Since a stable solution requires $|\phi(t + \Delta t)/\phi(t)| \leqq 1$, Eq. 4.61 can be substituted into the inequality

$$\left[\frac{\phi(t + \Delta t)}{\phi(t)}\right]^2 \leqq 1 \qquad (4.62)$$

to find what, if any, conditions must be satisfied for stability. The requirement, Eq. 4.62, that the disturbance amplitude not grow with time leads to the inequality

$$[2T(\cos \omega_1\Delta + \cos \omega_2\Delta) - (4T + \frac{K'}{B'}\Delta^2)]\,[2T(\cos \omega_1\Delta + \cos \omega_2\Delta)$$

$$- (4T + \frac{K'}{B'}\Delta^2) + 2\frac{S\Delta^2}{\Delta t}] \leqq 0 \qquad (4.63)$$

The factor enclosed in the first set of brackets is always less than, or equal, to zero. Thus, the factor enclosed in the second set of brackets must always be greater than, or equal, to zero.

$$2T(\cos \omega_1\Delta + \cos \omega_2\Delta) - (4T + \frac{K'}{B'}\Delta^2) + 2\frac{S\Delta^2}{\Delta t} \geqq 0 \qquad (4.64)$$

The worst possible case occurs when $\cos \omega_1\Delta = \cos \omega_2\Delta = -1$, and this result substituted into Eq. 4.64 leads to the final result

$$\frac{S\Delta^2}{T\Delta t} \geqq 4 + \frac{1}{2}\left(\frac{K'}{B'}\right)\frac{\Delta^2}{T} \qquad (4.65)$$

The stability requirement in Eq. 4.65 has the effect of limiting the time step. For a numerical example, suppose that an unconfined aquifer analysis has $S = .1$, $T = .1\ m^2/s$, $\Delta = 1$ km and $K'/B' = 0$. Then the time step must satisfy the requirement

$$\Delta t \leqq 2.9 \text{ days} \qquad (4.66)$$

On the other, if the aquifer is a confined aquifer with $S = 10^{-4}$,

$$\Delta t \leqq .0029 \text{ days} \qquad (4.67)$$

Thus, if the aquifer is unconfined and if calculations are to be carried out for a total time period of hours, days or even months, Eq. 4.66 is not overly restrictive. On the other hand, if calculations are to be carried out over a much longer time period, or if the aquifer is confined, then satisfying Eq. 4.67 at each time step could make the calculations prohibitively expensive. In this latter case, it would be better to use either Eq. 4.55 or Eq. 4.56.

A similar calculation for Eq. 4.55 gives the equation

$$\phi(t)\ [2T(\cos \omega_1\Delta + \cos \omega_2\Delta) - (4T + \frac{K'}{B'}\Delta^2)]$$
$$= S\Delta^2 \frac{\phi(t) - \phi(t - \Delta t)}{\Delta t} \qquad (4.68)$$

This gives the following expression for the ratio of disturbance amplitudes:

$$\frac{\phi(t)}{\phi(t - \Delta t)} = \frac{\frac{S\Delta^2}{\Delta t}}{\frac{S\Delta^2}{\Delta t} - 2T(\cos \omega_1\Delta + \cos \omega_2\Delta) + (4T + \frac{K'}{B'}\Delta^2)} \qquad (4.69)$$

The variable t can be replaced with t + Δt in Eq. 4.69 and the result substituted into Eq. 4.62 to give

$$0 \leqq [(4T + \frac{K'}{B'}\Delta^2) - 2T(\cos \omega_1\Delta + \cos \omega_2\Delta)][(4T + \frac{K'}{B'}\Delta^2)$$
$$- 2T(\cos \omega_1\Delta + \cos \omega_2\Delta) + 2\frac{S\Delta^2}{\Delta t}] \qquad (4.70)$$

But each of the two factors in brackets on the right side of Eq. 4.70 is always greater than, or equal, to zero. Thus, Eq. 4.70 is always satisfied, and Eq. 4.55 is unconditionally stable when T is constant.

Substituting Eqs. 4.58 and 4.59 into Eq. 4.56 gives

$$\frac{1}{2}[\phi(t) + \phi(t - \Delta t)]\ [2T(\cos \omega_1\Delta + \cos \omega_2\Delta) - (4T + \frac{K'}{B'}\Delta^2)]$$
$$= S\Delta^2 \frac{\phi(t) - \phi(t - \Delta t)}{\Delta t} \qquad (4.71)$$

The solution of Eq. 4.71 for the ratio of disturbance amplitudes at times t and t - Δt gives

$$\frac{\phi(t)}{\phi(t - \Delta t)} = \frac{2\left(\frac{S\Delta^2}{\Delta t}\right) + 2T(\cos \omega_1\Delta + \cos \omega_2\Delta) - (4T + \frac{K'}{B'}\Delta^2)}{2\left(\frac{S\Delta^2}{\Delta t}\right) - 2T(\cos \omega_1\Delta + \cos \omega_2\Delta) + (4T + \frac{K'}{B'}\Delta^2)} \qquad (4.72)$$

Finally, setting the square of the right side of Eq. 4.72 less than, or equal, to unity leads to the requirement

$$2T(\cos \omega_1\Delta + \cos \omega_2\Delta) \leqq (4T + \frac{K'}{B'}\Delta^2) \qquad (4.73)$$

But Eq. 4.73 is always true, so that Eq. 4.56 is also unconditionally stable for constant values of T.

The main results of this section can be summarized by listing the strong and weak points of Eqs. 4.54 - 4.56. Truncation errors in time are smallest for Eq. 4.56 and of larger, but equal, magnitude for Eqs. 4.54 and 4.55. Thus, solutions of comparable accuracy require a larger number of shorter time steps when using either Eq. 4.54 or 4.55 than when using Eq. 4.56. Equation 4.54 is the simplest of the three approximations to program on a computer, and Eq. 4.56 is the most complicated. Stability requirements, however, can make a numerical solution of Eq. 4.54 relatively inefficient, while Eqs. 4.55 and 4.56 are unconditionally stable (for constant values of T). Since space limitations require that we use only one of these three approximations, we will choose the backward difference approximation, Eq. 4.55, as being a reasonable compromise between the conflicting requirements of accuracy, simplicity and stability.

19. Computer Coding

The computer coding of Eq. 4.55 is, in many respects, very similar to the coding for the steady flow problem. It does, however, require the use of the following, additional variables:

S(I) = storage coefficient at node I;

HLAST(I) = value of H(I) from the last time step;

TIME(K) = value of t at the end of each time step, with TIME(1) = 0;

NT = number of values of TIME(K); and

DT = time step = TIME(K) - TIME(K - 1).

The coding sets the initial values of h, F_p, h', R and Q equal to zero, and changes or perturbations from these zero values during any time step must be inserted into the program as the calculations proceed. Thus, the governing equation is assumed to be linear, and the program calculates changes in h that are caused by changes in F_p, h', R or Q.

A flow-chart for an unsteady flow program is shown in Figs. 4.10 and 4.11. The flow-chart shown in Fig. 4.10 reads in data, prints out the same data, computes the coefficients in Eqs. 4.20 and 4.55 and sets values of h, F_p, h', R and Q equal to zero. It differs from the flow-chart in Fig. 4.7 for steady flow in the following ways:

1. The integer NT, which equals the number of time steps plus one, is read in statement 1.
2. Values of F(I), HP(I), R(I) and Q(I) are not read in statements 5 and 16 but, instead, are set equal to zero in statements 12 and 20-22.
3. A storage coefficient, S(I), is read in for each node in statement 17.
4. Values of h at t = 0, HLAST(I), are set equal to zero in statements 11 and 19.
5. Calculation of the coefficient B0(I) = B_0 in Eqs. 4.53 and 4.55 is delayed until statement 41 in the second half of the progam (Fig. 4.11) since it must be recomputed at the beginning of every time step.

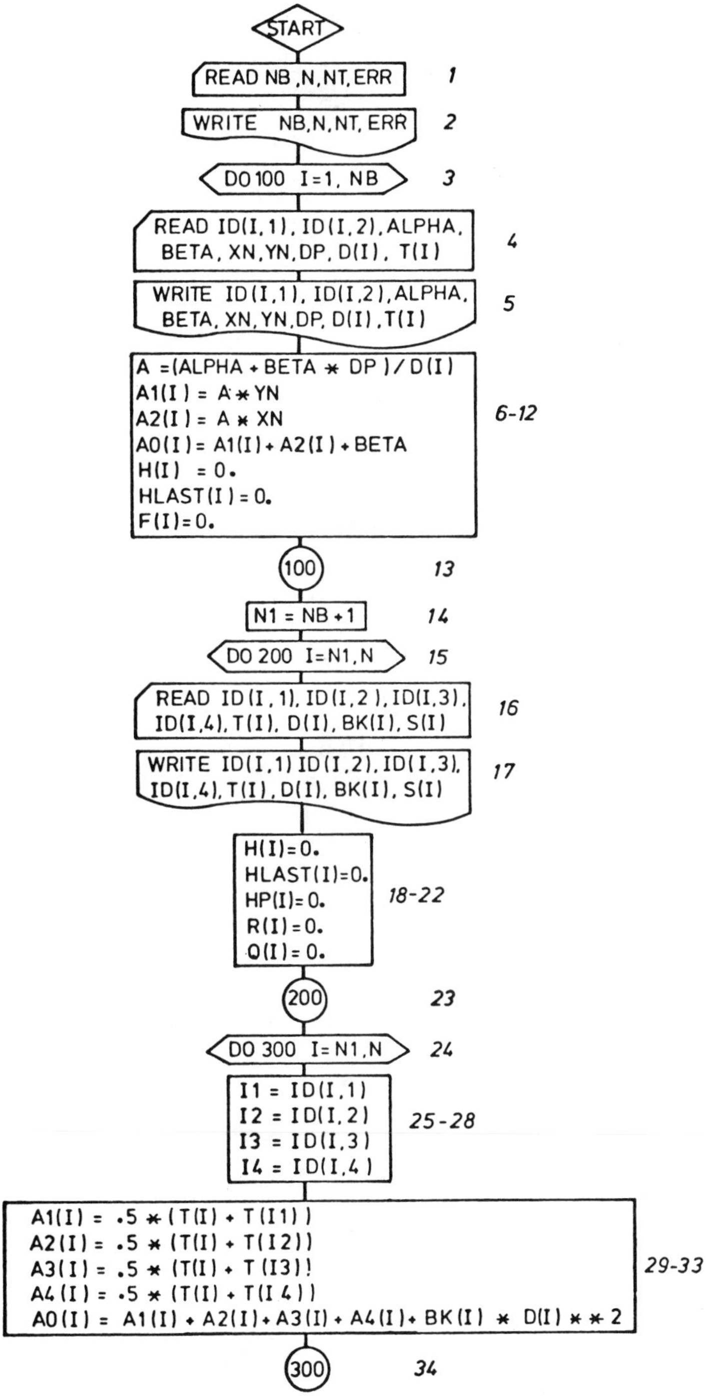

Figure 4.10 - A flowchart for the first half of a computer program for unsteady flow.

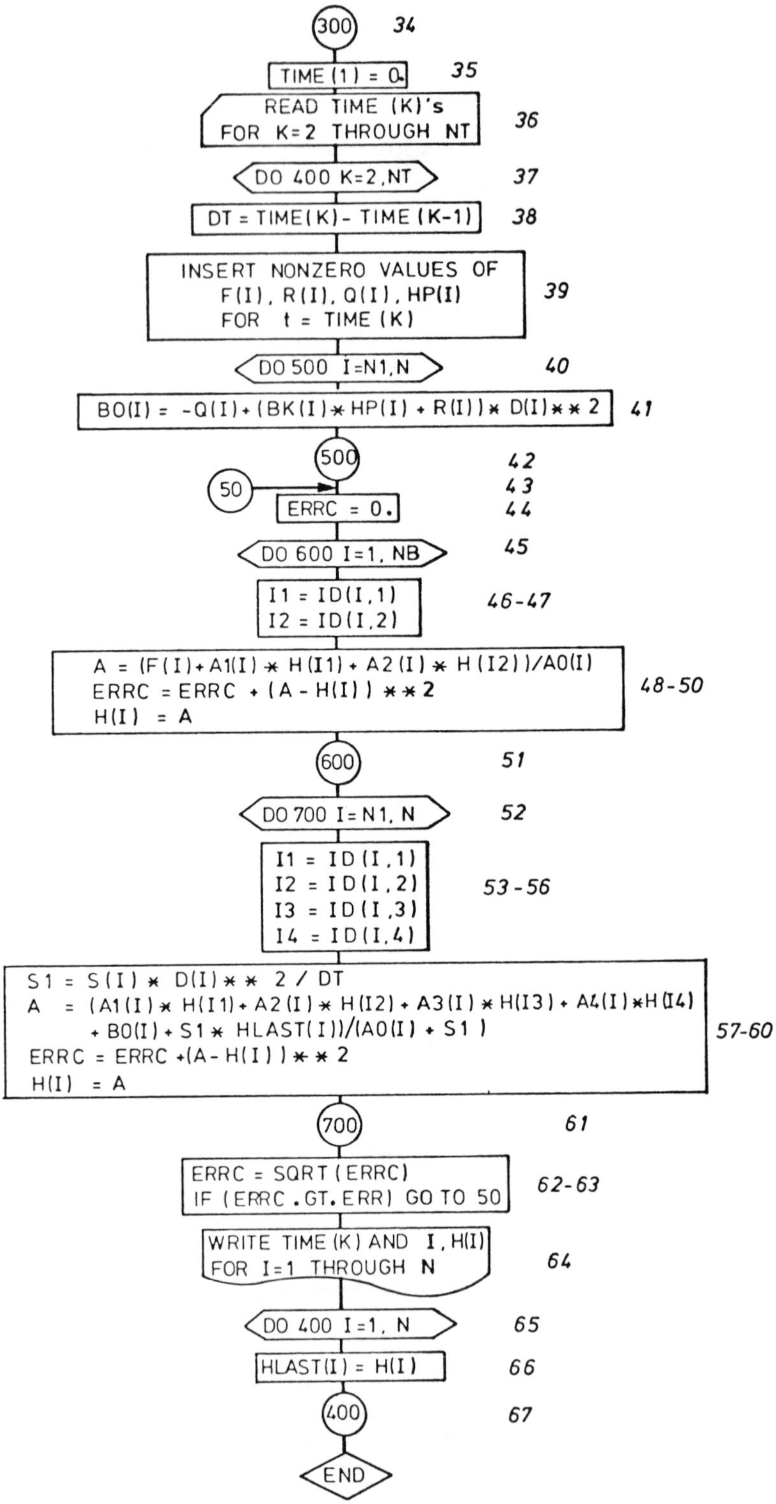

Figure 4.11 - A flowchart for the second half of a computer program for unsteady flow.

The second half of the flow-chart, shown in Fig. 4.11, uses the Gauss-Seidel iteration to solve Eqs. 4.20 and 4.55 at the end of every time step. A more detailed discussion of this flow-chart follows:

Statements 35-36: Values of t = TIME(K) at the end of every time step are read in, with TIME(1) = 0. The time step, DT = TIME(K) - TIME(K - 1), need not be constant for all values of K.

Statement 39: Nonzero values of F_p, R, Q and h' are inserted at the beginning of each time step.

Statement 41: Values of R, Q and h' inserted in statement 39 are used to compute B_0 in Eqs. 4.53 and 4.55 at the beginning of each time step.

Statments 43-63: The Gauss-Seidel iteration is used to compute the solution of Eqs. 4.20 and 4.55 at the end of every time step.

Statement 66: Values of h computed at the end of a time step are placed in HLAST(I) for use as initial values during the next time step.

The flow-charts can be modified to use the method of successive over-relaxation, SOR, merely by replacing statements 48 and 58, respectively, with Eq. 4.46 and

```
A = H(I) + W*(A1(I)*H(I1) + A2(I)*H(I2) + A3(I)*H(I3)
  + A4(I)*H(I4) + BO(I) + S1*HLAST(I) - (AO(I) + S1)*
   H(I))/(AO(I) + S1)                                    (4.74)
```

Since the optimum value for the relaxation factor W, depends only upon the coefficient matrix of the finite-difference equations, and since the terms in this coefficient matrix will change as Δt changes from one cycle to the next, there is no single optimum value for W that will hold for all time steps in any given calculation or from one calculation to the next for a particular aquifer. The author has never used SOR methods with unsteady flow calculations and, therefore, is unable to comment upon the sensitivity of W to changes in Δt.

Finally, it should be pointed out that a variable time step and superposition can sometimes be used to reduce the expense of calculations. For example, assume that changes in h are to be calculated from the linearized equation

$$\overline{\nabla}.(T\overline{\nabla}h) = S\frac{\partial h}{\partial t} + \left(\frac{K'}{B'}\right) h + Q_0\delta(x - x_0)\ \delta(y - y_0) \qquad (4.75)$$

Once values of h have been computed for a given Q_0, then multiplying both sides by the ratio Q_1/Q_0 shows that values of h when Q_0 is replaced with Q_1 can be obtained from the first solution merely by multiplying the first solution for h by the ratio Q_1/Q_0. Thus, a single calculation for any value of Q is sufficient to calculate the particular value of Q that will create a specified drawdown at a specified location and time. As a second example, a single solution of Eq. 4.75 for the pumping schedule shown in Fig. 4.12a can be used to obtain a solution for the more complex pumping schedule in Fig. 4.12b simply by displacing the first solution for h by an amount $t = \tau$ and subtracting these values of h from the first solution for h. The solution for Fig. 4.12b can then be superimposed to model more complicated pumping schedules, like the one shown in Fig. 4.12c. The savings in computer time can be considerable in this instance, because time steps for the first problem, shown in Fig. 4.12a, would need to be small only near the first discontinuity in Q, at t = 0, whereas Δt would have to be small in the neighbourhood of each of the discontinuities in Q that are shown in Figs. 4.12b and c if superposition is not used.

20. Other Numerical Techniques

Two other numerical techniques, the finite element method and the integral equation method, are sometimes used to obtain approximate solutions to groundwater problems. Unsteady flow solutions of Eq. 4.51 are usually obtained with the finite element method by replacing the time derivative with a first-order, backward-difference approximation. Then the dependent variable, h, is approximated in local sub-regions with a finite-degree polynomial, whose coefficients are written in terms of unknown values of the dependent variable at surrounding nodes. Finally, an algebraic approximation to the solution of the partial differential equation is obtained for each node by inserting the polynomial approximation for h into a definite integral that is equivalent to the partial differential equation. The integral formulation of the partial differential equation is often obtained from a variational theorem, a Galerkin approximation or a least-squares residual approximation. The resulting algegraic equations are usually, but not always, solved with direct techniques.

Probably the biggest advantage of the finite element method is its flexibility in the use of variable and irregular grids and node locations. It is more flexible in this regard than the finite-difference method. On the other hand, this increasing flexibility is achieved at the cost of longer, more involved computer codings and large increases in computer storage requirements when the approximating set of algebraic equations is solved with direct techniques. A comprehensive introduction to finite element methods is given by Norrie and de Vries (1978).

The integral equation method provides a fundamentally different way to obtain approximate solutions. This method uses a boundary integral to distribute singular solutions of the partial differential equation along the solution domain boundary, and an approximate solution for an

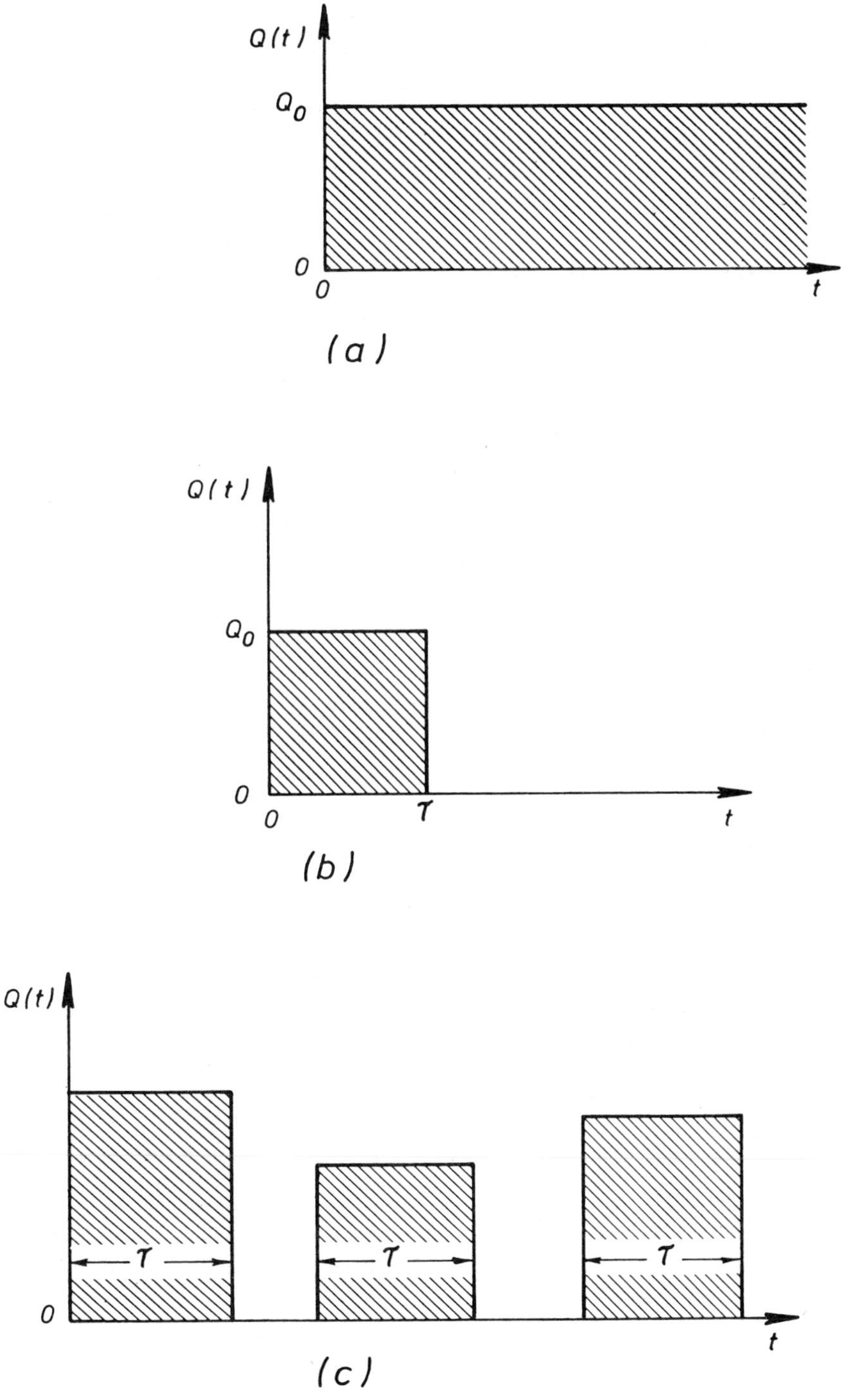

Figure 4.12 - The use of a simple solution and superposition to model complex pumping schedules.

unknown weighting function in the integrand is found by requiring boundary conditions to be satisfied at nodes along the boundary. The resulting set of simultaneous equations is usually, but not always, solved by direct techniques. The coefficient matrix of the simultaneous equations is not sparse, but it often has a strong main diagonal that makes it relatively easy to obtain numerical solutions. Furthermore, the unknowns are located only along the solution domain boundary and are, therefore, far fewer in number than the unknowns in either the finite-difference or finite element methods.

The chief advantage of the integral equation method is the relatively small amount of input data that is required and the ease of data preparation. This makes the method particularly useful for obtaining solutions to problems with free surfaces, where the position and shape of a free surface must be altered by trial and error until two boundary conditions are satisfied simultaneously. The smaller number of unknowns also makes it feasible to obtain solutions to problems with three spatial dimensions. On the other hand, since the method superimposes known, singular solutions of the governing partial differential equation, the method is, in practice, limited to the solution of linear partial differential equations with constant coefficients. Illustrative applications of this method to groundwater flow problems have been given by Liggett (1977) and Isaacs and Hunt (1981), and a more general introduction to the method is given by Brebbia and Walker (1979).

21. Hele-Shaw Experiments

In earlier years before digital computers became so readily available, experimental analogs furnished the only way in which approximate solutions could be obtained for the more difficult problems. In recent years, most of these analog experiments have given way to numerical solutions. One of the analogs, however, the vertical Hele-Shaw experiment, will probably continue to be used well into the future. This is because a Hele-Shaw experiment locates directly, without trial and error, a free surface or an interface between fluids with different densities.

The Hele-Shaw analog makes use of the fact that the laminar flow of a viscous fluid between two parallel, plane boundaries is described approximately by a set of equations that is virtually identical with the equations of groundwater flow. We can show this by choosing the (x, y) plane to be parallel to the vertical, plane boundaries and midway between them. Since the flow is laminar and incompressible, it is described by the following form of the Navier-Stokes equations:

$$-g\overline{\nabla}h + \nu\nabla^2\overline{u} = \overline{a} \tag{4.76}$$

The piezometric head, $h = \frac{p}{\rho g} - \hat{g}.\bar{r}$, has the same meaning as in groundwater flow problems, ν = kinematic viscosity, g = gravitational constant, $\bar{u}$ = fluid velocity and $\bar{a}$ = fluid acceleration. Because the flow is "creeping," we can set $\bar{a} \simeq 0$. Because the spacing of the vertical, plane boundaries is small compared with the dimensions of the region in the (x, y) plane, it is permissible to make the boundary-layer approximation that $\bar{u}$ changes much more rapidly in a direction normal to the plane boundaries than in directions parallel to the x and y coordinate axes. Thus, Eq. 4.76 is approximated with

$$-g\bar{\nabla}h + \nu \frac{\partial^2 \bar{u}}{\partial z^2} = 0 \qquad (4.77)$$

in which $\bar{\nabla}$ and $\bar{u}$ only have x and y components. Finally, since a uniform flow between two plane boundaries has a velocity distribution given by a second-degree polynomial for its exact solution, it is reasonable to assume that $\bar{u}$ takes the following form:

$$\bar{u} = \frac{3}{2}\bar{U}(x, y, t)\ [1 - \left(\frac{z}{B/2}\right)^2] \qquad (4.78)$$

The two-dimensional vector, $\bar{U}(x, y, t)$, is a flux velocity ($= \frac{2}{3}$ maximum velocity), and B = spacing of the vertical, plane boundaries. The substitution of Eq. 4.78 into Eq. 4.77 gives the result

$$\bar{U} = -K\bar{\nabla}h, \quad \left(K = \frac{gB^2}{12\ \nu}\right) \qquad (4.79)$$

Equation 4.79 is, of course, analogous to Darcy's law in two dimensions, and the analogy is completed by writing a conservation of volume statement for the incompressible flow.

$$\bar{\nabla}.\bar{U} = 0 \qquad (4.80)$$

Because the piezometric head, h, has the same meaning in the Hele-Shaw experiment and in groundwater flow, boundary conditions along any free surface are identical with Eqs. 2.56 and 2.62 provided that the porosity, σ, is taken as unity in the Hele-Shaw experiment.

The approximate nature of Eqs. 4.79 and 4.80 is easily seen when it is noted that an exact solution for the Hele-Shaw experiment requires $\bar{u}$ to vanish along any fixed, solid boundary in the (x, y) plane. But the unique solution of Eqs. 4.79 and 4.80 only requires the normal velocity component to vanish along such a boundary, which means that solutions of Eqs. 4.79 and 4.80 will allow the fluid to slip along fixed boundaries. Experience has shown, though, that the effect of a boundary layer along these fixed boundaries is usually relatively small and that Hele-Shaw experiments are capable of producing very accurate experimental solutions for most problems.

The vertical, plane boundaries for a Hele-Shaw experiment are often constructed from clear plastic since plastic is more easily machined than glass. Furthermore, Eq. 4.79 shows that the permeability is proportional to the square of the plate spacing, and the flexible nature of plastic permits spacers (sleeves or washers held in place with small bolts) to be inserted at regular intervals to insure a uniform boundary spacing. Boundaries of constant piezometric head are modelled by suddenly increasing B to form a relatively wide reservoir of fluid. Oil is often used in vertical Hele-Shaw models with a free surface because its viscous nature allows the flow to remain laminar for wider spacings of the plate boundaries. If water is used, then the plate spacing must be reduced to keep the flow laminar, and small plate spacings can create a relatively large capillary "climb" along a free surface.

Discontinuities in permeability are modelled routinely in Hele-Shaw experiments by inserting discontinuous changes in B. Thus, Eq. 4.79 shows that the ratio of B for any two regions will be equal to the square root of the ratio of permeabilities. Anisotropic aquifers can also be modelled by cutting grooves in the vertical plate boundaries in the direction of the largest principal value of the permeability tensor. The size and spacing of these grooves cannot be calculated, however, but must be found with an experimental trial and error procedure.

The permeability of a Hele-Shaw experiment can, in theory, be calculated with Eq. 4.79. In practice, however, it is usually more satisfactory to construct a permeameter. For example, a rectangular "aquifer," either horizontal or vertical, can be bounded at each end with reservoirs, and the piezometric gradient between the two reservoirs, the flow rate and the cross sectional area of the aquifer can be used to calculate K from Darcy's law. This permeability can then be used to calculate the permeability for another experiment by multiplying by the ratio of the square of the plate spacings.

Unsteady flow through the embankment shown in Fig. 2.1 will be used to illustrate a method for scaling results from a Hele-Shaw experiment. A photograph of an experiment for this type of problem is shown in Fig. 4.13. The problem is described by Eqs. 3.46 - 3.52, and a list of the dimensionless variables governing the experiment could be found simply by rewriting these equations in dimensionless variables and noting the various dimensionless variables and parameters that appear in the result. Alternatively, we can note from Eqs. 3.46 - 3.52 and the geometry of the solution domain that the solution for h will depend upon the following variables and parameters:

$$h = f\left(x,\ y,\ t,\ H,\ L,\ \frac{K}{\sigma},\ \frac{R}{\sigma}\right) \tag{4.81}$$

The embankment base width, L, is needed to characterize the solution domain geometry, and the ratios K/σ and R/σ are obtained by dividing

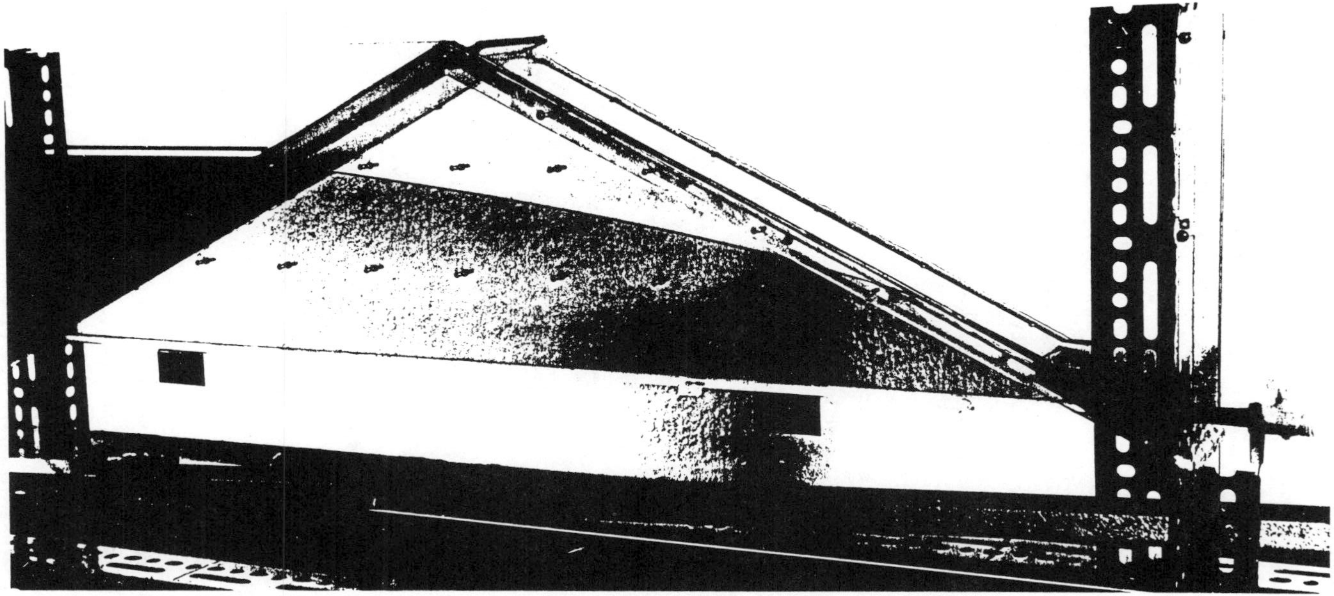

Figure 4.13 - A Hele-Shaw experiment modelling flow through an embankment.

both sides of Eq. 3.52 by σ. It is also assumed that the embankment slopes will be identical for model and prototype and that the permeability, K, is everywhere constant.

Equation 4.81 has eight variables containing the two basic dimensions of length and time. Thus, as shown in problem 11, two repeating variables can be used in a dimensional analysis to obtain six dimensionless variables:

$$\frac{h}{L} = f\left(\frac{x}{L}, \frac{y}{L}, \frac{Kt}{\sigma L}, \frac{H}{L}, \frac{R}{K}\right) \tag{4.82}$$

The use of Eq. 3.51 in Eq. 4.82 shows that the free surface coordinates can be plotted according to the functional relationship

$$\frac{y}{L} = f\left(\frac{x}{L}, \frac{Kt}{\sigma L}, \frac{H}{L}, \frac{R}{K}\right) \tag{4.83}$$

The experimental results can be scaled to a prototype by either using the experimental data plotted according to Eq. 4.83 or by transferring the experimental results point by point (i.e. - by noting from Eq. 4.83 that y/L will be identical for model and prototype when x/L, $Kt/(\sigma L)$, H/L and R/K are all simultaneously identical for model and prototype).

The special case of steady flow with no recharge can be obtained from Eqs. 4.81 - 4.83 by letting $t \to \infty$ and $R \to 0$ to obtain

$$\frac{h}{L} = f\left(\frac{x}{L}, \frac{y}{L}, \frac{H}{L}\right) \tag{4.84}$$

$$\frac{y}{L} = f\left(\frac{x}{L}, \frac{H}{L}\right) \tag{4.85}$$

Since the steady flow rate, q, is obtained by integrating a flux velocity calculated with Darcy's law along a fixed path,

$$q = f(K, H, L) \tag{4.86}$$

Thus, a dimensional analysis yields

$$\frac{q}{KH} = f\left(\frac{H}{L}\right) \tag{4.87}$$

in which it has been assumed that q is a two-dimensional flow rate with dimensions of length^2/time.

Finally, it should be noted that the form of the dimensionless variables in Eqs. 4.82 - 4.85 and 4.87 is not unique. In fact, it is always possible to replace any dimensionless variable with the result obtained by multiplying that dimensionless variable with any combination of powers of the other dimensionless variables. The only inviolable rule for this manipulation is that it must not decrease the total number of dimensionless variables. As examples, q/(KH) could be replaced with q/(KL) in Eq. 4.87, or $Kt/(\sigma L)$ could be replaced with $Rt/(\sigma H)$ in Eqs. 4.82 and 4.83. In general, although the final form of a dimensional analysis is not unique, it is possible to derive all other combinations from any particular dimensionless grouping. The usual procedure in most experimental work is to choose final forms for the dimensionless variables that can be controlled easily in an experiment and that have well-established physical meanings.

REFERENCES

Brebbia, C.A. and Walker, S. 1979. Boundary Element Techniques in Engineering, Newnes-Butterworths, London.

Hunt, B. 1978. "Finite Difference Approximation of Boundary Conditions along Irregular Boundaries," International Journal for Numerical Methods in Engineering, Vol. 12, pp. 229-235.

Hunt, B. and Isaacs, L.T. 1981. "Integral Equation Formulation for Groundwater Flow," ASCE Journal of the Hydraulics Division, Vol. 107, No. HY10, October, pp. 1197-1209.

Liggett, J.A. 1977. "Location of Free Surface in Porous Media," Journal of the Hydraulics Division, ASCE, Vol. 103, No. HY4, April, pp. 353-365.

Norrie, D.H. and de Vries, G. 1978. An Introduction to Finite Element Analysis, Academic Press, New York.

Smith, G.D. 1965. Numerical Solution of Partial Differential Equations, first edition, Oxford University Press, London, pp. 149-151.

Smith, G.D. 1978. Numerical Solution of Partial Differential Equations: Finite Difference Methods, second edition, Clarendon Press, Oxford, pp. 230-264.

Varga, R.S. 1962. Matrix Iterative Analysis, Prentice-Hall, Englewood Cliffs, p. 73.

PROBLEMS

1. Substitute Eqs. 4.6 - 4.9 into Eq. 4.1. Then differentiate Eq. 4.1 once with respect to x and y to obtain the backward-difference approximations

$$\frac{\partial h_1}{\partial x} = \frac{h_3 + 3h_1 - 4h_0}{2\Delta} + 0(\Delta^2)$$

$$\frac{\partial h_2}{\partial y} = \frac{h_4 + 3h_2 - 4h_0}{2\Delta} + 0(\Delta^2)$$

and the forward-difference approximations

$$\frac{\partial h_3}{\partial x} = - \frac{h_1 + 3h_3 - 4h_0}{2\Delta} + O(\Delta^2)$$

$$\frac{\partial h_4}{\partial y} = - \frac{h_2 + 3h_4 - 4h_0}{2\Delta} + O(\Delta^2)$$

2. The truncation error, $O(\Delta^2)$, in Eq. 4.6 indicates that the approximation becomes exact if h_0 is a polynomial of degree three or lower. Show that this is true for the particular case

$$h = 1 - 2x + 3x^2 + x^3, \ \Delta = 1$$

Do this by calculating h_0, h_1 and h_3 and substituting these numerical values into Eq. 4.6. Then compare the result with the exact result

$$\frac{\partial^2 h_0}{\partial x^2} = (6 + 6x)_{x=0} = 6$$

3. Substitute Eqs. 4.6 and 4.8 into Eq. 4.1 for the case when $h = h(x)$ only. Then integrate the result to obtain Simpson's rule for intergration:

$$\int_{-\Delta}^{\Delta} h(x)\,dx = \frac{\Delta}{3}(h_1 + 4h_0 + h_3) + O(\Delta^5)$$

Note that approximation of a function with Eq. 4.1 leads to second-order approximations for derivatives at nodes 0 but gives a fifth-order approximation for a definite integral. This result can be interpreted geometrically by noting that the approximate and exact curves for h(x) agree exactly at the nodes only. Thus, any oscillations between nodes will cause errors to cancel in the integration, which computes the area beneath the curve, while slopes of the exact and approximating curves can have substantial differences at the nodes.

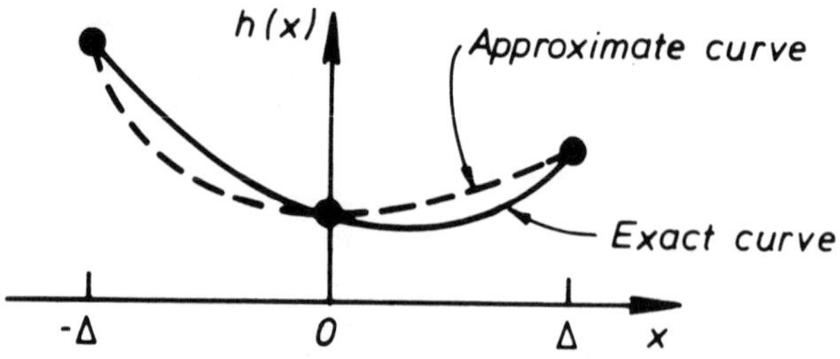

PROB. 3

4. Expand the derivatives in the following equation

$$\overline{\nabla}.(T\overline{\nabla}h) = 0$$

to obtain

$$T\nabla^2 h + \overline{\nabla}T.\overline{\nabla}h = 0$$

Obtain a second-order approximation to this equation by replacing the derivatives with the central-difference approximations given by Eqs. 4.6 - 4.9. Then notice that the resulting, algebraic approximation is not diagonally dominant and that off-diagonal terms in the coefficient matrix for h can become relatively large in regions where T changes rapidly.

5. Use the Gauss-Seidel iteration to attempt to calculate the solution to the following set of equations:

$$h_1 + 10\ h_2 = 1$$

$$6\ h_1 + h_2 = 2$$

Do the calculations by first keeping the equations in their given order. Then interchange the order of the two equations and try again.

6. Discuss briefly the changes that would have to be made in the flow-charts shown in Figs. 4.7 and 4.8 for a nonlinear, free-surface flow in which T varies with h. Note that the linearized equations should only be applied to free-surface flows in which computed changes in h are small compared with the saturated aquifer thicknesses.

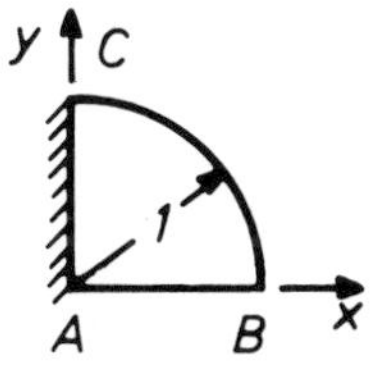

PROB. 7

7. Superimpose upon the upper right quadrant of a unit circle a uniform, finite-difference grid with a spacing of $\Delta = 0.1$ Use a computer program written from the flow-charts in Figs. 4.7 and 4.8 to solve the following problem within the quarter circle:

$$\overline{\nabla}.(T\overline{\nabla}h) = 0 \text{ within ABC}$$

$$\frac{dh}{dn} = 0 \text{ along CA}$$

$$h = \frac{1}{2}(x^2 - y) + 1 \text{ along ABC}$$

Assume that T = exp(2y) at each node. Compare the computed values of h at each node with the exact solution, which is given by

$$h = \frac{1}{2}(x^2 - y) + 1 \text{ within ABC}$$

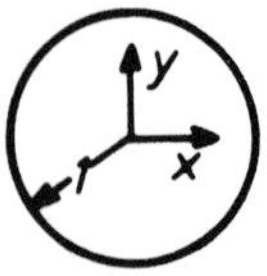

PROB.8

8. Superimpose upon a unit circle, with its center at the coordinate origin, a finite-difference grid with a uniform spacing of $\Delta = 0.2$. Then set T = Q = 1 to solve the following problem:

$$\overline{\nabla}.(T\overline{\nabla}h) = Q\delta(x)\,\delta(y),\ [\delta(x)\,\delta(y) \simeq \frac{1}{\Delta^2} \text{ at } x = y = 0]$$

$$h = 0 \text{ for } x, y \text{ on } x^2 + y^2 = 1$$

Compare numerical values of h at each node with the exact solution

$$h = \frac{1}{2\pi} \ln (x^2 + y^2)^{1/2}$$

Finally, introduce a graded mesh, similar to the one shown in Fig. 4.6, around the origin to see if this improves the solution near the well.

9. Use an explicit, forward-difference approximation in time to obtain a finite-difference representation for

$$T\frac{\partial^2 h}{\partial x^2} = S\frac{\partial h}{\partial t}$$

in which T and S are constant. Investigate the finite-difference equation for stability and show that a stable calculation requires

$$\frac{T\Delta t}{S\Delta^2} \leqq \frac{1}{2}$$

Note the difference between this result and the corresponding result given by Eq. 4.64 for two spatial dimensions.

$$\frac{T\Delta t}{S\Delta^2} \leqq \frac{1}{4}$$

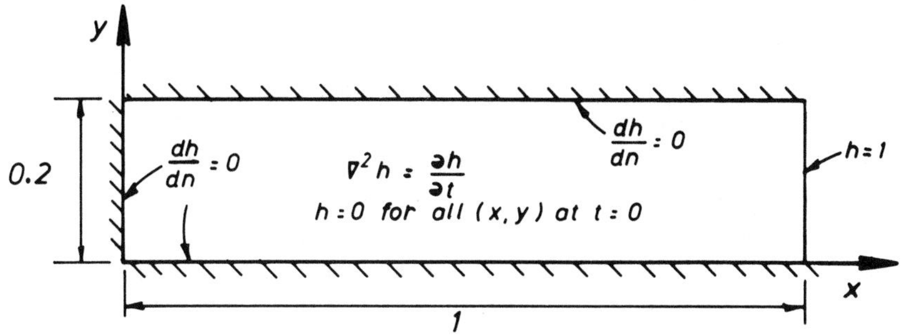

PROB. 10

10. Use a node spacing of $\Delta = 0.1$ and a computer program written from the flow-charts in Figs. 4.9 and 4.10 to obtain a numerical solution to the problem shown in the sketch. Compare the numerical solution along $x = 0$ with the exact solution

$$h(0, y, t) = 1 + \frac{4}{\pi}\sum_{n=1}^{\infty}\frac{(-1)^n}{2n-1}\exp[-(2n-1)^2\pi^2 t/4]$$

11. A fairly formal introduction to dimensional analysis is given in most introductory fluid mechanics texts. This approach usually includes a statement of the Buckingham π theorem and then solves one set of simultaneous equations for each dimensionless variable that is obtained. A much simpler procedure, however, can be used to obtain the same results. A description of this procedure follows:

(a) List the variables involved, as in Eq. 4.80, making sure there is only one dependent variable.

(b) Choose as many repeating variables as fundamental dimensions involved. Length and time are the two fundamental dimensions represented in Eq. 4.80, and the repeating variables were chosen as L and K/σ. Mass and temperature are examples of other fundamental dimensions that can appear in problems. The choice of repeating variables is not unique but must satisfy the following three conditions:

1. There are as many repeating variables as fundamental dimensions.
2. The repeating variables must contain, between them, all of the fundamental dimensions.
3. It must not be possible to form a dimensionless variable by using only the repeating variables.

As an example of 2 and 3, we could not choose H and L together as our repeating variables in Eq. 4.80. On the other hand we could choose H and t, L and t, L and K/σ, H and R/σ, L and R/σ, etc.

(c) Combine the repeating variables with each of the remaining variables to form one dimensionless variable with each of these remaining variables. Frequently this process can be carried out by inspection. For example, combining the repeating variables L and K/σ with x can only result in the dimensionless ratio of the two lengths, x/L. In more difficult instances, it may be easier to remove the basic dimensions, one at a time, by multiplying or dividing by powers of the individual repeating variables. For example, in combining t with the repeating variables L and K/σ, the fundamental dimension of time can be removed by forming the product

$$\frac{Kt}{\sigma}$$

This result has dimensions of length, which can be removed next by dividing with L to obtain $Kt/(\sigma L)$.

(d) Once completed, the final form of a dimensional analysis can be altered by multiplying any dimensionless variable with any combination of powers of the other dimensionless variables provided that the total number of dimensionless variables remains unchanged.

Apply the procedure just described to find a group of dimensionless variables that can be used to describe the solution for the critical flow rate, Q_{crit}, that could be calculated from Eqs. 3.41 - 3.46. Assume that K and R are constant in this sea water intrusion problem.

12. Steady flow through a homogeneous, isotropic (K = scalar constant) embankment is modelled with a Hele-Shaw experiment. The experiment gives

$$q = 2\ \text{cm}^3/\text{s},\ K = 1\ \text{cm/s},\ H = 5\ \text{cm},\ L = 20\ \text{cm}$$

Use Eq. 4.86 to calculate the corresponding prototype values for q and H if K and L have values of

$$K = 10^{-5}\ \text{m/s},\ L = 30\ \text{m}$$

5 The Inverse Problem

22. Transmissivity Calculations in Steady Flow

Chapter IV considered the direct problem of calculating the piezometric head, h, when the aquifer parameters T, S and K'/B' are known. This chapter will consider the inverse problem of calculating T, S and K'/B' from field measurements of h. The inverse problem is important because, in general, it must be solved before a solution can be attempted for the direct problem. Unfortunately, though, solution methods for the inverse problem are not as satisfactory as solution methods for the direct problem.

The theory of calculating transmissivities from steady flow distributions of h was first worked out by Nelson (1960, 1961) and is relatively complete. A steady flow with no wells or leakage satisfies the following equation:

$$\bar{\nabla}.(T\bar{\nabla}h) = 0 \tag{5.1}$$

An expansion of the derivatives in Eq. 5.1 shows that T satisfies a first-order, partial differential equation when h is a known, measured function.

$$\bar{\nabla}T.\bar{\nabla}h + T\nabla^2 h = 0 \tag{5.2}$$

Equation 5.2 is easily put into characteristic form by dividing with $|\bar{\nabla}h|$ to obtain

$$\frac{\bar{\nabla}h}{|\bar{\nabla}h|}.\bar{\nabla}T + T\frac{\nabla^2 h}{|\bar{\nabla}h|} = 0 \tag{5.3}$$

But $\bar{\nabla}h/|\bar{\nabla}h|$ is shown by Eq. 1.11 to be a unit vector that is normal to the contours of constant h and that points in the direction of increasing h. In other words, $\bar{\nabla}h/|\bar{\nabla}h|$ is the unit tangent vector to the steady flow streamlines (which are lines that are everywhere tangent to the velocity vector). Thus, the directional derivative given in Eq. 1.8 allows Eq. 5.3 to be written as the ordinary differential equation

$$\frac{dT}{ds} + \frac{\nabla^2 h}{|\bar{\nabla}h|}T = 0 \tag{5.4}$$

along the characteristic curves

$$\frac{d\bar{r}}{ds} = \frac{\bar{\nabla}h}{|\bar{\nabla}h|} \tag{5.5}$$

Since $\nabla^2 h/|\bar{\nabla}h|$ in Eqs. 5.4 - 5.5 is, in theory, a known function, Eq. 5.4 is readily integrated to obtain

$$\frac{T(s)}{T(0)} = \exp\left\{-\int_0^s \frac{\nabla^2 h}{|\bar{\nabla}h|}\,ds\right\} \tag{5.6}$$

in which T(0) is an initial value for T(s) at one, arbitrary point on a streamline. The equation of each streamline is found by integrating the two scalar equations in the vector Eq. 5.5.

Equations 5.5 and 5.6 appear, at first glance, to provide a complete and simple solution to an inverse problem. Unfortunately, the practical use of Eq. 5.6 is severely limited because of the great difficulty in calculating values of $\nabla^2 h$ from field measurements of h. This is because field measurements always contain errors in measurements of h and because Eq. 5.1 is actually only an approximation to reality, and these errors are magnified tremendously when they are differentiated twice with finite-difference approximations. (The comments in problem 3 of the previous chapter have some relevance to this observation.) The difficulty might also have been anticipated from slightly different considerations. The solution of Eq. 5.1 for h is known to be relatively insensitive to errors, or perturbations, in T. Thus, small perturbations in h must cause relatively large errors in the solution for T. At any rate, Eqs. 5.5 and 5.6 probably make their most valuable contribution by showing that a known distribution of h is not sufficient to calculate a unique distribution for T. In particular, we must also know one, and only one, initial value for T along each and every streamline. Equation 2.37 shows that this requirement for uniqueness is completely equivalent to knowing the flow rate through each stream tube (i.e. - a relatively thin tube bounded with streamlines along each side).

The remark at the end of the last paragraph lays the foundation for what is probably the most satisfactory method for obtaining solutions to this particular inverse problem. Field measurements of h can be used to draw contours of constant h. The resulting piezometric contour map, in turn, permits the construction of streamlines since the streamlines are orthogonal to contours of constant h. (In fact, Eq. 5.5 shows that each streamline passing through any given point has a unique geometry.) A portion of such a flow net is shown in Fig. 5.1, and it is worth remarking that the flow net elements are, in general, curvilinear rectangles with differing ratios of length to width. Finally, once the solution of the inverse problem has been made unique by specifying the flow rate, Q, through each stream tube, a second-order, central-difference approximation allows T to be computed at the central point of each flow net element from

$$Q = T|\Delta h|\left(\frac{\Delta n}{\Delta s}\right) \tag{5.7}$$

in which Δn and Δs are the average width and length, respectively, of a stream tube element. Equation 5.7 shows, incidentally, that the elements along any stream tube will all have the same ratio of $(\Delta n/\Delta s)$ only if T

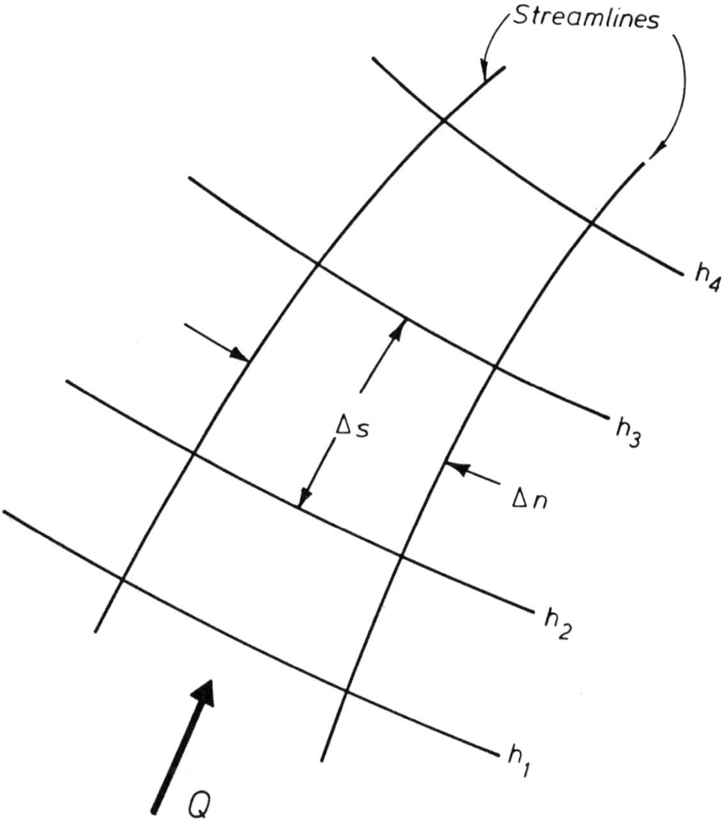

Fig. 5.1 - A portion of a flow net used to calculate the transmissivity, T.

and the contour interval, $|\Delta h|$, are the same for each element.

The method just described was used successfully by Nelson (1961). More recently, Hunt (1976) and Day and Hunt (1977) applied the method to several regional problems in New Zealand. In both of these latter studies, the computed transmissivities were used in finite-difference models to compute distributions for h, and Figs. 5.2 and 5.3 show comparisons between the computed contours of h and the measured contours of h that were used to compute T. The comparisons are good, which leads the author to suggest two reasons for the relative success of this simple calculation: first, the construction of piezometric contours and streamlines creates a smoothing of the piezometric field data, and, second, Eq. 5.7 only requires the calculation of first-order derivatives from the field data rather than the second-order derivatives that must be obtained for the use of Eq. 5.6. The calculation, of course, makes use of the same curvilinear, characteristic coordinates that appear in the formulation of Eqs. 5.5 and 5.6. The flow nets used for the transmissivity calculations by Hunt (1976) and Day and Hunt (1977) are shown in Figs. 5.4 and 5.5, respectively. It is also worth noting that the calculations of Day and Hunt (1977) included the effect of abstractions from wells and springs simply by changing the value of Q at appropriate points along each stream tube.

The calculations of T from any set of piezometric contours should always be checked by using the calculated distribution of T in a numerical solution for h and seeing how closely the computed and measured distributions of h agree. As noted before, this process is illustrated in Figs. 5.2 and 5.3. A close agreement indicates that calculations of T along each stream tube were carried out with acceptable accuracy. On the other hand, it does not prove that the assumed distribution of Q (or one initial value for T along each stream tube, which is equivalent to an assumed distribution of Q) is correct. The agreement between measured and calculated distributions of h can always be improved by using trial and error adjustments of T in the numerical solution for h. These calculations seem to be reasonably efficient and straightforward provided that the original distribution of T is calculated fairly accurately. The guiding principle in making these trial and error adjustments is that increasing transmissivities in a local region usually has the effect of increasing the spacing of piezometric contours in that region.

23. Disturbance Speeds in Groundwater Flow

Disturbance speeds can sometimes be measured in groundwater flow and used to calculate the ratios K/σ or T/S. For example, a patch of dye or pollutant or a groundwater mound of relatively small height in an unconfined aquifer is carried along with a speed, dx/dt, given by

$$\frac{dx}{dt} = -\left(\frac{K}{\sigma}\right)\frac{\partial h}{\partial x} \tag{5.8}$$

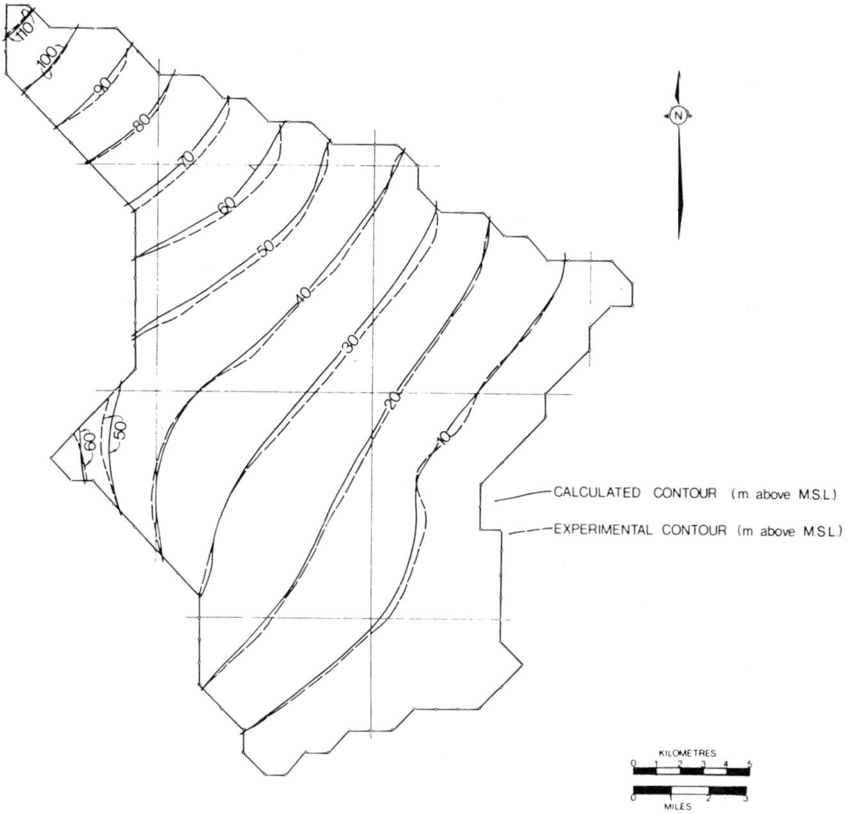

Fig. 5.2 - A comparison between measured piezometric contours and contours computed from a transmissivity distribution calculated by Hunt (1976) with Eq. 5.7. The flow net that was used for this calculation is shown in Fig. 5.4.

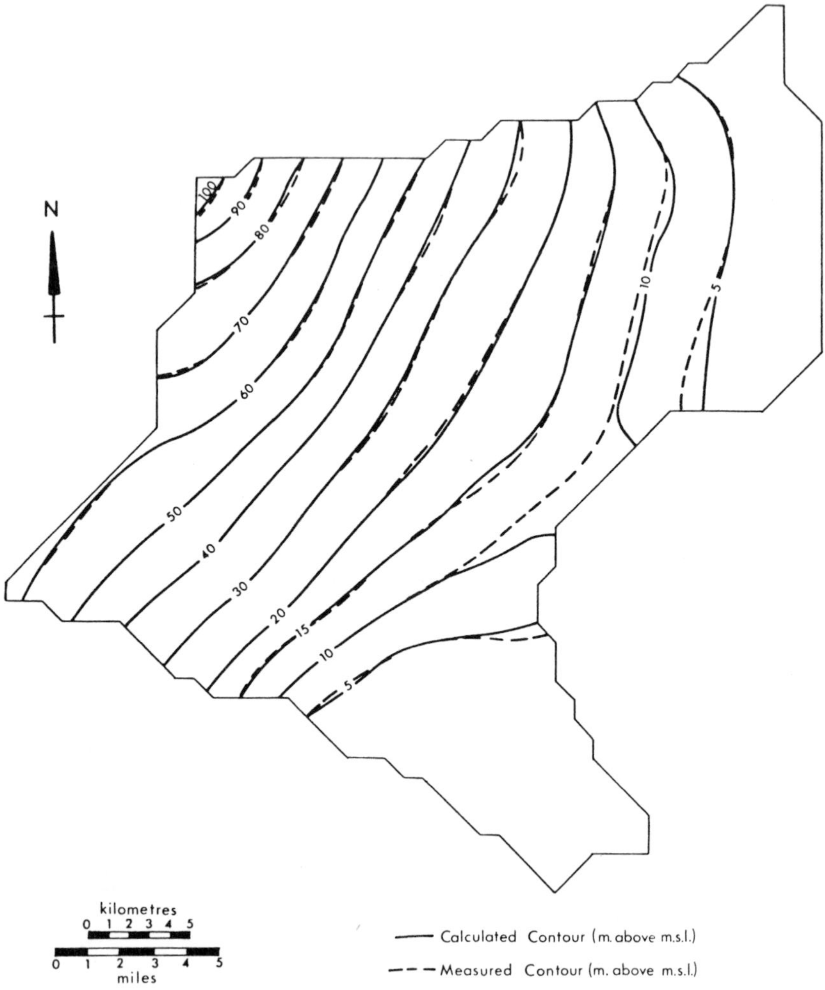

Fig. 5.3 - A comparison between measured piezometric contours and contours computed from a transmissivity distribution calculated by Day and Hunt (1977) with Eq. 5.7. The flow net that was used for this calculation is shown in Fig. 5.5.

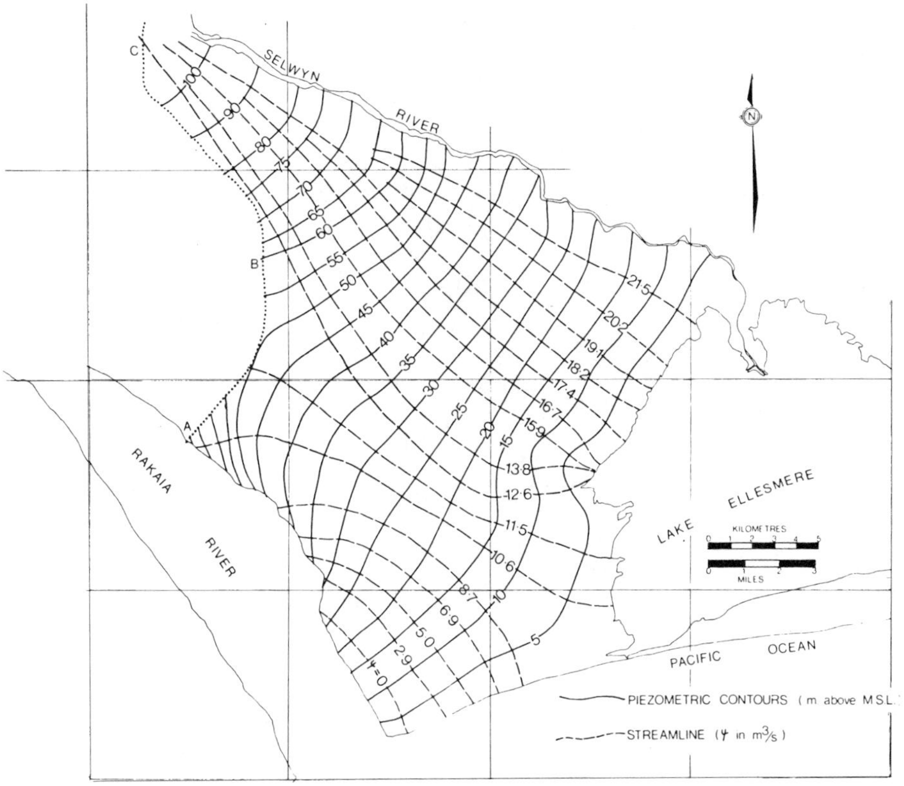

Fig. 5.4 - The flow net used in the calculations by Hunt (1976).

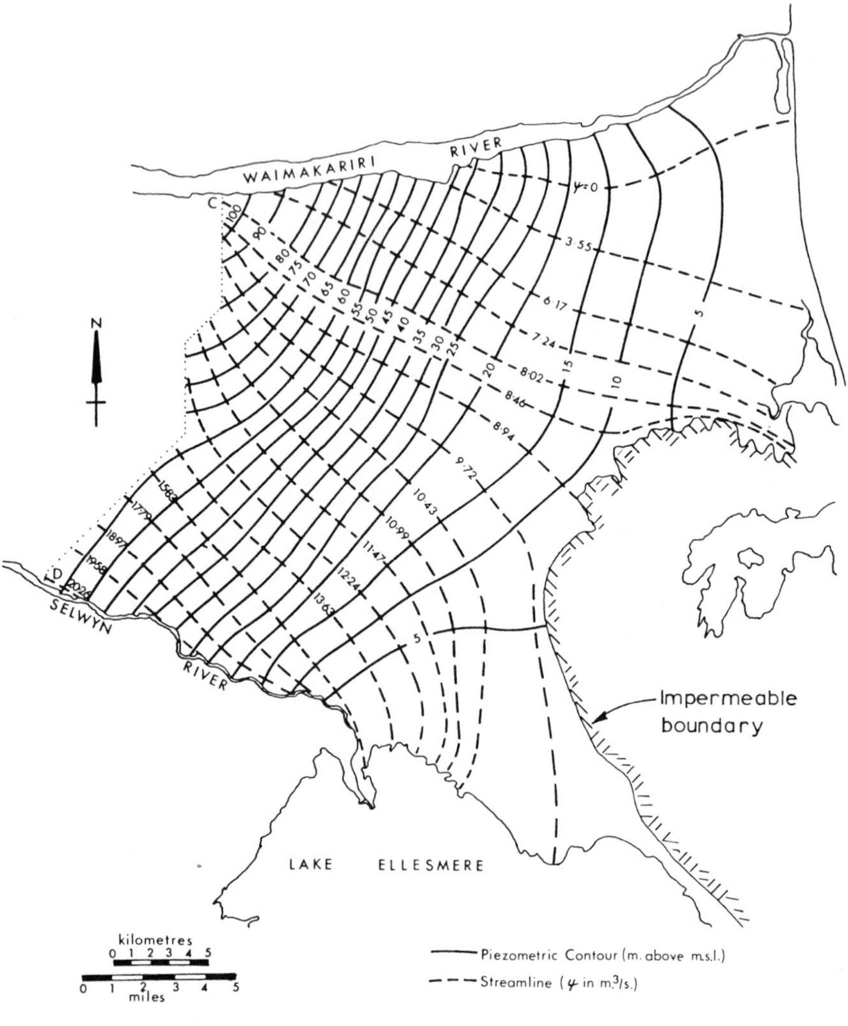

Fig. 5.5 - The flow net used in the calculations by Day and Hunt (1977).

Thus measurements of the disturbance speed and piezometric gradients permit the calculation of K/σ. The porosity, σ, can be set equal to S for an unconfined flow, and the transmissivity can be approximated with $T \simeq KB$ in which B = saturated aquifer thickness. However, care must be taken to insure that dx/dt and $\partial h/\partial x$ are measured over a distance which is large compared with the scale of the larger aquifer heterogeneities. The reasons for this are discussed in section 4.

The disturbance speed of waves can also be used to calculate the ratio T/S in some confined aquifers. As an example, the rise and fall of a tide often creates a sinusoidal, piezometric wave within an aquifer. This disturbance in h satisfies the equation

$$T \frac{\partial^2 h}{\partial x^2} = S \frac{\partial h}{\partial t} \tag{5.9}$$

in which T and S are assumed constant and x is measured in the direction of wave propagation. Substituting the expression

$$h = h_0 \exp(i\omega t + \lambda x) \tag{5.10}$$

into Eq. 5.9 shows that

$$\lambda = \pm \sqrt{\frac{S\omega}{T}} \exp\left(i \frac{\pi}{4}\right) = \pm \sqrt{\frac{S\omega}{2T}} (1 + i) \tag{5.11}$$

Thus, a solution to Eq. 5.9 is given by the imaginary part of Eq. 5.10 in the form

$$h = h_0 \exp\left(- x \sqrt{\frac{S\omega}{2T}}\right) \sin \left(\omega t - x \sqrt{\frac{S\omega}{2T}}\right) \tag{5.12}$$

Equation 5.12 shows that the wave is sinusoidal with an amplitude of h_0 at $x = 0$ and that the wave amplitude decays exponentially with x. This decay, or damping, is so rapid in unconfined aquifers that the wave is seldom observed. However, confined aquifers have very small values of S, and the waves are often observed at relatively large distances from the seacoast. Since the speed of a "fixed" point on this wave, such as the point $h = 0$, is found by keeping the argument of the sin function constant, the wave is seen to propagate with the constant speed.

$$\frac{dx}{dt} = \sqrt{2\omega \left(\frac{T}{S}\right)} \tag{5.13}$$

Equation 5.13 gives a relatively easy way to calculate T/S for a confined, coastal aquifer.

24. Pump Tests

A pump test is a controlled field experiment in which measured values of h are compared with a mathematical solution for the purpose of computing T, S and K'/B'. The experiment is most often conducted by pumping a constant, measured flow rate from a test well and measuring drawdowns, or changes in h, in nearby observation wells. An extensive and detailed description of various kinds of pump tests and testing procedures is given by Kruseman and de Ridder (1979).

Well-conducted pump tests are expensive and time consuming to carry out. Furthermore, like any experiment in an uncontrolled environment, pump tests do not always yield usable results. Thus, it is important to carry out only those pump tests that are necessary and to carry these out as carefully as possible. A number of important considerations follow:

1. Well-conducted pump tests measure drawdowns in small-diameter, observation wells rather than the large-diameter, pumped well. The use of drawdown data from a pumped well is relatively inaccurate because of head losses, well-storage effects and three-dimensional effects that are not included in the theoretical solution that is matched to the experimental data. The small-diameter observation bores are less expensive to construct than large diameter bores and have smaller well-storage effects.

2. There should be, under ideal conditions, at least three or four observation bores. These bores need not be screened, but they should be "developed" by pumping water out, pumping compressed air in or by surging with a plate on the end of a rod. This well development prevents some, but not necessarily all, of the observation bores from clogging, and using a number of bores helps insure that good data will be obtained if some of the observation bores do clog. In addition, data from a number of observation bores usually results in better average values being calculated for T, S and K'/B'.

3. Since pumped wells are often screened over only a portion of the total aquifer thickness, observation wells should not be located closer to the pumped well than about one aquifer thickness. This is to make sure that vertical velocity components, which are usually neglected in the theoretical solution, are negligibly small near the observation wells.

4. Observation wells can, and should, be placed further from the pumped well in a confined aquifer than in an unconfined aquifer. This is because the celerity of a piezometric wave is considerably greater in a confined aquifer than in an unconfined aquifer as a result of the differences in magnitude of S. (Drawdowns may never be observed throughout a test in a well that is placed too far away from a pumped well in an unconfined aquifer.) The best way to determine observation well spacing is to use estimated values for T, S and K'/B' in a theoretical solution.

Drawdown measurements in the observation wells can be carried out with a number of different devices, ranging from an automatic water level recorder to a string with a weight on the end. The most satisfactory device known to the writer consists of lead weights attached to the end of a length of television aerial wire. The other end of the wire is connected to a small battery and galvanometer. When the bared ends of the weighted end of the wire touch the free surface inside the well, the circuit is completed and the galvanometer is deflected. The entire length of wire can be wound compactly upon a small reel, with the battery and galvanometer mounted on the side or within the center of the reel. Flow rates from the pumped well are usually measured with an orifice meter.

The theoretical solution for flow to a completely penetrating well in an infinite, isotropic and homogeneous aquifer is obtained by solving the following boundary-value problem:

$$T \frac{1}{r}\frac{\partial}{\partial r}\left(r \frac{\partial h}{\partial r}\right) = S \frac{\partial h}{\partial t} + \left(\frac{K'}{B'}\right) h, \quad (0 < r < \infty,\ 0 < t < \infty) \tag{5.14}$$

$$h(r, 0) = 0, \quad (0 < r < \infty) \tag{5.15}$$

$$h(\infty, t) = 0, \quad (0 < t < \infty) \tag{5.16}$$

$$\lim_{r \to 0} \left(r \frac{\partial h}{\partial r}\right) = \frac{Q}{2\pi T}, \quad (0 < t < \infty) \tag{5.17}$$

Equation 5.14 is the particular form of Eq. 4.50 that results when $h' = Q = R = 0$, T = constant, S = constant and $h = h(r, t)$. Equation 5.15 is an initial condition that requires a zero drawdown at $t = 0$, Eq. 5.16 requires a zero drawdown for all values of t when the radial coordinate, r, becomes infinite and Eq. 5.17 prescribes the strength of the singularity at the well so that the flow rate, Q, to the well is a constant. The special case of flow with zero leakage will be obtained from the final result by setting $K'/B' = 0$.

The solution of Eqs. 5.14 - 5.17 can be found by using the zero-order, Fourier-Bessel transform

$$\phi(\alpha, t) = \int_0^\infty r\, h(r, t)\, J_0(\alpha r)\, dr \tag{5.18a}$$

$$h(r, t) = \int_0^\infty \alpha\, \phi(\alpha, t)\, J_0(\alpha r)\, d\alpha \tag{5.18b}$$

in which J_0 is the zero-order Bessel function of the first kind. Thus, multiplying Eqs. 5.14 - 5.17 by $r\, J_0(\alpha r)$ and integrating from $r = 0$ to $r = \infty$ allows the problem for $h(r, t)$ to be replaced with a simpler problem for $\phi(\alpha, t)$. This gives, after several integrations by parts and the use of Eqs. 5.16 and 5.17, the following result:

$$\frac{\partial \phi(\alpha,\ t)}{\partial t} + \left(\frac{T\alpha^2}{S} + \frac{K'}{SB'}\right) \phi(\alpha,\ t) = - \frac{Q}{2\pi S} \tag{5.19}$$

$$\phi(\alpha,\ 0) = 0 \tag{5.20}$$

The solution of Eqs. 5.19 - 5.20 is given by

$$Q(\alpha,\ t) = - \frac{Q}{2\pi S} \int_0^t \exp\left[- \left(\frac{T\alpha^2}{S} + \frac{K'}{SB'}\right) (t - \tau)\right] d\tau \tag{5.21}$$

Finally, substituting Eq. 5.21 into Eq. 5.18b, interchanging the order of integration and evaluating the inside integral leads to the result

$$h(r,\ t) = - \frac{Q}{4\pi T} \int_{\frac{Sr^2}{4Tt}}^{\infty} \exp\left(- x - \frac{K'r^2}{4B'Tx}\right) \frac{dx}{x} \tag{5.22}$$

Since the drawdown is given by $|h|$, Eq. 5.22 can be written in the form

$$|h| = \frac{Q}{4\pi T} W\left(u,\ \frac{r}{L}\right) \tag{5.23}$$

in which

$$u = \frac{Sr^2}{4Tt} \tag{5.24}$$

$$\frac{r}{L} = r\sqrt{\frac{K'}{B'T}} \tag{5.25}$$

$$W\left(u,\ \frac{r}{L}\right) = \int_u^{\infty} \exp\left(- x - \frac{r^2}{4L^2x}\right) \frac{dx}{x} \tag{5.26}$$

The variable $L = \sqrt{B'T/K'}$ has dimensions of length but is not a measurable, geometric length in the problem, and the function W is the "well function for leaky aquifers" that was first derived by Hantush and Jacob (1955).

Series expansions for $W\left(u,\ \frac{r}{L}\right)$ can be obtained in the following forms:

$$W\left(u,\ \frac{r}{L}\right) = \sum_{n=0}^{\infty} \frac{E_{n+1}(u)}{n!} \left(- \frac{r^2}{4uL^2}\right)^n,\ \left(0 \leqq \frac{r^2}{4uL^2} < \infty\right) \tag{5.27}$$

$$W\left(u,\ \frac{r}{L}\right) = 2K_0\left(\frac{r}{L}\right) - \sum_{n=0}^{\infty} \frac{(-u)^n}{n!} E_{n+1}\left(\frac{r^2}{4uL^2}\right),$$

$$\left(0 < \frac{r}{L} \leqq \infty,\ 0 \leqq u < \infty\right) \tag{5.28}$$

The function K_0 is the zero-order, modified Bessel function of the second kind, and $E_n(u)$ is the exponential integral

$$E_n(u) = \int_1^{\infty} \exp(- ux) \frac{dx}{x^n} \tag{5.29}$$

In practice, it is only necessary to calculate E_1 since other values of E_n can be calculated from the recurrence formula

$$E_{n+1}(u) = \frac{1}{n} [\exp(-u) - uE_n(u)], \quad (n = 1, 2, \ldots) \qquad (5.30)$$

Equation 5.30 is derived from Eq. 5.29 by integrating once, by parts. Abramowitz and Stegun (1964) show how E_1 and K_0 can be calculated from infinite series expansions, asymptotic expansions, approximating polynomials and tables. The infinite series in Eqs. 5.27 and 5.28 are absolutely convergent and alternating, which means that the truncation error is always less in magnitude than the first neglected term provided that this neglected term has a smaller magnitude than the previous term. An aquifer that is not leaky ($K'/B' = 0$) is seen from Eqs. 5.23 - 5.27 to have the solution

$$|h| = \frac{Q}{4\pi T} E_1(u) \qquad (5.31)$$

The exponential integral, $E_1(u)$, is often written as $W(u)$ and called the "well function" in books on groundwater hydrology. Some values of $E_1(x)$ are given in Table 2, and values of $K_0(x)$ are given in Table 3. A tabulation of $W(u, r/L)$ that was obtained by Hantush (1964) is given in Table 4, and a derivation of the series expansions in Eqs. 5.27 and 5.28 will be given in section 30 of chapter 6.

25. Analysis of Pump Test Data

Curves plotted from Eq. 5.23 on semi-log paper are shown in Fig. 5.6. There are two important characteristics of these curves that are extremely useful in analyzing field data: first, drawdowns for a leaky aquifer ($r/L > 0$) approach horizontal asymptotes as t becomes large, and, second, drawdowns for an aquifer without leakage ($r/L = 0$) are asymptotic to a straight line that is not horizontal as t becomes large. These two characteristics are easily used to determine from field data whether an aquifer should be analyzed as a leaky or nonleaky aquifer. A plot of experimental values of $|h|$ = drawdown versus t/r^2 for all observation wells is made on semi-log paper, as shown in Fig. 5.7. If data from all observation wells tends to cluster along a line similar to the curve marked A in Fig. 5.7, then the data should be analyzed by assuming zero leakage ($r/L = 0$). If data from each well approaches a horizontal line for large t with a different horizontal asymptote for each different value of r, as shown by the curves marked B in Fig. 5.7, then the data should be analyzed by assuming a finite leakage ($r/L > 0$).

There are two standard methods that are often used to analyze data for nonleaky ($r/L = 0$) aquifers. The first method is known as Jacob's method and can be used if measurements have been carried out over a long enough time period to allow the drawdown data in the semi-log plot of

- Table 2 -

Values of $E_1(x)$

x	$E_1(x)$
5	0.0011
3	0.0130
2	0.0489
1	0.2194
0.8	0.3106
0.6	0.4544
0.4	0.7024
0.2	1.2227
0.1	1.8229

For $x \to 0$:

$$E_1(x) = \ln\left(\frac{0.56145948}{x}\right) + x - \frac{1}{4}x^2 + \frac{1}{18}x^3 - \frac{1}{96}x^4 + \ldots\ldots$$

For $x >> 1$:

$$E_1(x) \sim \frac{\exp(-x)}{x}\left(1 - \frac{1}{x} + \frac{2}{x^2} - \frac{6}{x^3} + \ldots\ldots\right)$$

- Table 3 -

Values of $K_0(x)$

x	$K_0(x)$
3	0.0347
2	0.1139
1	0.4210
0.8	0.5653
0.6	0.7775
0.4	1.1145
0.2	1.7527
0.1	2.4271

For $x \to 0$:

$$K_0(x) = \left(1 + \frac{1}{4}x^2 + \frac{1}{64}x^4 + \ldots\right)\ln\left(\frac{1.12291897}{x}\right) + \frac{1}{4}x^2 + \frac{3}{128}x^4 + \ldots\ldots$$

For $x >> 1$:

$$K_0(x) \sim \exp(-x)\sqrt{\frac{\pi}{2x}}\left(1 - \frac{1}{8x} + \frac{9}{128\,x^2} - \frac{75}{1024\,x^3} + \ldots\ldots\right)$$

- Table 4 -

Values of W(u, β)*

u \ β	0	.001	.002	.005	.01	.02	.05	.1	.2	.5	1.0
.000001	13.24	13.00	12.44	10.83	9.44	8.06	6.23	4.85	3.51	1.85	0.84
.000002	12.55	12.42	12.10	10.82	9.44	8.06	6.23	4.85	3.51	1.85	0.84
.000005	11.63	11.58	11.44	10.68	9.44	8.06	6.23	4.85	3.51	1.85	0.84
.00001	10.94	10.91	10.84	10.40	9.42	8.06	6.23	4.85	3.51	1.85	0.84
.00002	10.24	10.23	10.19	9.95	9.30	8.06	6.23	4.85	3.51	1.85	0.84
.00005	9.33	9.32	9.31	9.21	8.88	8.01	6.23	4.85	3.51	1.85	0.84
.0001	8.63	8.63	8.62	8.57	8.40	7.84	6.23	4.85	3.51	1.85	0.84
.0002	7.94	7.94	7.94	7.91	7.82	7.50	6.22	4.85	3.51	1.85	0.84
.0005	7.02	7.02	7.02	7.01	6.98	6.83	6.08	4.85	3.51	1.85	0.84
.001	6.33	6.33	6.33	6.33	6.31	6.23	5.80	4.83	3.51	1.85	0.84
.002	5.64	5.64	5.64	5.64	5.63	5.59	5.35	4.71	3.50	1.85	0.84
.005	4.73	4.73	4.73	4.72	4.72	4.71	4.61	4.30	3.46	1.85	0.84
.01	4.04	4.04	4.04	4.04	4.04	4.03	3.98	3.82	3.29	1.85	0.84
.02	3.35	3.35	3.35	3.35	3.35	3.35	3.33	3.24	2.95	1.84	0.84
.05	2.47	2.47	2.47	2.47	2.47	2.47	2.46	2.43	2.31	1.71	0.84
.1	1.82	1.82	1.82	1.82	1.82	1.82	1.82	1.81	1.75	1.44	0.82
.2	1.22	1.22	1.22	1.22	1.22	1.22	1.22	1.22	1.19	1.06	0.71
.5	0.56	0.56	0.56	0.56	0.56	0.56	0.56	0.56	0.55	0.52	0.42
1.0	0.22	0.22	0.22	0.22	0.22	0.22	0.22	0.22	0.22	0.21	0.19

*A more extensive tabulation to four decimal places is given by Hantush (1964).

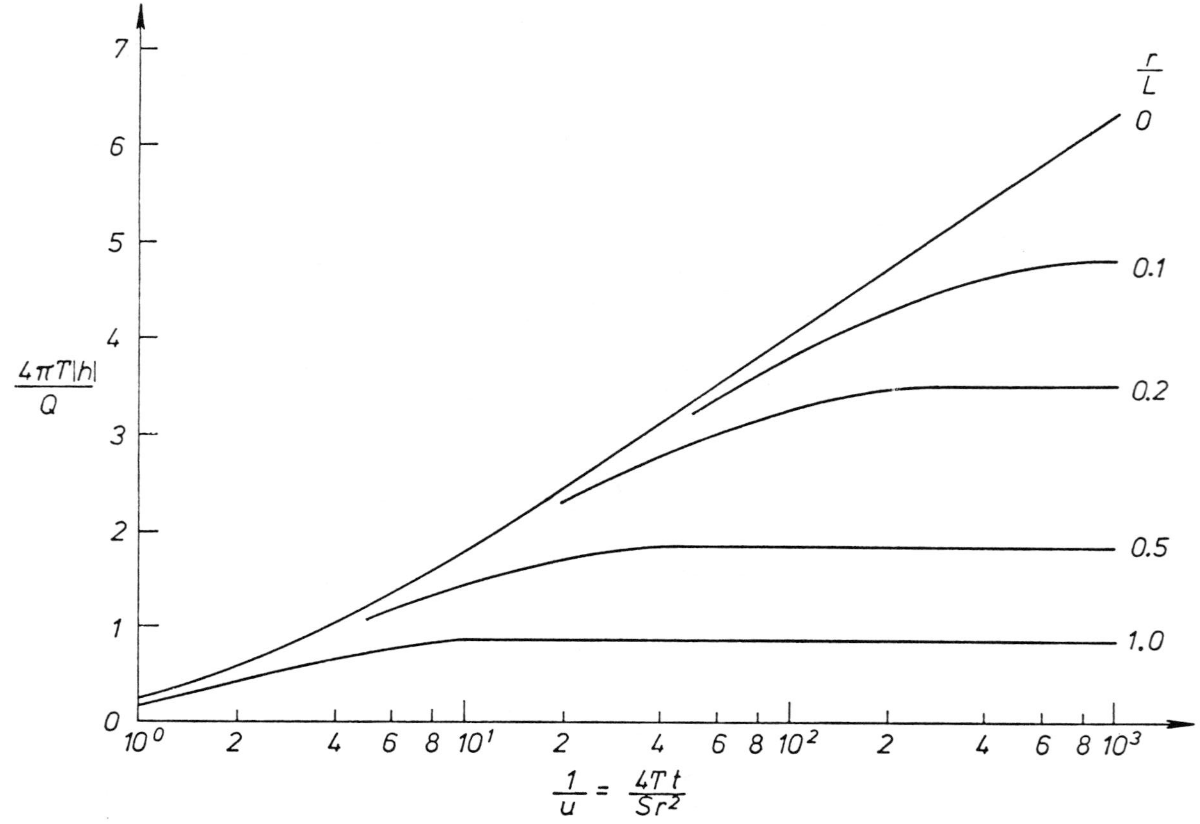

Fig. 5.6 - Curves plotted from Eq. 5.23 on semi-log paper.

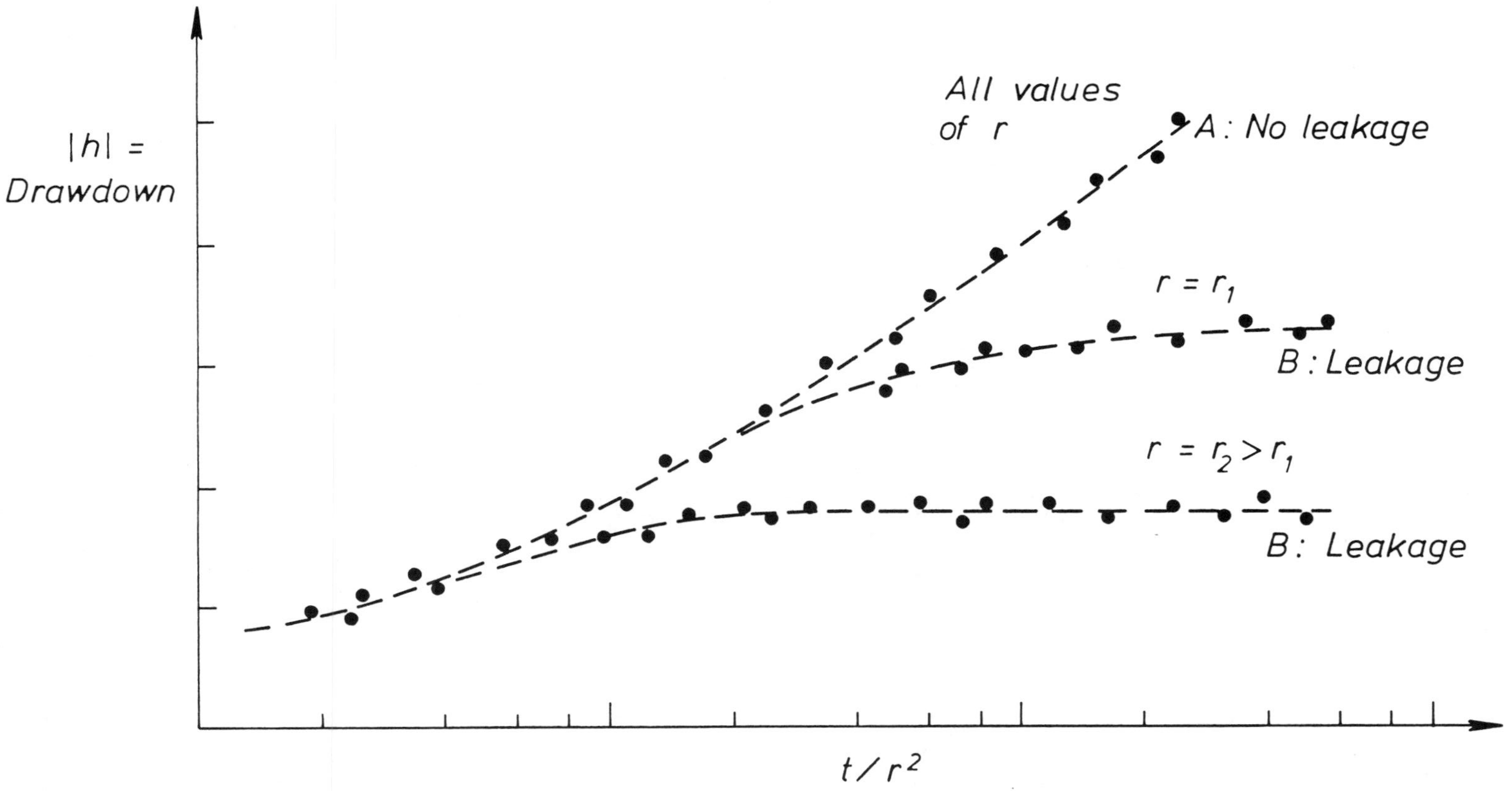

Fig. 5.7 - Observation-well data plotted on semi-log paper.

$|h|$ versus t/r^2 to become asymptotic to a straight line. Under these conditions, a straight line can be averaged through the data for all observation wells, and the values of $|h|$ and t/r^2 for any two points on this straight line can be substituted into the approximation

$$\frac{4\pi T|h|}{Q} = E_1(u) \simeq \ln\left(\frac{0.56145948}{u}\right), \quad \left(u = \frac{Sr^2}{4Tt}\right) \tag{5.32}$$

Equation 5.32 results from using the first term in the expansion of $E_1(u)$ for small u, as shown in Table 2, and putting two sets of values for $|h|$ and t/r^2 in Eq. 5.32 gives two equations that can be solved simultaneously for T and S. For example, subtracting the two equations leads to the result

$$T = \frac{Q}{4\pi(|h_1| - |h_2|)} \ln \frac{(t/r^2)_1}{(t/r^2)_2} \tag{5.33}$$

Since Q is a known number, T can be calculated directly from Eq. 5.33. Then the calculated value of T can be used to calculate S from

$$S = 0.56145948(4T)\left(\frac{t}{r^2}\right) \exp\left(-\frac{4\pi T|h|}{Q}\right) \tag{5.34}$$

in which values of t/r^2 and $|h|$ are given by the coordinates of either of the two points on the straight line.

The second method of analyzing data for nonleaky ($r/L = 0$) aquifers is known as either the Theis or match-point method. This method makes use of a plot of Eq. 5.31 on log-log paper, as shown in Fig. 5.8, and an experimental, log-log plot of $|h|$ = drawdown versus t/r^2, as shown in Fig. 5.9. All of the experimental drawdown data, for all values of r, should plot along the same smooth curve, and the amount of scatter present in this experimental plot is a measure of the departure from the assumed condition of an infinite, homogeneous, isotropic aquifer. Because of the property $\log(ab) = \log(a) + \log(b)$, the ordinates and abscissas of Figs. 5.8 and 5.9 will differ only by additive constants. Thus, the two plots can be superimposed upon a light table and translated relative to each other, taking care to keep the coordinate axes parallel, until the theoretical curve in Fig. 5.8 forms a satisfactory fit to the experimental data in Fig. 5.9. Any two points lying above and below each other on these two plots give the four numbers a, b, c and d:

$$|h| = a \tag{5.35a}$$

$$\frac{t}{r^2} = b \tag{5.35b}$$

$$\frac{4\pi T|h|}{Q} = c \tag{5.35c}$$

$$\frac{4Tt}{Sr^2} = d \tag{5.35d}$$

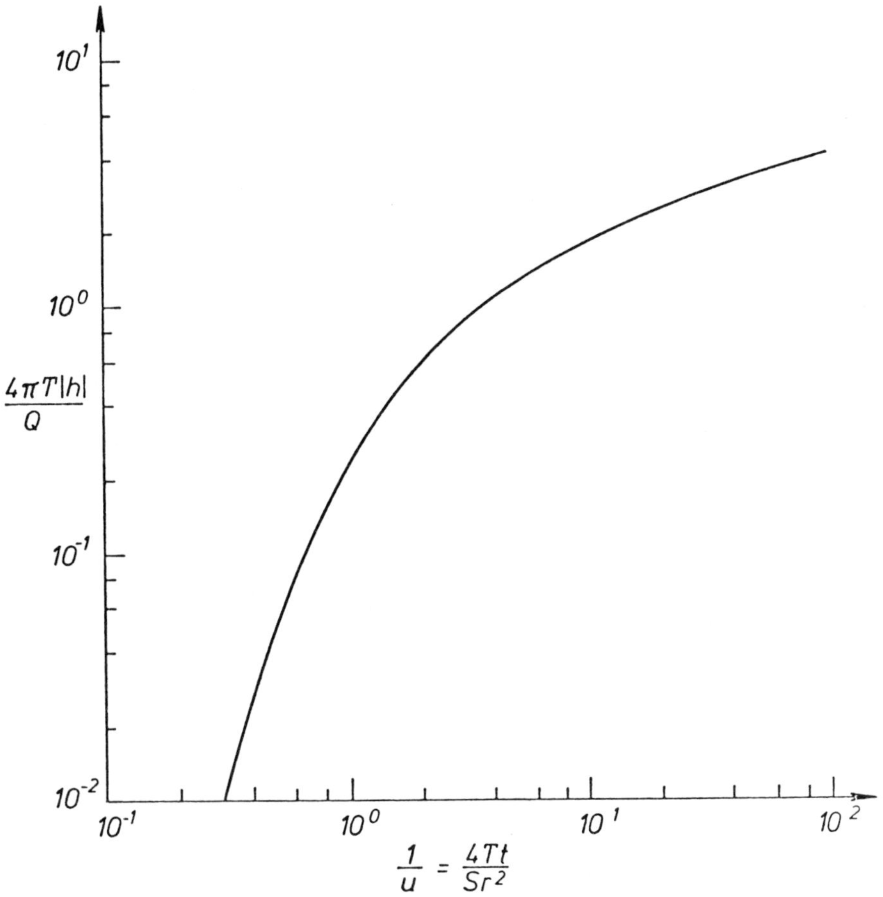

Fig. 5.8 - Equation 5.31 plotted on log-log paper.

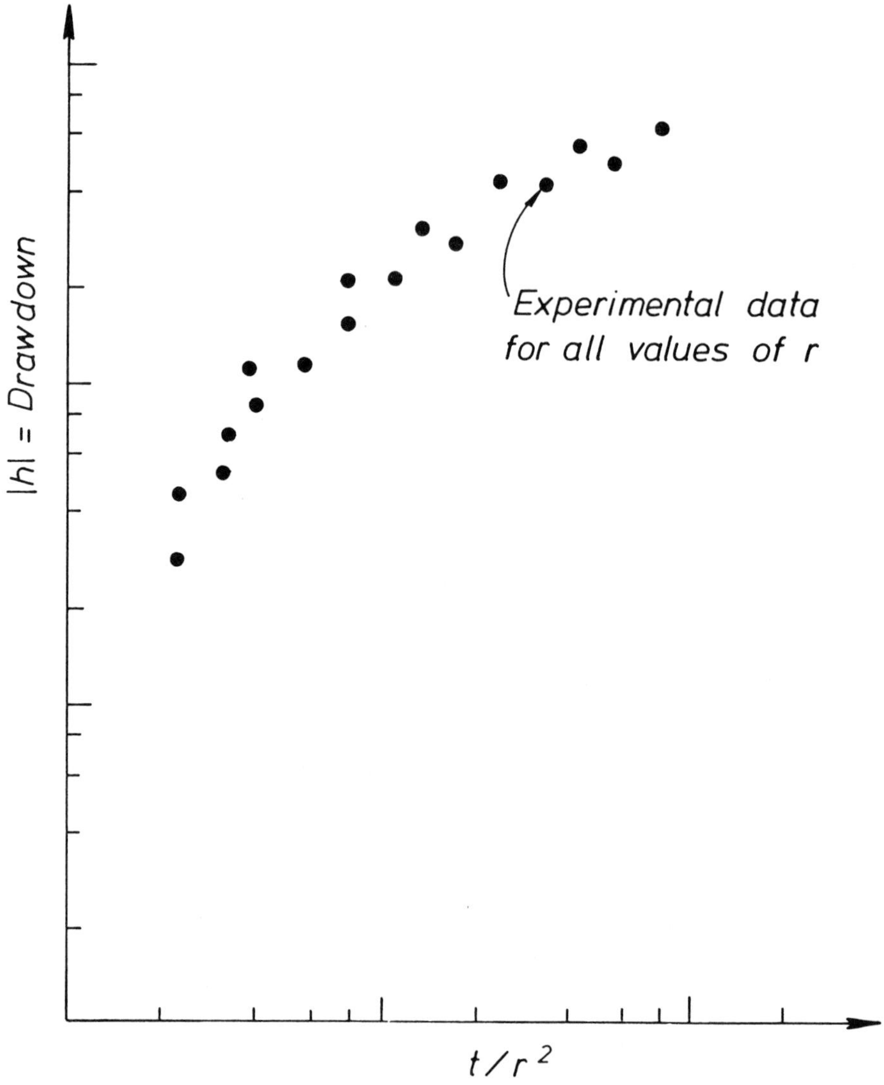

Fig. 5.9 - Observation-well data plotted on log-log paper.

Since a, b, c, d and Q are all known numbers, the elimination of $|h|$ and t/r^2 allows T and S to be calculated from

$$T = \frac{Qc}{4\pi a} \tag{5.36a}$$

$$S = \frac{Qbc}{\pi ad} \tag{5.36b}$$

Finally, the calculated values of T and S can be put into Eq. 5.31 and the theoretical curve can be plotted in Fig. 5.9 to show the quality of agreement between theory and experiment.

The reader should be warned that the amount of scatter in this curve-fitting process tends to increase as the number of observation wells increases. (The smallest amount of scatter, as well as the least accurate result, can be obtained by using one observation well). As an example, Fig. 5.10 shows the experimental data and theoretical curve that resulted in an analysis carried out by Hunt (1980). In this case, data was used from twelve observation wells, and the writer considers the amount of scatter present in Fig. 5.10 to be typical for pump tests using a number of observation wells.

The analysis of pump test data for leaky ($r/L > 0$) aquifers is relatively difficult. If data is taken in at least two observation wells, and if the water levels in at least two of these observation wells have stabilized during the period of experimental measurement, then Eqs. 5.23, 5.24 and 5.28 show that

$$|h_1| = \frac{Q}{2\pi T} K_0\left(\frac{r_1}{L}\right) \tag{5.37a}$$

$$|h_2| = \frac{Q}{2\pi T} K_0\left(\frac{r_2}{L}\right) \tag{5.37b}$$

in which the subscripts 1 and 2 refer to the equilibrium values at wells 1 and 2. Eliminating T between Eqs. 5.37a and 5.37b gives a transcendental equation for L:

$$|h_1| K_0\left(\frac{r_2}{L}\right) = |h_2| K_0\left(\frac{r_1}{L}\right) \tag{5.38}$$

Once L has been calculated from Eq. 5.38, T can be calculated from either Eq. 5.37a or 5.37b. Finally, the storage coefficient, S, can be calculated by using the match-point method. To be more specific, using calculated values of L and T in Eq. 5.23 allows $|h|$ to be plotted for each of the two observation wells as a function of $\log [4Tt/(Sr^2)]$. Superimposing this theoretical, semi-log plot upon the experimental data in Fig. 5.7 and translating the two plots in the <u>horizontal</u> direction until the theoretical curves fit the experimental data gives the match-point data

$$\frac{t}{r^2} = a \tag{5.39a}$$

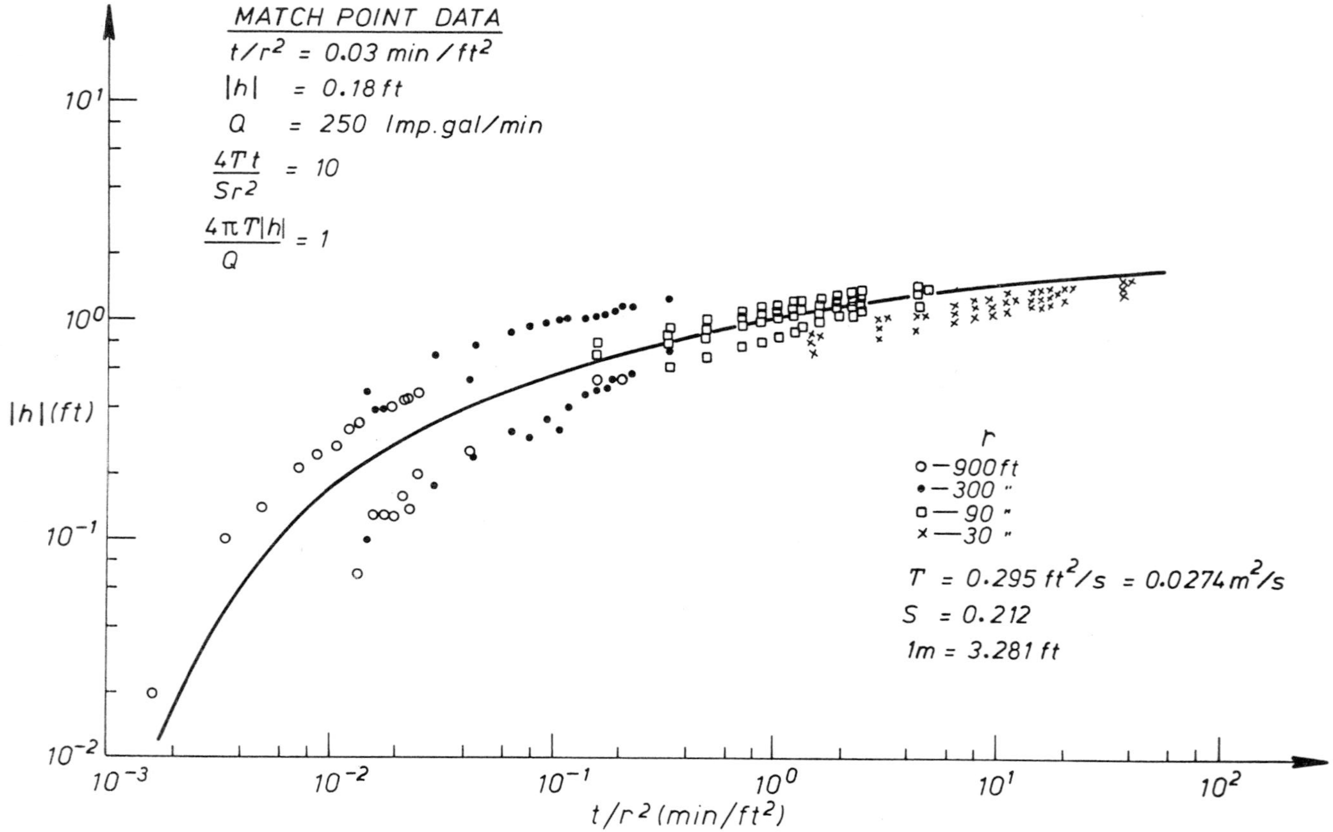

Fig. 5.10 - The result of a pump test analysis. The data was gathered by E.R. Garden and Partners, of Dunedin, New Zealand, and analyzed by Hunt (1980).

$$\frac{4Tt}{Sr^2} = b \tag{5.39b}$$

in which a and b are the horizontal coordinates of any two points that lie above and below each other. Thus, S can be calculated by eliminating t/r^2 from Eqs. 5.39a and 5.39b to obtain

$$S = 4T \frac{a}{b} \tag{5.40}$$

The leakage factor, K'/B', can be calculated for use in Eq. 4.50 from the definition of L:

$$\frac{K'}{B'} = \frac{T}{L^2} \tag{5.41}$$

An unpublished example of the use of this technique is shown in Fig. 5.11 and 5.12. Kruseman and de Ridder (1979) give, and illustrate with numerical examples, a number of other methods for analyzing pump test data for both leaky and nonleaky aquifers.

26. Some Additional Inverse Methods

The methods just described for calculating T, S and K'/B' can occasionally be supplemented with additional techniques. For example, geophysical methods, such as electrical resistivity or seismic measurements, can sometimes be used to map variations in saturated aquifer thicknesses. Then Eq. 2.38 shows that the ratio of transmissivities at any two points will be equal to the ratio of saturated aquifer thicknesses if the average permeabilities are identical.

River gagings at different stations along a river can sometimes be used to estimate distributions of K'/B' or T. If a river is spring-fed from deeper, confined aquifers, then measurements of h, h' and the increase in flow rate between any two gaging stations can be used in Eq. 2.43 to estimate an average value of K'/B' for an area contained between the gaging stations. If a river gains or loses water from an unconfined aquifer, then the change in river flow rates between gaging stations can sometimes be used with a flow net drawn from a measured piezometric contour map to estimate transmissivity values along each stream tube that intersects the river.

Relatively inexpensive specific capacity tests can also be used to estimate transmissivities. A specific capacity is defined as the ratio of a well flow rate divided by the drawdown in the pumped well, and a specific capacity test is simply a pump test in which drawdown measurements are taken within the pumped well. The dangers in using this kind of measurement are discussed in section 24. However, when observation wells can not be used, then a specific capacity test can still be used to estimate an order of magnitude for T. The procedure consists of substituting the well radius, flow rate, measured well drawdown at a

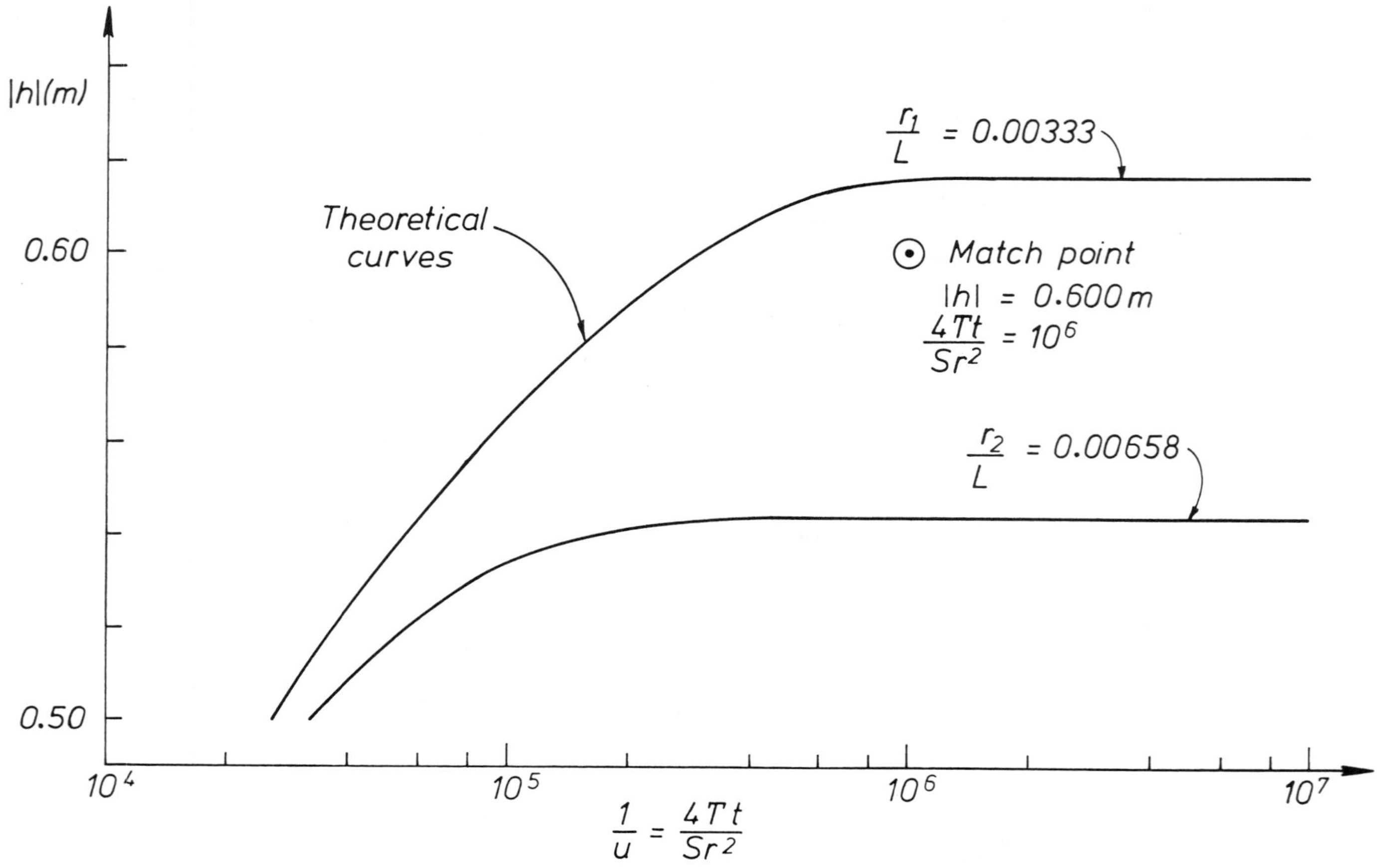

Fig. 5.11 - Theoretical curves for a leaky aquifer that were used to determine S in Fig. 5.12.

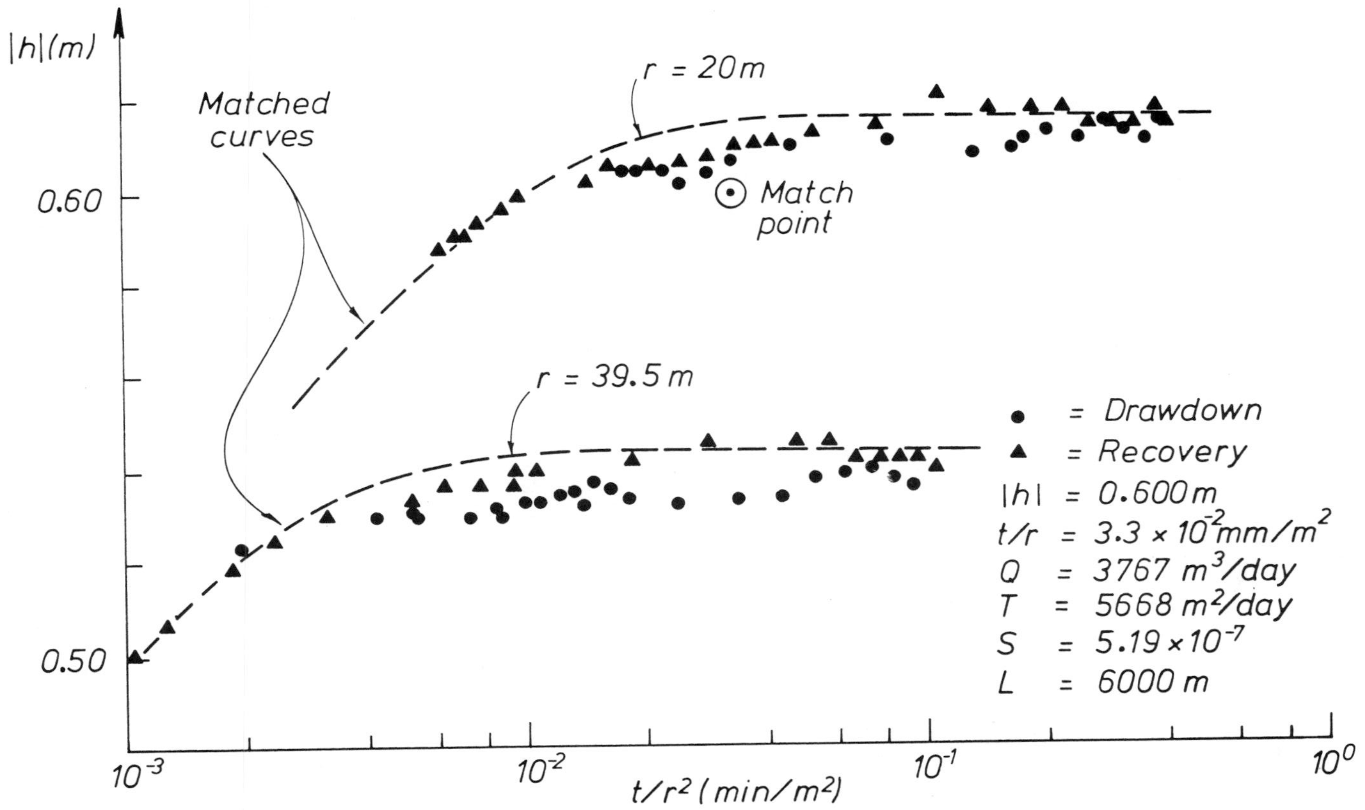

Fig. 5.12 - Experimental data and matched curves for a leaky aquifer. The data was gathered by the South Canterbury Catchment Board, of Timaru, New Zealand, and analyzed by the author in an unpublished report.

particular value of t and an estimate for S into Eq. 5.31 and solving Eq. 5.31 for T. The resulting value for T is relatively insensitive to the guess for S, which is probably the biggest reason why the test can give a reasonable order of magnitude for T. Walton (1970) gives a detailed discussion of specific capacity calculations.

A head loss that results when turbulent flow passes through a well screen and between a well casing and submerged well pump becomes one of the major sources of error when data measured within a pumped bore is used to estimate aquifer parameters. Jacob (1946) showed how a step-drawdown test can be used to calculate a well loss coefficient. This test is conducted by taking drawdown measurements within a pumped well while the flow is pumped at a number of different, constant rates over equal time intervals. Typical plots of drawdowns and flow rates as functions of time are shown in Fig. 5.13. The superposition principle discussed at the end of section 19 allows the drawdown at $t = t_n$ to be calculated from

$$|\Delta h_n| + |h_n| = \sum_{i=1}^{n} \frac{\Delta Q_i}{4\pi T} E_1\left[\frac{Sr^2}{4T(t_n - t_{i-1})}\right] + C(Q_n + \Delta Q_n)^2 \tag{5.42}$$

in which C = well loss coefficient. If $\Delta Q_n = 0$, Eq. 5.42 yields

$$|h_n| = \sum_{i=1}^{n-1} \frac{\Delta Q_i}{4\pi T} E_1\left[\frac{Sr^2}{4T(t_n - t_{i-1})}\right] + C\,Q_n^2 \tag{5.43}$$

Thus, subtracting Eqs. 5.43 and 5.42 and dividing the result by ΔQ_n gives

$$\frac{|\Delta h_n|}{\Delta Q_n} = \frac{1}{4\pi T} E_1\left[\frac{Sr^2}{4T(t_n - t_{n-1})}\right] + C(2Q_n + \Delta Q_n) \tag{5.44}$$

A similar equation can be obtained from Eq. 5.44 by replacing n with n + 1 to give

$$\frac{|\Delta h_{n+1}|}{\Delta Q_{n+1}} = \frac{1}{4\pi T} E_1\left[\frac{Sr^2}{4T(t_{n+1} - t_n)}\right] + C(2Q_{n+1} + \Delta Q_{n+1}) \tag{5.45}$$

Finally, since $t_n - t_{n-1} = t_{n+1} - t_n = \Delta t$ and $Q_{n+1} - Q_n = \Delta Q_n$, Eqs. 5.44 and 5.45 can be subtracted and solved for C in the form

$$C = \frac{\dfrac{|\Delta h_{n+1}|}{\Delta Q_{n+1}} - \dfrac{|\Delta h_n|}{\Delta Q_n}}{\Delta Q_{n+1} + \Delta Q_n} \tag{5.46}$$

Once C is obtained from Eq. 5.46, then an equation similar to Eq. 5.44 can be used to estimate T and S by using either Jacob's method or the match-point method:

$$|h'| = \frac{\Delta Q_n}{4\pi T} E_1\left[\frac{Sr^2}{4T(t - t_{n-1})}\right] \quad \text{for } t_{n-1} \leqq t \leqq t_n \tag{5.47}$$

in which the aquifer drawdown, $|h'|$, is given by

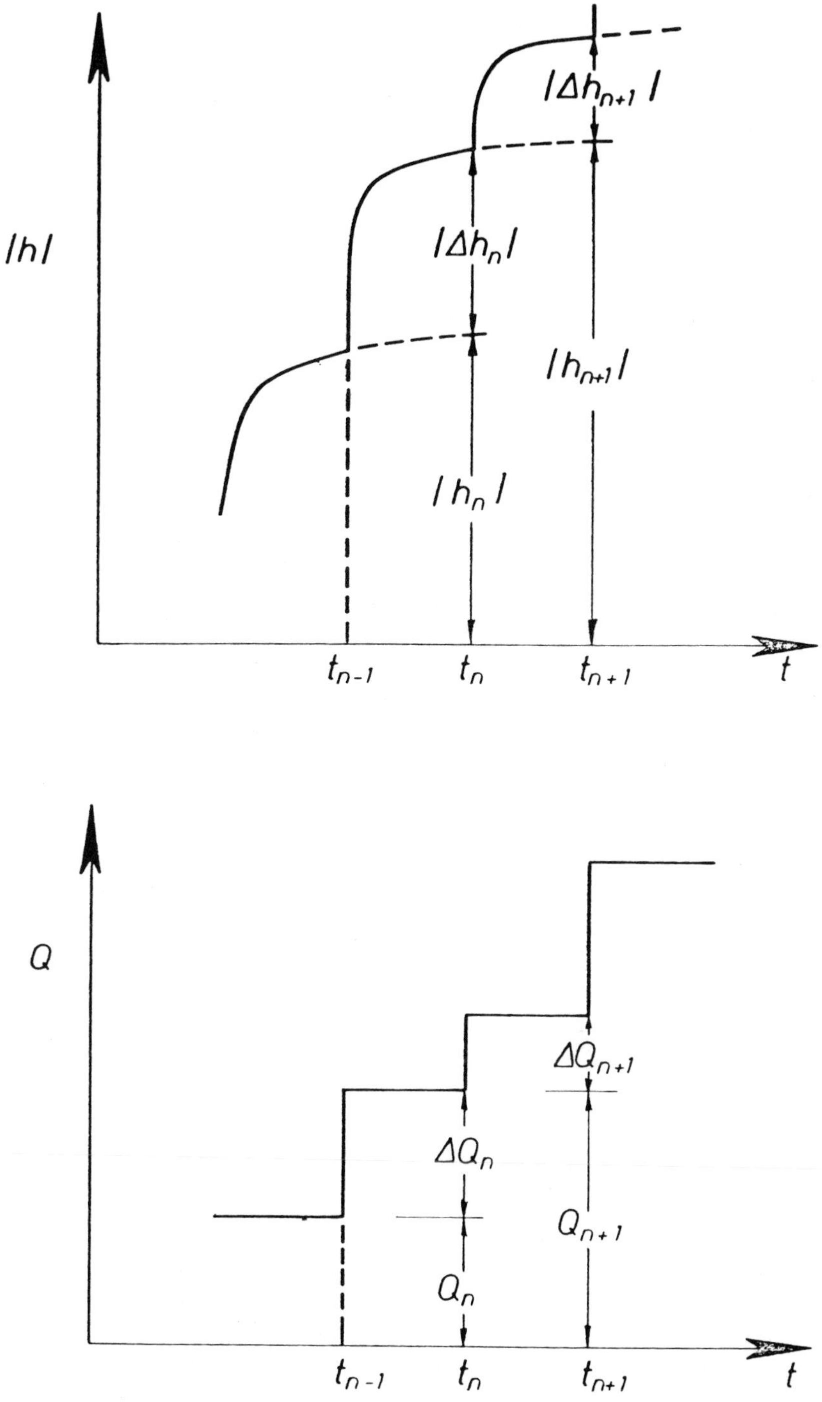

Fig. 5.13 - Drawdowns and flow rates for a step drawdown well test in which the time steps are equal.

$$|h'| = |\Delta h_n| - C(2Q_n + \Delta Q_n)\Delta Q_n \tag{5.48}$$

Sometimes r in Eq. 5.47 is set equal to a value slightly greater than the actual well radius, which makes a difference in the numerical value calculated for S but not for T. In this case, r is called the "effective" well radius and has the physical significance of being the radial distance to the point surrounding the well where the laminar flow becomes turbulent. It should also be pointed out that Eq. 5.46 holds under more general conditions than those assumed in its derivation. For example, Eq. 5.46 still holds if E_1 in Eq. 5.42 is replaced with the leaky aquifer function. In fact, Eq. 5.46 holds for any situation in which the governing equations are linear with coefficients that are independent of t.

Aquifer tests can also be carried out in a single well when either an instantaneous rise or drop in water level is created within the well. For example, if an aquifer is screened over its entire thickness, an instantaneous change in water level of H_0 within the well is shown in Chapter VIII to create water level changes within the well of

$$\frac{|H(t)|}{H_0} = \frac{8S}{\pi^2}\int_0^\infty \exp\left[-x^2\left(\frac{Tt}{Sr_0^2}\right)\right]\frac{dx}{x\Delta(x)} \tag{5.49}$$

in which r_0 = well radius, $H(t) = h(r_0, t)$, $H_0 = H(0)$ and

$$\Delta(x) = [xJ_0(x) - 2SJ_1(x)]^2 + [xY_0(x) - 2SY_1(x)]^2 \tag{5.50}$$

The symbols J_i and Y_i represent Bessel functions of the first and second kind, respectively. Numerical values of the right side of Eq. 5.49 have been tabulated by Cooper, Bredehoeft and Papadopulos (1967, 1973) and are plotted in Fig. 5.14. The recommended procedure is to drop a weighted float into the well and use Archimedes' buoyancy principle to calculate the resulting volume of displaced water. After water levels surrounding the well have stabilized, the float is suddenly withdrawn to give an instantaneous drawdown, H_0, which is equal to the volume of water displaced by the float divided by the inside area of the well pipe. Then the water level recovery within the well is measured, and a semi-log plot of $|H(t)|/H_0$ versus log t is made with the experimental data. Thus, superimposing this plot upon a plot similar to the one shown in Fig. 5.14 and translating the experimental plot in the horizontal direction allows the calculation of both T and S with the match-point method. Cooper et al. (1967, 1973) point out that calculated values of S are of questionable accuracy when S becomes small because the curves in Fig. 5.14 are nearly "parallel." An alternative experimental procedure, of course, is to create an instantaneous rise in water level by pouring water into the well.

A test that is analogous to the one just described consists of creating an instantaneous water level change in a pipe with an open end and measuring

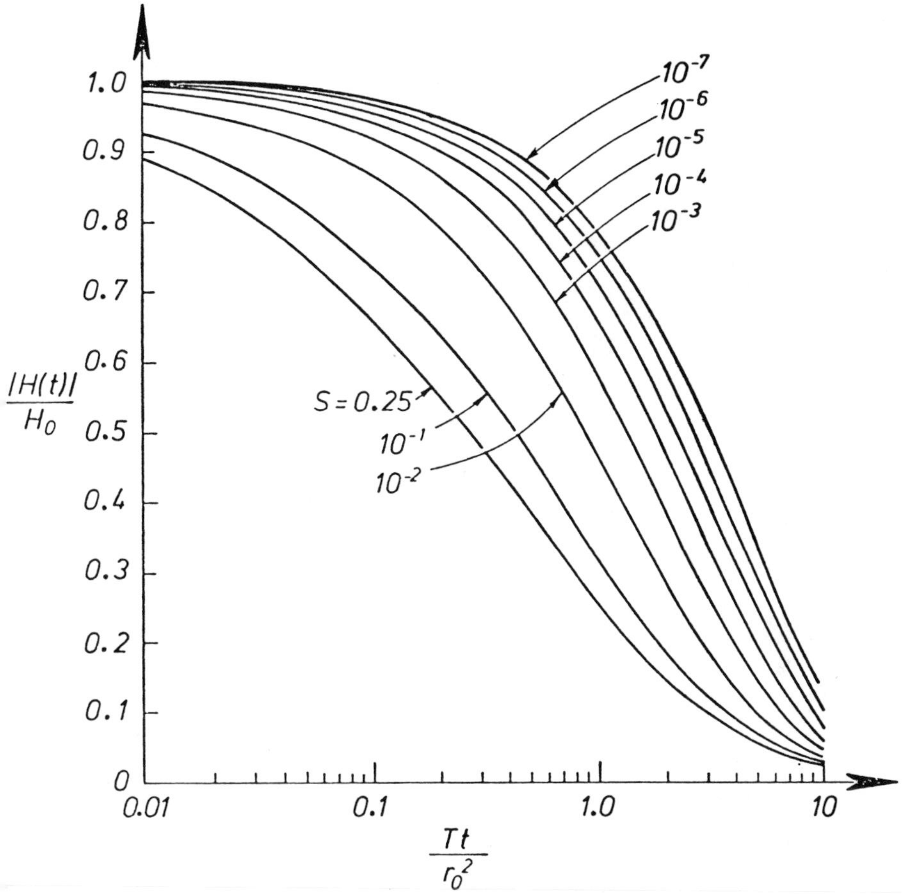

Fig. 5.14 - Water level changes within a well after an initial water level change of H_0.

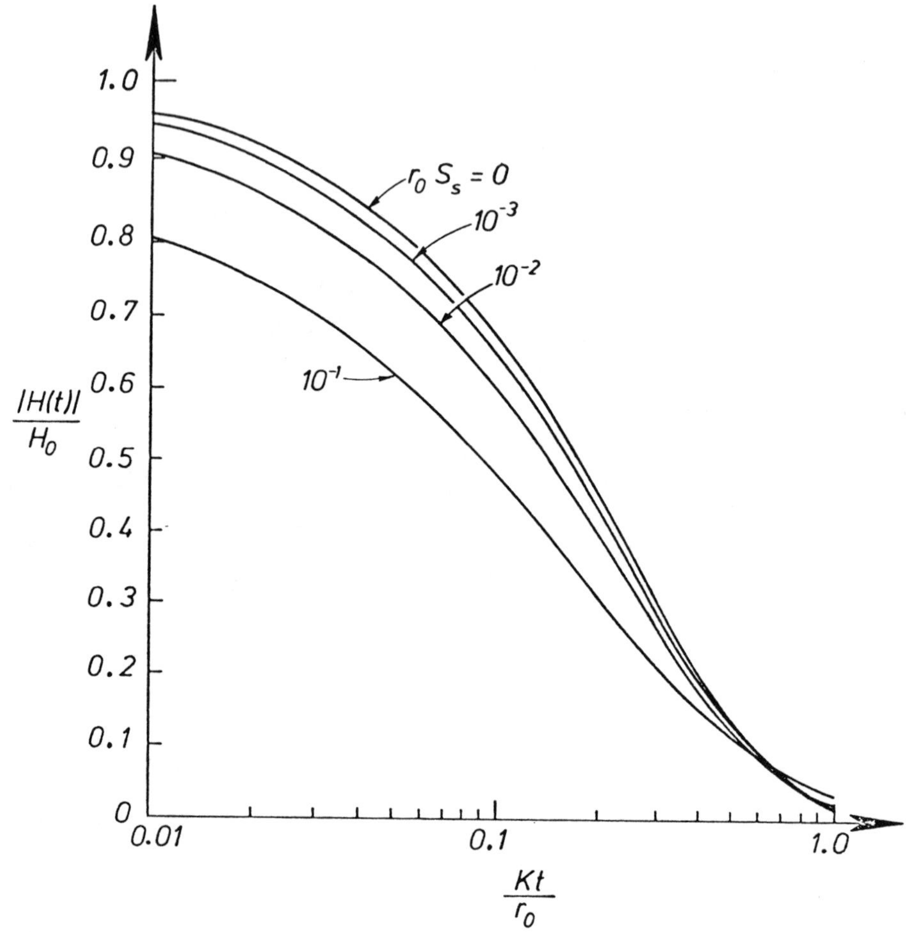

Fig. 5.15 - Water level changes within an open-end pipe after an initial water level change of H_0.

the water level recovery within the pipe. It is shown in Chapter VIII that the solution of this problem is

$$\frac{|H(t)|}{H_0} = \frac{4}{\pi} \int_0^\infty \frac{\exp\left[-x\left(\frac{Kt}{r_0}\right)\right] \sqrt{S_s r_0 x}}{(4 - x)^2 + 16\, S_s r_0 x}\, dx \qquad (5.51)$$

in which r_0 = pipe radius. The product of the specific storage, S_s, and the pipe radius, r_0, is normally a very small dimensionless number, and it is shown in Chapter VIII that

$$\frac{|H(t)|}{H_0} = e^{-4\tau}\,[1 - 8F(\tau)\sqrt{\frac{\varepsilon\tau}{\pi}} + 16\varepsilon\tau(1-2\tau) + \ldots] \qquad (5.52)$$

in which $\tau = Kt/r_0$, $\varepsilon = S_s r_0$ and the function, $F(\tau)$, is given by

$$F(\tau) = 1 - \sum_{n=1}^{\infty} \frac{(4\tau)^n}{n!(4n^2 - 1)}, \quad (0 \leqq \tau < \infty) \qquad (5.53)$$

A plot of this solution is shown in Fig. 5.15. The field test and calculation of K and S_s can, in theory, be carried out in the same way as described for the previous problem. However, accurate values for S_s will be extremely difficult to calculate in most applications since there is so little difference between the curves for $r_0 S_s \leqq 10^{-3}$.

REFERENCES

Abramowitz, M. and Stegun, I.A. (Editors), 1964. Handbook of Mathematical Functions, U.S. National Bureau of Standards, Applied Mathematics Series, No. 55, U.S. Government Printing Office, Washington, D.C., Chs. 5 and 9.

Cooper, H.H., Bredehoeft, J.D. and Papadopulos, I.S. 1967. "Response of a Finite-Diameter Well to an Instantaneous Charge of Water," Water Resources Research, Vol. 3, No. 1, pp. 263-269.

Day, M.C. and Hunt, B.W. 1977. Groundwater Transmissivites in North Canterbury (Note)," New Zealand Journal of Hydrology, Vol. 16, No. 2, pp. 158-163.

Hantush, M.S. and Jacob, C.E. 1955. "Non-steady Radial Flow in an Infinite Leaky Aquifer," Transactions of the American Geophysical Union, Vol. 36, pp. 95-100.

Hantush, M.S. 1964. "Hydraulics of Wells." In: V.T. Chow (Editor), Advances in Hydroscience, Academic Press, New York, Vol. 1, pp. 321-324.

Hunt, B. 1976. "An Aquifer Simulation Study in Northern Canterbury," New Zealand Journal of Science, Vol. 19, pp. 265-276.

Hunt, B. 1980. "Tiwai Peninsula: A Groundwater Resource Analysis," Transactions of the New Zealand Institution of Engineers Incorporated, Vol. 7, No. 2/CE, July, pp. 77-84.

Jacob, C.E. 1946. "Drawdown Test to Determine Effective Radius of Artesian Well," Proceedings of the American Society of Civil Engineers, Vol. 72, No. 5, pp. 629-646.

Kruseman, G.P. and de Ridder, N.A. 1979. Analysis and Evaluation of Pumping Test Data, Fourth Edition, Bulletin 11, International Institute for Land Reclamation and Improvement, P.O. Box 45, Wageningen, The Netherlands.

Nelson, R.W. 1960. "In-place Measurement of Permeability in Heterogeneous Media.1. Theory of a Proposed Method," Journal of Geophysical Research, Vol. 65, No. 6, pp. 1753-1758.

Nelson, R.W. 1961. "In-place Measurement of Permeability in Heterogeneous Media.2. Experimental and Computational Considerations," Journal of Geophysical Research, Vol. 66, No. 8, pp. 2469-2478.

Papadopulos, I.S., Bredehoeft, J.D. and Hilton, H.C. 1973. "On the Analysis of "Slug Test" Data," Water Resources Research, Vol. 9, No. 4, pp. 1087-1089.

Walton, W.C. 1970. Groundwater Resource Evaluation, McGraw-Hill Book Co. New York, pp. 311-321.

PROBLEMS

1. The piezometric distribution in problem 7 of Chapter IV

$$h = \frac{1}{2}(x^2 - y) + 1$$

has the family of streamlines, ψ = constant, given by

$$\psi = \frac{1}{2} x \exp(2y)$$

Prove that the lines ψ = constant and h = constant are orthogonal by showing that $\bar{\nabla} h . \bar{\nabla}\psi = 0$. Next, solve the second equation for y and plot the streamlines $\psi = 0.3$ and $\psi = 0.6$. Then solve the first equation for y and plot the portions of the piezometric contours h = .5, .6, .7, .8, .9 and 1.0 which intersect both streamlines. Finally, use Eq. 5.7 to calculate values of T as multiples of one, arbitrarily chosen reference value of T at some point along the stream tube and compare the result with the exact solution

$$T = \exp(2y)$$

2. Use Eq. 5.13 and a tidal period of 12 hours ($\omega = 2\pi$/period) to calculate the wave celerity in a confined aquifer for which $T = 0.1\ m^2/s$ and $S = 10^{-4}$. Then compare this celerity with the celerity for an unconfined aquifer in which S = 0.1. Finally, use Eq. 5.12 to compute the value of x for each aquifer at which $Max\{h/h_0\} = 0.01$. Note that the wave celerities calculated from Eq. 5.13 are considerably greater than pore velocities.

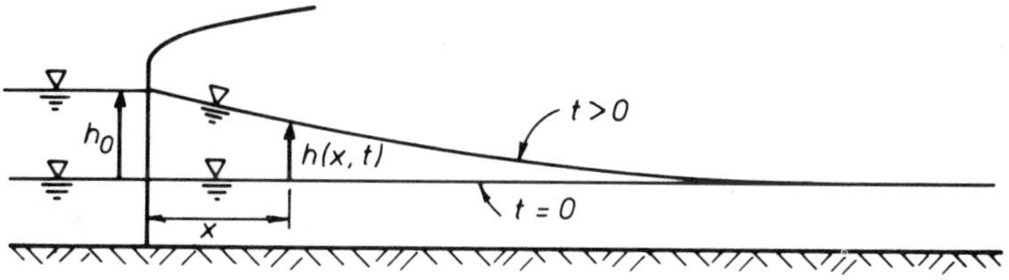

PROB. 3

3. The river level at one end of a semi-infinite aquifer suddenly rises a distance of h_0. The theoretical solution of this problem is given by

$$h(x, t) = h_0 \ \mathrm{erfc}\left(\frac{x}{2}\sqrt{\frac{S}{Tt}}\right)$$

in which erfc = complimentary error function defined as

$$\mathrm{erfc}\left(\frac{x}{2}\sqrt{\frac{S}{Tt}}\right) = \frac{2}{\sqrt{\pi}} \int_{\frac{x}{2}\sqrt{\frac{S}{Tt}}}^{\infty} \exp(-y^2)\ dy$$

A number of observation wells, located at known distances from the river, permit measurements to be taken of h as a function of t. The rise in river level, h_0, is also known. Devise a matchpoint method, using semi-log paper, that would allow the calculation of S/T from the experimental data.

4. Use Jacob's method to obtain S and T from the following data. This data was obtained by the TSE Group Consultants Ltd. of Auckland, New Zealand in 1979 and was analyzed by the author as part of a water supply investigation for a meat freezing company in Blenheim, New Zealand.

Observation Well No. 1 (r = 228 m)

$\|h\|$ (m)	t/r^2 (s/m^2)	$\|h\|$ (m)	t/r^2 (s/m^2)
.04	2.31×10^{-3}	.46	5.77×10^{-2}
.14	5.77×10^{-3}	.52	8.66×10^{-2}
.24	1.15×10^{-2}	.58	1.39×10^{-1}
.29	1.73×10^{-2}	.65	2.25×10^{-1}
.33	2.31×10^{-2}		
.39	3.46×10^{-2}		

Observation Well No. 2 (r = 210 m)

$\|h\|$ (m)	t/r^2 (s/m^2)	$\|h\|$ (m)	t/r^2 (s/m^2)
.08	2.72×10^{-3}	.34	4.08×10^{-2}
.17	6.80×10^{-3}	.38	6.80×10^{-2}
.23	1.36×10^{-2}	.42	1.02×10^{-2}
.27	2.04×10^{-2}	.46	1.63×10^{-1}
.30	2.72×10^{-2}	.51	2.65×10^{-1}

Observation Well No. 3 (r = 296 m)

$\|h\|$ (m)	t/r^2 (s/m^2)	$\|h\|$ (m)	t/r^2 (s/m^2)
.10	1.37×10^{-3}	.39	2.05×10^{-2}
.20	3.42×10^{-3}	.44	3.42×10^{-2}
.27	6.85×10^{-3}	.48	5.14×10^{-2}
.31	1.03×10^{-2}	.53	8.22×10^{-2}
.34	1.37×10^{-2}	.58	1.34×10^{-1}

Observation Well No. 4 (r = 497 m)

$\|h\|$ (m)	t/r^2 (s/m^2)	$\|h\|$ (m)	t/r^2 (s/m^2)
.10	1.21×10^{-3}	.28	1.21×10^{-2}
.15	2.43×10^{-3}	.31	1.82×10^{-2}
.18	3.64×10^{-3}	.35	2.91×10^{-2}
.20	4.86×10^{-3}	.38	4.74×10^{-2}
.24	7.29×10^{-3}		

Observation Well No. 5 (r = 938 m)

|h| (m)	t/r^2 (s/m^2)	|h| (m)	t/r^2 (s/m^2)
.03	1.36×10^{-4}	.13	2.05×10^{-3}
.06	3.41×10^{-4}	.17	3.41×10^{-3}
.07	6.82×10^{-4}	.19	5.11×10^{-3}
.09	1.02×10^{-3}	.22	8.18×10^{-3}
.11	1.36×10^{-3}	.25	1.33×10^{-2}

Observation Well No. 6 (r = 709 m)

|h| (m)	t/r^2 (s/m^2)	|h| (m)	t/r^2 (s/m^2)
.01	5.97×10^{-4}	.12	5.97×10^{-3}
.03	1.19×10^{-3}	.15	8.95×10^{-3}
.05	1.79×10^{-3}	.18	1.43×10^{-2}
.07	2.39×10^{-3}	.22	2.33×10^{-2}
.08	3.58×10^{-3}		

Observation Well No. 7 (r = 460 m)

|h| (m)	t/r^2 (s/m^2)	|h| (m)	t/r^2 (s/m^2)
.01	5.67×10^{-4}	.16	8.51×10^{-3}
.04	1.42×10^{-3}	.21	1.42×10^{-2}
.08	2.84×10^{-3}	.24	2.13×10^{-2}
.11	4.25×10^{-3}	.28	3.40×10^{-2}
.13	5.67×10^{-3}	.32	5.53×10^{-2}

The flow rate from the pumped well was $Q = 0.03825\ m^3/s$, and the measurements given above spanned a time period of 195 minutes.

Work problem 4 by using the Theis or match-point method.

6 Groundwater Pollution

27. Diffusion and Dispersion

In this chapter we will be concerned with the scattering of a pollutant in an aquifer. However, the subject is most easily introduced by first considering molecular diffusion in a fluid that is not surrounded by a porous medium. If a bit of pollutant, such as dye or salt water, is injected at a point in a container of motionless fluid, the pollutant will initially be concentrated in a small region surrounding the injection point. Molecular movement, though, will cause the pollutant particles to migrate away from each other even though the fluid has an average velocity of zero everywhere, and this scattering process will continue until the pollutant particles are distributed uniformly throughout the fluid. Injecting this same pollutant in a moving, laminar flow merely results in the superposition of this molecular scattering process upon the scattering and transport created by the moving fluid particles as they carry the pollutant particles downstream.

Considerations mentioned in the previous paragraph suggest that a conservation of mass statement can be written for the pollutant particles in the form

$$\int_S c\bar{u}_p \cdot \hat{e}_n \, dS + \frac{\partial}{\partial t}\int_{\forall} c \, d\forall = \int_{\forall} I \, d\forall \qquad (6.1)$$

in which c = pollutant concentration per unit volume of mixture (units of mass per unit volume, parts per million, etc.), $\bar{u}_p$ = flux velocity of the pollutant particles and I = injection rate of c within $\forall$ (units of c per unit time per unit volume). Thus, Eq. 6.1 states that the mass transport of c out through S added to the rate of mass increase of c within $\forall$ equals the mass injection rate of c within $\forall$. An application of the divergence theorem to the surface integral in Eq. 6.1 gives

$$\int_{\forall} [\bar{\nabla}\cdot(c\bar{u}_p) + \frac{\partial c}{\partial t}] \, d\forall = \int_{\forall} I \, d\forall \qquad (6.2)$$

Finally, since Eq. 6.2 must hold for all arbitrary choices of $\forall$, the following equation must be satisfied at all points in the fluid:

$$\bar{\nabla}\cdot(c\bar{u}_p) + \frac{\partial c}{\partial t} = I \qquad (6.3)$$

Equation 6.3 contains the unknown vector $\bar{u}_p$, which must be related to the other unknowns in the problem in order to obtain as many equations as unknowns. This is done by introducing an empirical equation known as Fick's first law:

$$c(\bar{u}_p - \bar{u}) = -D\bar{\nabla}c \qquad (6.4)$$

The scalar D is an experimentally measured constant known as the diffusion coefficient (with units of length ^{2}/time). Thus, Eq. 6.4 states that the pollutant particles move relative to the fluid particles in a direction that is orthogonal to surfaces of constant c and in the direction of decreasing c. Equation 6.4 is analogous to Darcy's law, Fourier's law of heat conduction and Ohm's law.

The elimination of $\bar{u}_p$ from Eqs. 6.3 and 6.4 gives

$$D\nabla^2 c = \bar{\nabla}.(\bar{u}c) + \frac{\partial c}{\partial t} - I \qquad (6.5)$$

Finally, since we will restrict our attention to incompressible flows,

$$\bar{\nabla}.\bar{u} = 0 \qquad (6.6)$$

so that Eq. 6.5 reduces to

$$D\nabla^2 c = \bar{u}.\bar{\nabla}c + \frac{\partial c}{\partial t} - I \qquad (6.7)$$

Equation 6.7 is the diffusion equation for an incompressible, laminar flow. Since $\bar{u} = d\bar{r}/dt$, Eq. 6.7 is more easily interpreted by writing it in the form

$$\frac{dc}{dt} = D\nabla^2 c + I \qquad (6.8)$$

in which dc/dt has the physical meaning of the time rate of increase of c for a fluid particle moving with the velocity $\bar{u} = d\bar{r}/dt$. (dc/dt is often written as Dc/Dt in books on fluid mechanics and is called the material derivative, substantial derivative or convective derivative). Thus, Eq. 6.8 states that the pollutant concentration of a fluid particle moving with a flow will change as the result of molecular diffusion or internal production of c. The case of radioactive (exponential) decay is described with Eq. 6.7 by setting

$$I = -\lambda c \qquad (6.9)$$

in which λ is a decay constant that is equal to the reciprocal of the mean lifetime of a radioactive pollutant.

Since scattering in a laminar fluid motion appears superficially to resemble scattering in laminar groundwater flow, it might appear reasonable to use a slightly modified form of Eq. 6.7 to describe diffusion in groundwater flow.

$$D\nabla^2 c = \frac{\bar{u}}{\sigma}.\bar{\nabla}c + \frac{\partial c}{\partial t} - I \qquad (6.10)$$

The porosity, σ, has been inserted in Eq. 6.10 as a result of Eq. 2.2 so that the first two terms on the right side of Eq. 6.10 retain the same physical meaning as in Eq. 6.7:

$$\frac{\bar{u}}{\sigma}.\bar{\nabla}c + \frac{\partial c}{\partial t} = \frac{d\bar{r}}{dt}.\bar{\nabla}c + \frac{\partial c}{\partial t} = \frac{dc}{dt} \tag{6.11}$$

Solutions of Eq. 6.10 for the case of uniform flow ($\bar{u}$ = constant vector everywhere) state that contours of constant c which are initially circular will remain circular with a radius that increases with time as the pollutant is transported downstream. [See Fig. 6.1(a)]. Experimental evidence, however, shows that contours of constant c which are initially circular become elliptical in shape as they are transported downstream, with the major axis of the elliptical contours remaining parallel to the velocity vector. [See Fig. 6.1(b)]. Furthermore, the lengths of the major and minor axes of the ellipses are observed to change when the magnitude of the uniform velocity is changed. This suggests that D is a function of the velocity magnitude and direction, with a larger value in the direction of $\bar{u}$ and a smaller value in any direction orthogonal to $\bar{u}$. Thus, scattering in uniform groundwater flow is described by

$$\frac{\partial}{\partial x}\left(D_1 \frac{\partial c}{\partial x}\right) + \frac{\partial}{\partial y}\left(D_2 \frac{\partial c}{\partial y}\right) + \frac{\partial}{\partial z}\left(D_2 \frac{\partial c}{\partial z}\right)$$
$$= \frac{u}{\sigma}\frac{\partial c}{\partial x} + \frac{\partial c}{\partial t} - I \tag{6.12}$$

in which $D_1(u)$ is greater than $D_2(u)$. Laboratory measurements, some of which are shown by Bear (1972), indicate that

$$D_1 \simeq D_m \text{ for } \frac{ud}{\sigma D_m} \leqq 1$$
$$\simeq d\,\frac{u}{\sigma} \text{ for } \frac{ud}{\sigma D_m} \geqq 1 \tag{6.13}$$

in which D_m = molecular diffusion coefficient, D_1 = longitudinal dispersion coefficient and d = mean grain diameter of the porous matrix. Less numerous measurements of D_2, the lateral dispersion coefficient, suggest that D_2 is about an order of magnitude smaller than D_1 when $ud/(\sigma D_m) > 1$.

The physical reason for the difference between scattering in laminar fluid motion and scattering in groundwater flow is the presence, or absence, of a porous matrix. Molecular diffusion is always present in groundwater flow, but its presence in flows with higher velocities is overshadowed by the mechanical splitting created by the porous matrix. In particular, two pollutant particles starting at two slightly different points in the same passageway of a porous matrix will, in general, travel two entirely different paths to arrive at the same downstream cross section. Thus, these two particles will arrive at this cross section at two different times (longitudinal dispersion) and at two different points in the cross section (lateral dispersion). On the other hand, very low velocity flows are dominated by molecular diffusion. In fact, the porous matrix actually slows diffusion in these flows, so that $D_1 = \beta D_m$ in which $\beta(u) \leqq 1$. Thus, choosing $\beta = 1$ in Eq. 6.13 is really an oversimplification that is only justified by the relative insensitivity of solutions for c to errors in

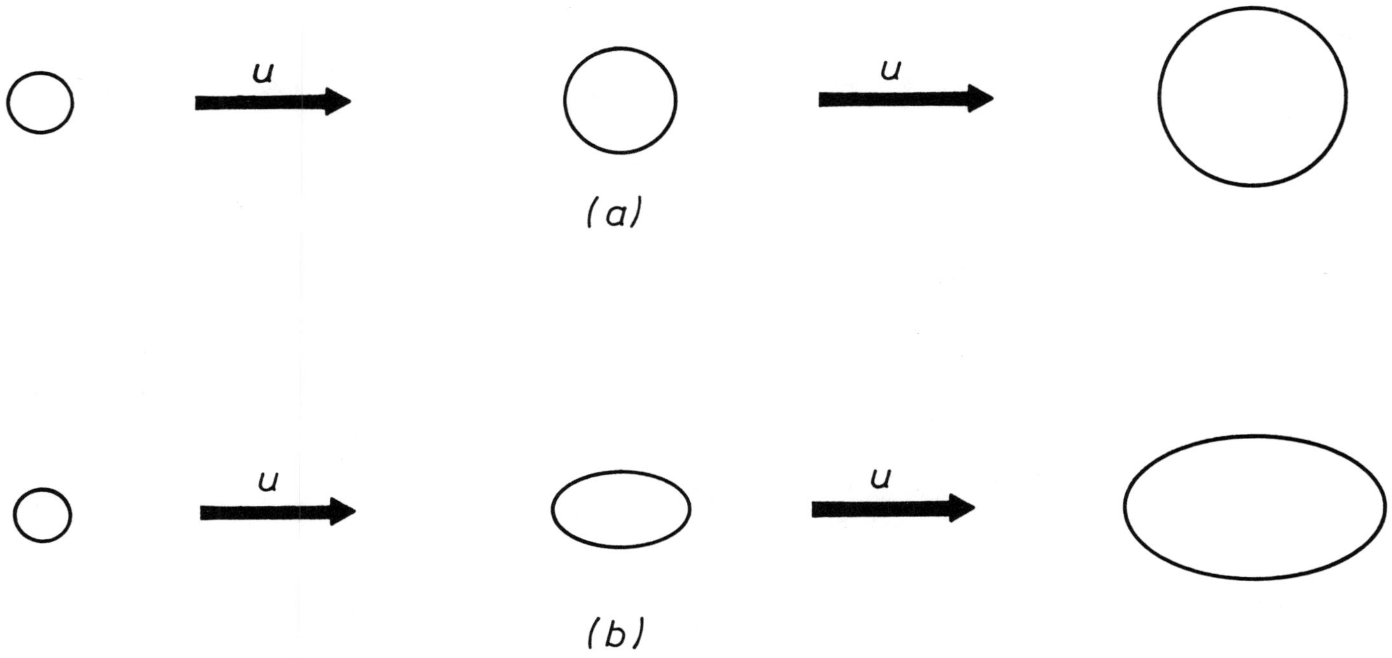

Fig. 6.1 - Contours of constant concentration in (a) uniform, laminar fluid motion and (b) uniform groundwater motion.

D_1 and D_2.

Most field applications are concerned with flows in which $ud/(\sigma D_m) >> 1$, so that the scattering process is dominated by mechanical dispersion rather than molecular diffusion. However, values of d obtained by comparing solutions of Eq. 6.12 with field measurements tend to be orders of magnitude larger than the diameter of the porous matrix particles. Thus, it is customary to write

$$D_1 = \alpha_1 \frac{u}{\sigma} \qquad (6.14)$$

$$D_2 = \alpha_2 \frac{u}{\sigma} \qquad (6.15)$$

in which the longitudinal and lateral dispersitivities, α_1 and α_2, respectively, have units of length and magnitudes that are characteristic scales for the larger aquifer heterogeneities. These dispersivities must always be determined by comparing field measurements with analytical or numerical solutions of the dispersion equation. Furthermore, since use of the dispersion equation implies the assumption of a continuum, the scale of the polluted region in the field experiments must be large compared to the scale of the larger aquifer heterogeneities.

28. The Dispersion Tensor

Equation 6.12 implies that D is a second order tensor with the principal values D_1 and D_2 in directions parallel and normal, respectively, to $\overline{u}$. Thus, Eq. 6.12 must be written for more general flows in the form

$$\overline{\nabla}.(\underline{D}.\overline{\nabla}c) = \frac{\overline{u}}{\sigma}.\overline{\nabla}c + \frac{\partial c}{\partial t} - I \qquad (6.16)$$

in which the dispersion tensor, $\underline{D}$, can be determined by requiring that Eq. 6.16 reduce to Eq. 6.12 for uniform flow. In particular, since $\hat{i} = \overline{u}/u$ and $\hat{j}\hat{j} + \hat{k}\hat{k} = \underline{1} - \hat{i}\hat{i} = \underline{1} - \overline{u}\overline{u}/u^2$, the tensor

$$\underline{D} = D_1 \frac{\overline{u}\overline{u}}{u^2} + D_2 \left(\underline{1} - \frac{\overline{u}\overline{u}}{u^2}\right) \qquad (6.17)$$

can be inserted into Eq. 6.16 to yield Eq. 6.12 for the uniform flow $\overline{u} = u\hat{i}$. (The identity tensor, $\underline{1}$, was introduced in problem 4 of chapter II.) But Eq. 6.17 is written in a form that can be calculated for any coordinate system, which means that it is the general expression for $\underline{D}$.

Writing Eqs. 6.16 and 6.17 in different coordinate systems readily shows the generality of these two equations. Setting $\overline{u} = u_1\hat{e}_1 + u_2\hat{e}_2 + u_e\hat{e}_3$ and computing

$$\begin{aligned}\overline{u}\overline{u} = {} & u_1^2\hat{e}_1\hat{e}_1 + u_1u_2\hat{e}_1\hat{e}_2 + u_1u_3\hat{e}_1\hat{e}_3 \\ & + u_2u_1\hat{e}_2\hat{e}_1 + u_2^2\hat{e}_2\hat{e}_2 + u_2u_3\hat{e}_2\hat{e}_3 \\ & + u_3u_1\hat{e}_3\hat{e}_1 + u_3u_2\hat{e}_3\hat{e}_2 + u_3^2\hat{e}_3\hat{e}_3\end{aligned} \qquad (6.18)$$

leads to the general expression for D_{ij}:

$$D_{ij} = D_2 \delta_{ij} + (D_1 - D_2) \frac{u_i u_j}{u^2} \tag{6.19}$$

The Kronecker delta, δ_{ij}, and velocity magnitude, u, are defined by

$$\begin{aligned} \delta_{ij} &= 1 \text{ for } i = j \\ &= 0 \text{ for } i \neq j \end{aligned} \tag{6.20}$$

$$u^2 = u_1^2 + u_2^2 + u_3^2 \tag{6.21}$$

Equation 6.19 holds for any right-handed, orthogonal coordinate system. Care must be taken, however, when inserting 6.17 into Eq. 6.16 since the unit base vectors, $\hat{e}_i$, of some coordinate systems are not constant vectors.

Equation 6.16 is particularly easy to write in scalar form for Cartesian coordinates since all unit base vectors are constant in the Cartesian system. Thus, the various terms in Eq. 6.16 are

$$\frac{\bar{u}}{\sigma}.\bar{\nabla}c = \frac{1}{\sigma}\left\{u_x \frac{\partial c}{\partial x} + u_y \frac{\partial c}{\partial y} + u_z \frac{\partial c}{\partial z}\right\} \tag{6.22}$$

$$\begin{aligned} \bar{\nabla}.(\underline{D}.\bar{\nabla}c) = {} & \frac{\partial}{\partial x}\left\{D_{xx}\frac{\partial c}{\partial x} + D_{xy}\frac{\partial c}{\partial y} + D_{xz}\frac{\partial c}{\partial z}\right\} \\ & + \frac{\partial}{\partial y}\left\{D_{yx}\frac{\partial c}{\partial x} + D_{yy}\frac{\partial c}{\partial y} + D_{yz}\frac{\partial c}{\partial z}\right\} \\ & + \frac{\partial}{\partial z}\left\{D_{zx}\frac{\partial c}{\partial x} + D_{zy}\frac{\partial c}{\partial y} + D_{zz}\frac{\partial c}{\partial z}\right\} \end{aligned} \tag{6.23}$$

$$[D_{ij}] = D_2 \begin{bmatrix} 1 & 0 & 0 \\ 0 & 1 & 0 \\ 0 & 0 & 1 \end{bmatrix} + \frac{(D_1 - D_2)}{u^2} \begin{bmatrix} u_x^2 & u_x u_y & u_x u_z \\ u_y u_x & u_y^2 & u_y u_z \\ u_z u_x & u_z u_y & u_z^2 \end{bmatrix} \tag{6.24}$$

Equations 6.23 and 6.24 are easily seen to reduce to the correct expressions for uniform flow. For example, setting $u_y = u_z = 0$ gives $D_{xx} = D_1$, $D_{yy} = D_{zz} = D_2$ and $D_{ij} = 0$ for $i \neq j$. Likewise, setting $u_x = u_z = 0$ gives $D_{xx} = D_{zz} = D_2$, $D_{yy} = D_1$ and $D_{ij} = 0$ for $i \neq j$.

In the cylindrical coordinate system (r, θ, z) shown in Fig. 6.2(a), general expressions for the gradient of a scalar and divergence of a vector [Hildebrand (1962)] are given, respectively, by

$$\bar{\nabla}c = \hat{e}_r \frac{\partial c}{\partial r} + \hat{e}_\theta \frac{1}{r}\frac{\partial c}{\partial \theta} + \hat{e}_z \frac{\partial c}{\partial z} \tag{6.25}$$

$$\bar{\nabla}.\bar{F} = \frac{1}{r}\frac{\partial}{\partial r}(rF_r) + \frac{1}{r}\frac{\partial F_\theta}{\partial \theta} + \frac{\partial F_z}{\partial z} \tag{6.26}$$

Thus, the convective terms in Eq. 6.16 are given by

$$\frac{\bar{u}}{\sigma}.\bar{\nabla}c = \frac{1}{\sigma}\left\{u_r \frac{\partial c}{\partial r} + \frac{u_\theta}{r}\frac{\partial c}{\partial \theta} + u_z \frac{\partial c}{\partial z}\right\} \tag{6.27}$$

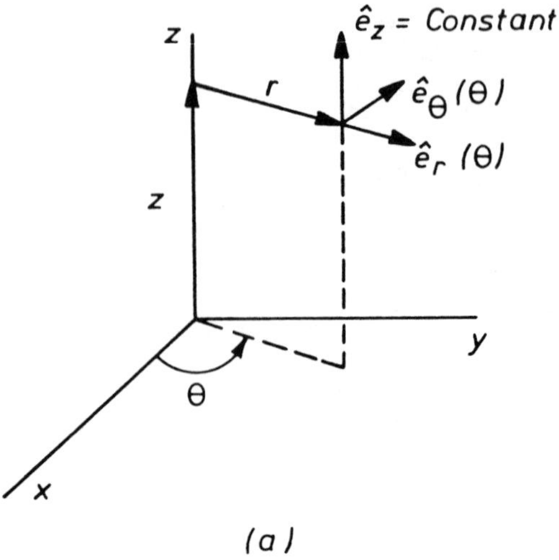

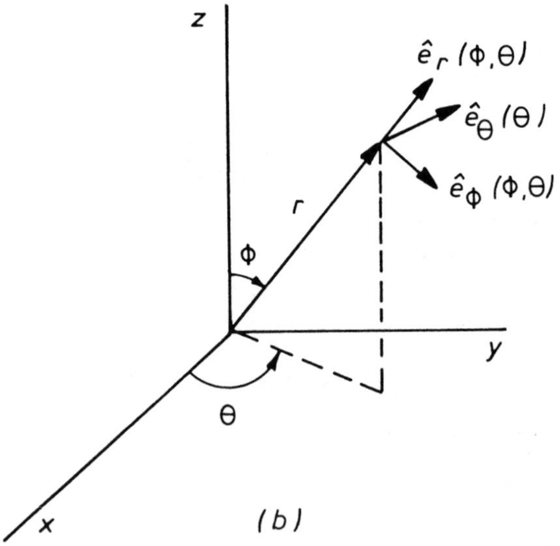

Fig. 6.2 - Coordinates and unit base vectors in (a) the cylindrical coordinate system and (b) the spherical coordinate system.

The left side of Eq. 6.16 can be obtained by first calculating the vector

$$\underline{D}.\overline{\nabla}c = \hat{e}_r\left(D_{rr}\frac{\partial c}{\partial r} + \frac{D_{r\theta}}{r}\frac{\partial c}{\partial \theta} + D_{rz}\frac{\partial c}{\partial z}\right) + \hat{e}_\theta\left(D_{\theta r}\frac{\partial c}{\partial r} + \frac{D_{\theta\theta}}{r}\frac{\partial c}{\partial \theta} + D_{\theta z}\frac{\partial c}{\partial z}\right) + \hat{e}_z\left(D_{zr}\frac{\partial c}{\partial r} + \frac{D_{z\theta}}{r}\frac{\partial c}{\partial \theta} + D_{zz}\frac{\partial c}{\partial z}\right) \quad (6.28)$$

Thus, substituting the three components of Eq. 6.28 for $\overline{F}$ in Eq. 6.26 gives

$$\overline{\nabla}.(\underline{D}.\overline{\nabla}c) = \frac{1}{r}\frac{\partial}{\partial r}\left[r\left(D_{rr}\frac{\partial c}{\partial r} + \frac{D_{r\theta}}{r}\frac{\partial c}{\partial \theta} + D_{rz}\frac{\partial c}{\partial z}\right)\right] + \frac{1}{r}\frac{\partial}{\partial \theta}\left(D_{\theta r}\frac{\partial c}{\partial r} + \frac{D_{\theta\theta}}{r}\frac{\partial c}{\partial \theta} + D_{\theta z}\frac{\partial c}{\partial z}\right) + \frac{\partial}{\partial z}\left(D_{zr}\frac{\partial c}{\partial r} + \frac{D_{z\theta}}{r}\frac{\partial c}{\partial \theta} + D_{zz}\frac{\partial c}{\partial z}\right) \quad (6.29)$$

Components of $\underline{D}$ for use in Eq. 6.29 can be obtained from Eq. 6.24 simply by replacing the (x, y, z) subscripts with (r, θ, z), respectively.

In the spherical coordinate system (r, φ, θ) shown in Fig. 6.2(b), general expressions for the gradient and divergence [Hildebrand (1962)] are given, respectively, by

$$\overline{\nabla}c = \hat{e}_r\frac{\partial c}{\partial r} + \hat{e}_\phi\frac{1}{r}\frac{\partial c}{\partial \phi} + \hat{e}_\theta\frac{1}{r\sin\phi}\frac{\partial c}{\partial \theta} \quad (6.30)$$

$$\overline{\nabla}.\overline{F} = \frac{1}{r^2}\frac{\partial}{\partial r}\left(r^2F_r\right) + \frac{1}{r\sin\phi}\frac{\partial}{\partial \phi}(F_\phi\sin\phi) + \frac{1}{r\sin\phi}\frac{\partial F_\theta}{\partial \theta} \quad (6.31)$$

Thus, Eqs. 6.30 and 6.31 lead to the results

$$\frac{\overline{u}}{\sigma}.\overline{\nabla}c = \frac{1}{\sigma}\left(u_r\frac{\partial c}{\partial r} + \frac{u_\phi}{r}\frac{\partial c}{\partial \phi} + \frac{u_\theta}{r\sin\phi}\frac{\partial c}{\partial \theta}\right) \quad (6.32)$$

$$\overline{\nabla}.(\underline{D}.\overline{\nabla}c) = \frac{1}{r^2}\frac{\partial}{\partial r}\left[r^2\left(D_{rr}\frac{\partial c}{\partial r} + \frac{D_{r\phi}}{r}\frac{\partial c}{\partial \phi} + \frac{D_{r\theta}}{r\sin\phi}\frac{\partial c}{\partial \phi}\right)\right] + \frac{1}{r\sin\phi}\frac{\partial}{\partial \phi}\left[\left(D_{\phi r}\frac{\partial c}{\partial r} + \frac{D_{\phi\phi}}{r}\frac{\partial c}{\partial \phi} + \frac{D_{\phi\theta}}{r\sin\phi}\frac{\partial c}{\partial \theta}\right)\sin\phi\right] + \frac{1}{r\sin\phi}\frac{\partial}{\partial \theta}\left(D_{\theta r}\frac{\partial c}{\partial r} + \frac{D_{\theta\phi}}{r}\frac{\partial c}{\partial \phi} + \frac{D_{\theta\theta}}{r\sin\phi}\frac{\partial c}{\partial \theta}\right) \quad (6.33)$$

Once again, components of $\underline{D}$ for use on Eq. 6.33 can be obtained from Eq. 6.24 by replacing the subscripts (x, y, z) with (r, φ, θ), respectively.

29. Pollution Sources in Uniform Flow

The complicated nature of the dispersion tensor in Eq. 6.17 means that exact, analytical solutions of Eq. 6.16 can only be obtained for the greatly simplified case of uniform flow. In this case, if the constant velocity is taken to be in the x direction and if the pollutant is assumed to follow the radioactive decay law of Eq. 6.9, then Eq. 6.16 takes the relatively simple form

$$D_1 \frac{\partial^2 c}{\partial x^2} + D_2 \frac{\partial^2 c}{\partial y^2} + D_2 \frac{\partial^2 c}{\partial x^2} = \frac{u_x}{\sigma} \frac{\partial c}{\partial x} + \frac{\partial c}{\partial t} + \lambda c \tag{6.34}$$

in which D_1, D_2, u_x/σ and λ are all constants.

One of the most useful solutions of Eq. 6.34 is obtained for an instantaneous point source of pollutant. In this case, the solution of Eq. 6.34 satisfies the initial condition

$$c(x, y, z, 0) = \frac{M_3}{\sigma} \delta(x) \, \delta(y) \, \delta(z) \tag{6.35}$$

in which δ is the Dirac delta function that was introduced in Eqs. 2.44 and 2.45. A physical meaning for the constant, M_3, can be obtained by integrating both sides of Eq. 6.35.

$$M_3 = \iiint_{-\infty}^{\infty} c(x, y, z, 0) \, \sigma \, dx \, dy \, dz \tag{6.36}$$

Thus, since c is the mass of pollutant per unit volume of mixture, Eq. 6.36 shows that M_3 is the total mass of pollutant released at the coordinate origin at $t = 0$.

The solution of Eqs. 6.34 and 6.35 can be accomplished in a straightforward manner by using the Fourier transform theorem:

$$f(x) = \frac{1}{2\pi} \int_{-\infty}^{\infty} F(\alpha) \exp(i\alpha x) \, d\alpha \tag{6.37}$$

$$F(\alpha) = \int_{-\infty}^{\infty} f(x) \exp(-i\alpha x) \, dx \tag{6.38}$$

The form of Eqs. 6.37 - 6.38 suggests that a solution to Eqs. 6.34 - 6.35 might be sought in the form

$$c(x, y, z, t) = \frac{1}{(2\pi)^3} \iiint_{-\infty}^{\infty} F(\alpha_1, \alpha_2, \alpha_3) \exp[i(\alpha_1 x + \alpha_2 y + \alpha_3 z) + \gamma t] \, d\alpha_1 \, d\alpha_2 \, d\alpha_3 \tag{6.39}$$

Substituting Eq. 6.39 into Eq. 6.34 shows that γ is given by

$$\gamma = -D_1 \alpha_1^2 - D_2 \left(\alpha_2^2 + \alpha_3^2\right) - i \frac{u_x}{\sigma} \alpha_1 - \lambda \tag{6.40}$$

Finally, $F(\alpha_1, \alpha_2, \alpha_3)$ can be calculated by setting $t = 0$ on both sides of Eq. 6.39 and inserting Eq. 6.35 to obtain

$$\frac{M_3}{\sigma} \delta(x) \, \delta(y) \, \delta(z) = \frac{1}{(2\pi)^3} \iiint_{-\infty}^{\infty} F(\alpha_1, \alpha_2, \alpha_3) \exp[i(\alpha_1 x + \alpha_2 y + \alpha_3 z)] \, d\alpha_1 \, d\alpha_2 \, d\alpha_3 \tag{6.41}$$

Thus, applying Eq. 6.38 three times to Eq. 6.41 gives

$$F(\alpha_1, \alpha_2, \alpha_3) = \frac{M_3}{\sigma} \int\int\int_{-\infty}^{\infty} \delta(x)\ \delta(y)\ \delta(z)$$

$$\exp[-i(\alpha_1 x + \alpha_2 y + \alpha_3 z)]\ d\alpha_1\ d\alpha_2\ d\alpha_3 = \frac{M_3}{\sigma} \tag{6.42}$$

Substituting Eqs. 6.42 and 6.40 in Eq. 6.39 gives the solution

$$c(x, y, z, t) = \frac{M_3}{\sigma} \exp(-\lambda t)\ I\left(x - \frac{u_x}{\sigma} t, D_1 t\right) I(y, D_2 t)\ I(z, D_2 t) \tag{6.43}$$

in which

$$I(y, D_2 t) = \frac{1}{2\pi} \int_{-\infty}^{\infty} \exp(i\alpha_2 y - \alpha_2^2 D_2 t)\ d\alpha_2 \tag{6.44}$$

Since the integrand of Eq. 6.44 has real and imaginary parts that are even and odd functions of α_2, respectively,

$$I(y, D_2 t) = \frac{1}{\pi} \int_0^{\infty} \exp(-\alpha_2^2 D_2 t) \cos(\alpha_2 y)\ d\alpha_2 \tag{6.45}$$

The integral in Eq. 6.45 is given in many tabulations of definite integrals [for example, Gradshteyn and Ryzhik (1965)]:

$$I(y, D_2 t) = \frac{1}{2\sqrt{\pi D_2 t}} \exp\left(-\frac{y^2}{4D_2 t}\right) \tag{6.46}$$

Since the same integral appears three times with different arguments in Eq. 6.43, the repeated use of Eq. 6.46 in Eq. 6.43 gives the solution for an instantaneous, point source at the origin with radioactive decay:

$$c(x, y, z, t) = \frac{M_3}{8\sigma\sqrt{\pi^3 t^3 D_1 D_2^2}} \exp\left(-\lambda t - \frac{1}{4t}\left[\frac{\left(x - \frac{u_x}{\sigma} t\right)^2}{D_1} + \frac{y^2}{D_2} + \frac{z^2}{D_2}\right]\right) \tag{6.47}$$

Equation 6.47 shows that $c = 0$ at $t = 0$ everywhere except at $x = y = z = 0$. At the origin, $c \to \infty$ as $t \to 0$ at a rate that requires the spatial integral of c throughout any region that includes the origin to give the total mass, M_3, of injected pollutant. For any fixed value of t greater than zero, surfaces of constant c have the equation

$$\frac{\left(x - \frac{u_x}{\sigma} t\right)^2}{D_1} + \frac{y^2}{D_2} + \frac{z^2}{D_2} = \text{constant} \tag{6.48}$$

Thus, the surfaces of constant c are axisymmetric ellipsoids centered at $z = y = 0$, $x = (u_x/\sigma)t$ and with major and minor axes proportional to

$\sqrt{D_1}$ and $\sqrt{D_2}$, respectively. If, as seems to be the case in many applications, D_1 is about one order of magnitude greater than D_2, then the major axis is parallel to the uniform velocity vector and has a length that is about three times greater than the length of the two minor axes.

Solutions for instantaneous, point sources in one and two dimensions can be derived in a similar manner. However, it is probably more instructive for future applications to derive these solutions from Eq. 6.47. Since Eq. 6.34 is a linear equation with constant coefficients, any function of (x, y, z, t) that is a solution of Eq. 6.34 will also be a solution when (x, y, z, t) are replaced with (x-ρ, y-η, z-ζ, t-τ), in which (ρ, η, ζ, τ) are independent of (x, y, z, t). This process, when applied to Eq. 6.47, yields the solution for an instantaneous, point source inserted at $x = \rho$, $y = \eta$, $z = \zeta$ at the time $t = \tau$. Once again, since Eq. 6.47 is linear with constant coefficients, these sources can be distributed in space and time to yield other solutions. The distribution process can be carried out by using either finite summations or definite integrals.

As an example of this superposition process, Eq. 6.47 can be used to obtain the solution for an instantaneous, point source in two dimensions by distributing three-dimensional sources along a straight line extending from $z = -\infty$ to $z = \infty$. Thus, replacing M_3 in Eq. 6.47 with $M_2 d\zeta$ and integrating gives

$$c(x, y, t) = \frac{M_2}{8\sigma\sqrt{\pi^3 t^3 D_1 D_2^{\,2}}} \exp\left\{-\lambda t - \frac{1}{4t}\left[\frac{\left(x - \frac{u_x}{\sigma}t\right)^2}{D_1} + \frac{y^2}{D_2}\right]\right\} \int_{-\infty}^{\infty} \exp\left[-\frac{(z-\zeta)^2}{4D_2 t}\right] d\zeta \tag{6.49}$$

Calculating the integral in Eq. 6.49 gives the result

$$c(x, y, t) = \frac{M_2}{4\sigma\pi t\sqrt{D_1 D_2}} \exp\left\{-\lambda t - \frac{1}{4t}\left[\frac{\left(x - \frac{u_x}{\sigma}t\right)^2}{D_1} + \frac{y^2}{D_2}\right]\right\} \tag{6.50}$$

Since M_3 in Eq. 6.47 was replaced with $M_2 d\zeta$ to obtain Eq. 6.50, it is seen that M_2 is the pollutant mass per unit length along the line source at $t = 0$. In exactly the same way, distributing point sources over the entire plane $x = 0$ gives the solution for an instantaneous, point source in one dimension:

$$c(x, t) = \frac{M_1}{2\sigma\sqrt{\pi t D_1}} \exp\left[-\lambda t - \frac{\left(x - \frac{u_x}{\sigma}t\right)^2}{4tD_1}\right] \tag{6.51}$$

Since M_3 in Eq. 6.47 was replaced with $M_1\, d\eta d\zeta$ to obtain Eq. 6.51, M_1 is seen to be the pollutant mass per unit area over the plane $x = 0$ at $t = 0$.

Continuous sources, the result of injecting a pollutant with a constant mass flow rate at a fixed point in an aquifer, can be obtained by distributing instantaneous sources with respect to time. For example, replacing M_3 in Eq. 6.47 with $\dot{M}_3 d\tau$ and integrating gives

$$c(x, y, z, t) = \frac{\dot{M}_3}{8\sigma\sqrt{\pi^3 D_1 D_2^2}} \int_0^t (t - \tau)^{-3/2} \exp\left[-\lambda(t - \tau) - \frac{[x - \frac{u_x}{\sigma}(t - \tau)]^2}{4D_1(t - \tau)} - \frac{1}{4(t - \tau)}\left(\frac{y^2}{D_2} + \frac{z^2}{D_2}\right)\right] d\tau \qquad (6.52)$$

in which $\dot{M}_3$ is the flux of M_3 emitted from the source. Carrying out the rather difficult integration in Eq. 6.52 gives (see problem 6)

$$c(x, y, z, t) = \frac{\dot{M}_3 \exp\left(\frac{xu_x}{2\sigma D_1}\right)}{8\sigma\pi D_2 R}\left[\exp\left(-\frac{aR}{\sqrt{D_1}}\right) \operatorname{erfc}\left(\frac{R}{2\sqrt{D_1 t}} - a\sqrt{t}\right) + \exp\left(\frac{aR}{\sqrt{D_1}}\right) \operatorname{erfc}\left(\frac{R}{2\sqrt{D_1 t}} + a\sqrt{t}\right)\right] \qquad (6.53)$$

in which R and a are defined as

$$R = \sqrt{x^2 + \frac{D_1}{D_2}(y^2 + z^2)} \qquad (6.54)$$

$$a = \sqrt{\lambda + \frac{u_x^2}{4D_1\sigma^2}} \qquad (6.55)$$

The complementary error function is defined as

$$\operatorname{erfc}(x) = \frac{2}{\sqrt{\pi}} \int_x^\infty e^{-u^2} du = 1 - \frac{2x}{\sqrt{\pi}} \sum_{n=0}^\infty \frac{(-x^2)^n}{n!\,(2n + 1)} \qquad (6.56)$$

For large values of x, erfc has the asymptotic expansion

$$\operatorname{erfc}(x) \sim \frac{\exp(-x^2)}{x\sqrt{\pi}}\left[1 + \sum_{n=1}^\infty (-1)^n \frac{1.3\ldots(2n - 1)}{(2x^2)^n}\right] \qquad (6.57)$$

Numerical values of erfc are tabulated by Abramowitz and Stegun (1970).

Similar calculations with Eq. 6.51 give the one-dimensional solution for a continuous source:

$$c(x, t) = \frac{\dot{M}_1 \exp\left(\frac{xu_x}{2\sigma D_1}\right)}{4\sigma a\sqrt{D_1}}\left[\exp\left(-\frac{aR}{\sqrt{D_1}}\right) \operatorname{erfc}\left(\frac{R}{2\sqrt{D_1 t}} - a\sqrt{t}\right) - \exp\left(\frac{aR}{\sqrt{D_1}}\right) \operatorname{erfc}\left(\frac{R}{2\sqrt{D_1 t}} + a\sqrt{t}\right)\right] \qquad (6.58)$$

in which $\dot{M}_1$ is the flux of M_1 emitted by the source. The variable a is defined by Eq. 6.55, and the definition of R in Eq. 6.54 reduces to

$$R = \sqrt{x^2} = |x| \tag{6.59}$$

The corresponding two-dimensional solution for a continuous source can be obtained from Eq. 6.50 in the form

$$c(x, y, t) = \frac{\dot{M}_2 \exp\left(\frac{xu_x}{2\sigma D_1}\right)}{4\sigma\pi\sqrt{D_1 D_2}} W\left(\frac{R^2}{4D_1 t}, \frac{aR}{\sqrt{D_1}}\right) \tag{6.60}$$

in which $\dot{M}_2$ is the flux of M_2 emitted by the source and R is defined as

$$R = \sqrt{x^2 + \frac{D_1}{D_2} y^2} \tag{6.61}$$

The function W is the well function for leaky aquifers defined by Eq. 5.26.

Steady-state solutions for the continuous sources can be obtained simply by letting $t \to \infty$ in Eqs. 6.53, 6.58 and 6.60. This gives

$$c(x, y, z, \infty) = \frac{\dot{M}_3}{4\sigma\pi D_2 R} \exp\left(\frac{xu_x}{2\sigma D_1} - \frac{aR}{\sqrt{D_1}}\right) \tag{6.62}$$

$$c(x, \infty) = \frac{\dot{M}_1}{2\sigma a\sqrt{D_1}} \exp\left(\frac{xu_x}{2\sigma D_1} - \frac{aR}{\sqrt{D_1}}\right) \tag{6.63}$$

$$c(x, y, \infty) = \frac{\dot{M}_2}{2\sigma\pi\sqrt{D_1 D_2}} \exp\left(\frac{xu_x}{2\sigma D_1}\right) K_0\left(\frac{aR}{\sqrt{D_1}}\right) \tag{6.64}$$

The zero-order, modified Bessel function of the second kind, K_0, has been met previously in Eq. 5.28, and values of K_0 are given in Table 3 of Chapter 5.

Our calculations with pollution sources in uniform flow will be truncated with Eq. 6.64. However, the different possibilities that can be modelled by using either continuous or discrete distributions of these singular solutions is endless. Hunt (1978a) has also used these solutions to answer such basic questions as how long a continuous source must be in place before steady-state conditions at any given point are approached, how deep must an aquifer be in order that the influence of a bottom boundary upon a pollution source at the free surface become negligible and how much time is required for the solutions for an instantaneous point source and an instantaneous source of finite size to approach each other.

30. Series Expansions of the Well Function for Leaky Aquifers

Two infinite series expansions for the leaky aquifer well function were given in Chapter 5. However, the derivation of these expansions has been delayed until after this same function has been met in the study of groundwater pollution (Eq. 6.60). This is because Eqs. 5.27 and 5.28 converge too slowly for the efficient calculation of W in many pollution applications, and the derivation of Eqs. 5.27 and 5.28 will be supplemented in this section with a derivation of an asymptotic expansion for use when the second argument of W is large.

The well function for leaky aquifers is defined as

$$W(u, \beta) = \int_u^\infty \exp\left(-y - \frac{\beta^2}{4y}\right) \frac{dy}{y} \tag{6.65}$$

If we set $y = ux$, Eq. 6.65 takes the form

$$W(u, \beta) = \int_1^\infty \exp(-ux) \exp\left(-\frac{\beta^2}{4ux}\right) \frac{dx}{x} \tag{6.66}$$

A Taylor's series expansion yields

$$\exp\left(-\frac{\beta^2}{4ux}\right) = \sum_{n=0}^{\infty} \frac{1}{n!}\left(-\frac{\beta^2}{4ux}\right)^n \quad \text{for } 0 \leqq \frac{\beta^2}{4ux} < \infty \tag{6.67}$$

Since Eq. 6.67 is absolutely convergent over the entire integration interval in Eq. 6.66, and since the resulting integrals exist when Eq. 6.67 is substituted into Eq. 6.66 and the summation and integration orders are interchanged, we obtain

$$W(u, \beta) = \sum_{n=0}^{\infty} \frac{1}{n!}\left(-\frac{\beta^2}{4u}\right)^n \int_1^\infty \exp(-ux) \frac{dx}{x^{n+1}} \tag{6.68}$$

The integral in Eq. 6.68 is the exponential integral that was introduced in Eq. 5.29. Thus, the final result is

$$W(u, \beta) = \sum_{n=0}^{\infty} \frac{E_{n+1}(u)}{n!}\left(-\frac{\beta^2}{4u}\right)^n \tag{6.69}$$

which agrees with Eq. 5.27. The ratio test can be used to show that the expansion in Eq. 6.69 is absolutely convergent for $0 \leqq \beta^2/(4u) < \infty$. Furthermore, the series is alternating, which means that its truncation error magnitude is always less than the magnitude of the first neglected term <u>if</u> the neglected term has a magnitude that is smaller than the magnitude of the previous term.

Equation 6.65 can also be written in the form

$$W(u, \beta) = \int_0^\infty \exp\left(-y - \frac{\beta^2}{4y}\right) \frac{dy}{y} - \int_0^u \exp\left(-y - \frac{\beta^2}{4y}\right) \frac{dy}{y} \tag{6.70}$$

The first integral is given by Gradshteyn and Ryshik (1965), and use of the substitution $y = u/x$ in the second integral gives the result

$$W(u, \beta) = 2K_0(\beta) - \int_1^{\infty} \exp\left(-\frac{\beta^2}{4u} x\right) \exp\left(-\frac{u}{x}\right) \frac{dx}{x} \tag{6.71}$$

Use of the absolutely convergent expansion

$$\exp\left(-\frac{u}{x}\right) = \sum_{n=0}^{\infty} \frac{(-u/x)^n}{n!} \quad \text{for } 0 \leqq u/x < \infty \tag{6.72}$$

in Eq. 6.71 leads to the result

$$W(u, \beta) = 2K_0(\beta) - \sum_{n=0}^{\infty} \frac{(-u)^n}{n!} \int_1^{\infty} \exp\left(-\frac{\beta^2}{4u} x\right) \frac{dx}{x^{n+1}} \tag{6.73}$$

Thus, use of Eq. 5.29 gives the final result

$$W(u, \beta) = 2K_0(\beta) - \sum_{n=0}^{\infty} \frac{(-u)^n}{n!} E_{n+1}\left(\frac{\beta^2}{4u}\right) \tag{6.74}$$

Equation 6.74 agrees with Eq. 5.28, and the infinite series in Eq. 6.74 is an alternating series that is absolutely convergent for $0 \leqq u < \infty$.

Equation 6.69 was first obtained by Case and Addiego (1977), who did not give the radius of convergence for the infinite series. Equation 6.74 was first obtained by Hunt (1977), who also derived Eq. 6.69 and gave the radii of convergence for the infinite series in Eqs. 6.69 and 6.74.

As pointed out at the beginning of this section, $W(u, \beta)$ must often be calculated for use in Eq. 6.60 when β is large. An asymptotic expansion for this case can be obtained by setting $y = \beta x$ in Eq. 6.65 to obtain

$$W(u,\beta) = \int_{u/\beta}^{\infty} \exp\left[-\beta\left(x + \frac{1}{4x}\right)\right] \frac{dx}{x} \tag{6.75}$$

The exponential function in the integrand of Eq. 6.75 vanishes at $x = 0$ and $x = \infty$. Furthermore, it is positive on the entire integration interval and has one maximum at $x = 1/2$. Since $x + 1(4x) \sim 1 + 2(x - 1/2)^2$ near this maximum, the peak becomes more and more compressed about $x = 1/2$ as β in Eq. 6.75 becomes larger. This suggests making the substitution

$$x + \frac{1}{4x} = 1 + 2\tau^2 \tag{6.76}$$

in Eq. 6.75 to obtain

$$W(u, \beta) = 2\exp(-\beta) \int_{\gamma/\sqrt{2\beta}}^{\infty} \exp(-2\beta\tau^2) \frac{d\tau}{\sqrt{1 + \tau^2}} \tag{6.77}$$

in which

$$\gamma = \left(\frac{u}{\beta} - \frac{1}{2}\right) \quad \frac{\beta}{\sqrt{u}} \tag{6.78}$$

Equation 6.76 moves the maximum in the exponential function at $x = 1/2$ to $\tau = 0$.

Equation 6.77 is an exact expression for W, and it shows clearly that the peak in $\exp(-2\beta\tau^2)$ becomes more compressed about $\tau = 0$ as β becomes large. Since the integrand in Eq. 6.77 vanishes rapidly, for large β, as one moves away from $\tau = 0$, it is reasonable to use the infinite series expansion

$$\frac{1}{\sqrt{1+\tau^2}} = 1 + \sum_{n=1}^{\infty} (-1)^n \frac{1.3.5\ldots(2n-1)}{n!\,2^n} \tau^{2n} \tag{6.79}$$

The infinite series in Eq. 6.79 converges only for $|\tau| < 1$, but the exponential function in Eq. 6.77 vanishes rapidly enough as $|\tau|$ increases to insure the existence of the integral. Thus, substituting Eq. 6.79 into Eq. 6.77 and integrating gives the asymptotic series

$$W(u, \beta) \sim \sqrt{\frac{2}{\beta}} \exp(-\beta) \left[I_0 + \sum_{n=1}^{\infty} (-1)^n \frac{1.3.5\ldots(2n-1)}{n!} \frac{I_{2n}(\gamma)}{(4\beta)^n} \right] \tag{6.80}$$

in which

$$I_{2n}(\gamma) = \int_{\gamma}^{\infty} x^{2n} \exp(-x^2)\, dx \tag{6.81}$$

Values of $I_{2n}(\gamma)$ can be computed by noting that

$$I_0(\gamma) = \frac{\sqrt{\pi}}{2} \operatorname{erfc}(\gamma) \tag{6.82}$$

Thus, integrating Eq. 6.81 once, by parts, allows other values of $I_{2n}(\gamma)$ to be computed from Eq. 6.82 and the recurrence formula

$$I_{2n}(\gamma) = \left(\frac{2n-1}{2}\right) I_{2n-2}(\gamma) + \frac{\gamma^{2n-1}}{2} \exp(-\gamma^2) \tag{6.83}$$

The complementary error function, $\operatorname{erfc}(\gamma)$, can be computed from

$$\operatorname{erfc}(\gamma) = 1 - \operatorname{erf}(\gamma) \tag{6.84}$$

$$\operatorname{erf}(-\gamma) = -\operatorname{erf}(\gamma) \tag{6.85}$$

The error function, $\operatorname{erf}(\gamma)$, can be computed either by using infinite series, as shown in problems 9 and 10, or by using tables, as given by Abramowitz and Stegun (1970).

The infinite series in Eq. 6.80 does not converge because the integration interval in Eq. 6.77 extends beyond the radius of convergence of the infinite series in Eq. 6.79. This is the usual case with asymptotic expansions, and it means that the error of approximation will reach a minimum with a finite number of terms. Thus, the number of terms used in Eq. 6.80 should never extend past the point at which individual terms start to increase in magnitude. For this reason, the error of approximation in using an asymptotic expansion is usually assumed equal to the magnitude of the first neglected term in the expansion. On the other hand, two or three terms in Eq. 6.80 can sometimes be used to obtain an approximation for W with an accuracy that could only be duplicated by using one or two hundred terms in Eqs. 6.69 and 6.74.

An interesting application of Eq. 6.80 occurs for $u \to 0$. Since Eq. 6.74 shows that

$$W(0, \beta) = 2K_0(\beta) \tag{6.86}$$

and since $\gamma \to -\infty$ as $u \to 0$, Eqs. 6.80 - 6.85 give the result

$$K_0(\beta) \sim \sqrt{\frac{\pi}{2\beta}} \exp(-\beta) \left\{ 1 + \sum_{n=1}^{\infty} (-1)^n \frac{[1.3.5...(2n-1)]^2}{n! \, (8\beta)^n} \right\} \tag{6.87}$$

The first four terms of Eq. 6.87 are written out in Table 3 of Chapter 5.

The first two terms of Eq. 6.80 were obtained by Wilson and Miller (1979). The mathematically inclined reader who would like to learn more about asymptotic techniques will find a clearly written, general introduction to the subject in Carrier, Krook and Pearson (1966).

31. Pollution Calculations in Nonuniform Flow

In this section we will assume that $I = 0$ in Eq. 6.16 and that velocities are large enough to allow D_1 and D_2 to be approximated with Eqs. 6.14 and 6.15. A mathematical justification for the results discussed herein is given by Hunt (1978b), who also considers the more general case when Eqs. 6.14 - 6.15 do not apply and when I is given by Eq. 6.9. It will be assumed that the solution depends upon, at most, two spatial coordinates.

Let a pollutant, of concentration c_0, be released in a two-dimensional flow over a region that has a characteristic dimension of L. For example, Fig. 6.3(a) shows a pollutant that has been released instantaneously over a region of finite size, and Fig. 6.3(b) shows the result of a continuous release of pollutant along the line AB. When the largest principal value of the dispersivity tensor is small compared to L ($\alpha_1/L << 1$), the dispersion terms on the left side of Eq. 6.16 have a very small influence upon the pollutant movement for small to moderate distances downstream. (These distances, of course, are measured as multiples of L.) Thus, the solution can be approximated by solving

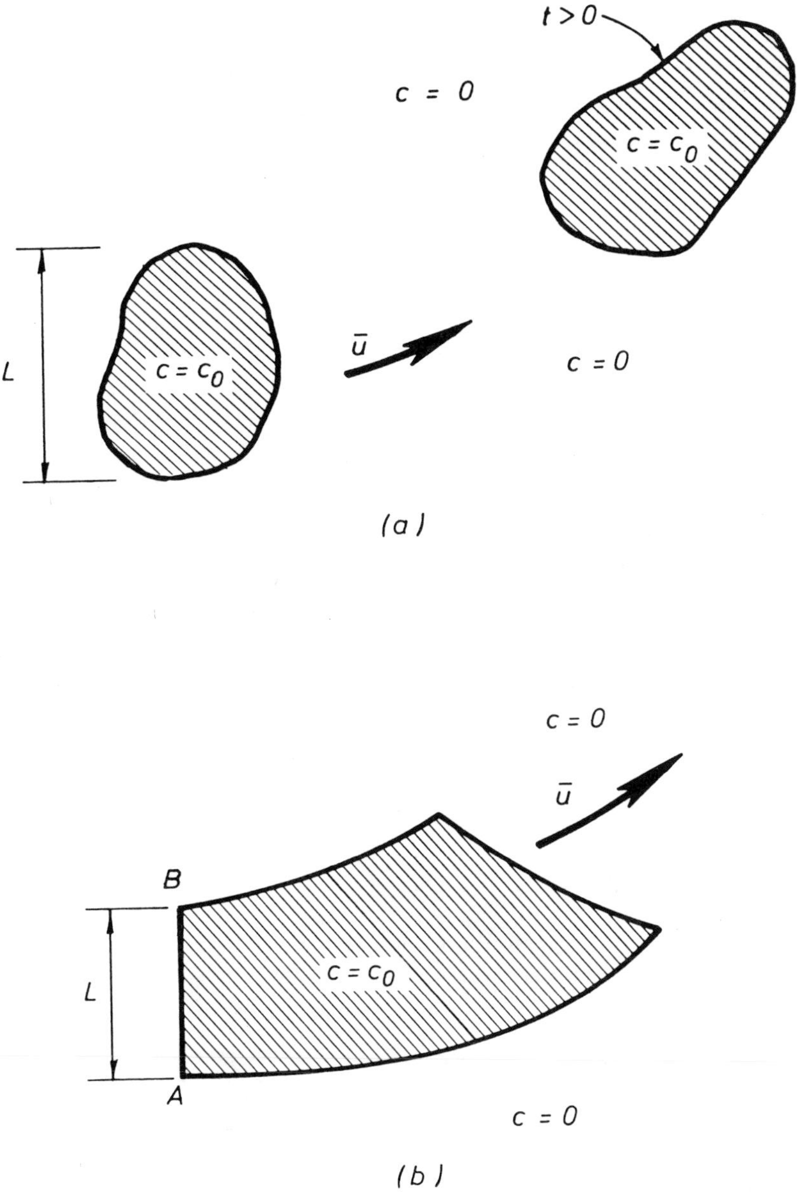

Fig. 6.3 - Pollution in nonuniform flow with (a) an instantaneous source of finite size and (b) a continuous source of finite size. Dispersion has been neglected.

$$0 = \frac{\bar{u}}{\sigma}.\bar{\nabla}c + \frac{\partial c}{\partial t} = \frac{dc}{dt} \qquad (6.88)$$

Equation 6.88 simply states that c will not change for a fluid particle moving with the velocity $d\bar{r}/dt = \bar{u}/\sigma$, and Fig. 6.3 shows the result of solving Eq. 6.88 for two particular problems. The boundaries of the cross-hatched regions in Fig. 6.3 are found by calculating the movement of individual fluid particles, a process that is carried out formally by integrating

$$\frac{d\bar{r}}{dt} = \frac{\bar{u}}{\sigma} \qquad (6.89)$$

to find $\bar{r} = \bar{r}(t)$ for a fluid particle that has been "tagged" with a concentration of c_0.

Including the dispersive terms on the left side of Eq. 6.16 will lead to a solution in which c varies continuously from c = 0 to $c = c_0$ along the boundaries of discontinuity in c that are shown in Fig. 6.3. However, when $\alpha_1/L << 1$, this "boundary-layer region" will be a very small fraction of L for small to moderate distances downstream, and concentration gradients will be very large within the boundary-layer region. Since numerical methods that might be used to solve Eq. 6.16 assume that c is approximated upon a grid of nodes with polynomials of finite degree, any attempt to obtain numerical solutions of Eq. 6.16 in regions where the boundary layer thickness is smaller than the node spacing will lead to great numerical difficulties. These numerical difficulties usually appear in the form of either numerical diffusion (an artificial dispersion that is not present in the physical experiment) or fairly violent oscillations in the solution that are somewhat similar in appearance to the oscillations shown for the unstable numerical solution in Fig. 4.9. Thus, numerical solutions of Eq. 6.16 should not be attempted until all boundary layer thicknesses in the solution domain are at least equal in magnitude to two or three node spacings.

The development of these pollution boundary layers is shown in Fig. 6.4 for the particular problems considered in Fig. 6.3. If s = arc length along the path of a fluid particle upon the boundary of discontinuity in c shown in Fig. 6.3, then the boundary layer thickness, δ, is given approximately by

$$\delta(\ell) = 6.58 \sqrt{\int_0^{\ell} [\alpha_2 + (\alpha_1 - \alpha_2)\cos^2\theta] \left[\frac{\Delta(\ell)}{\Delta(s)}\right]^2 ds} \qquad (6.90)$$

The symbols in Eq. 6.90 are ℓ = value of s at which δ is being calculated, α_1 and α_2 = longitudinal and lateral dispersivities, respectively, $\theta(s)$ = angle between $\bar{u}$ and the outward normal to the polluted region boundary in Fig. 6.3 and $\Delta(s)$ = normal spacing between the boundary of discontinuity in c shown in Fig. 6.3 and a line of fluid particles that moves "parallel" to this boundary of discontinuity. The normal distance $\Delta(s)$ appears as the result of approximating a normal derivative with finite

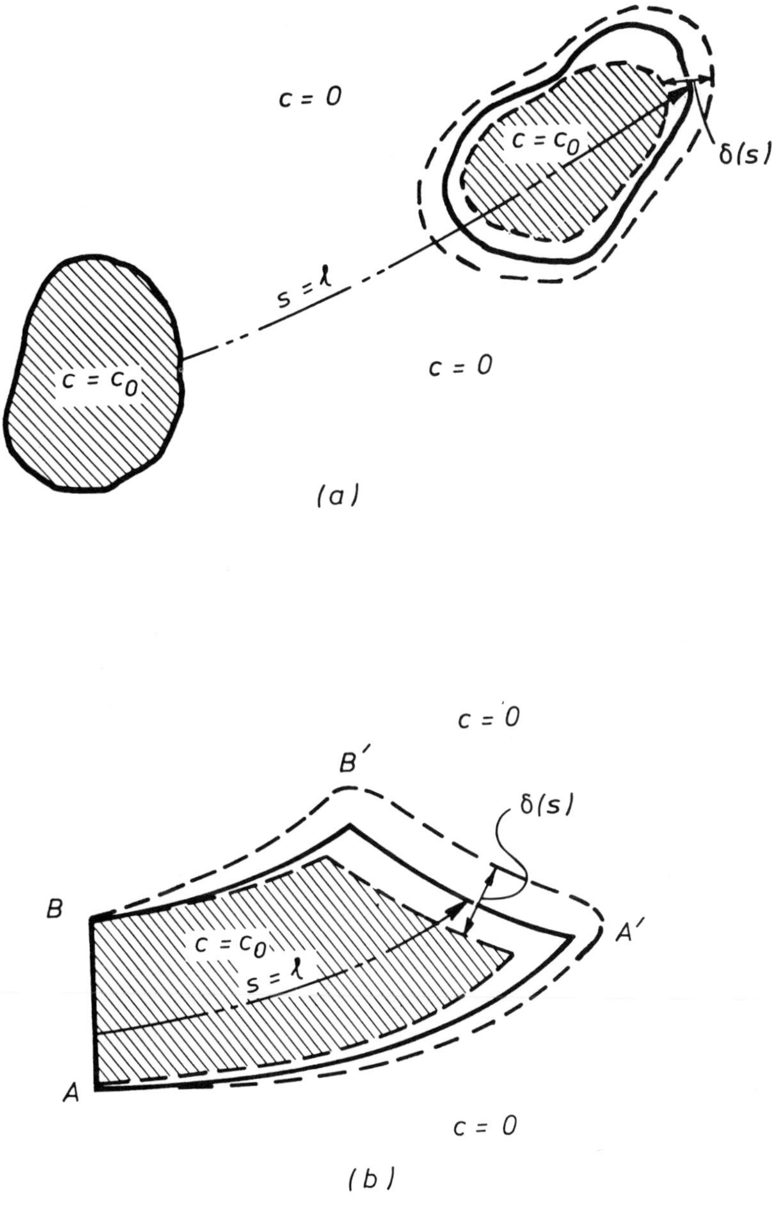

Fig. 6.4 - Development of boundary layers along the boundaries of discontinuity in c that are shown in Fig. 6.3.

differences, so that $\Delta(s)$ should be a small fraction of L throughout the calculations.

Since a streak line shows the instantaneous location of all fluid particles that have passed through a fixed point, and since time lines show the instantaneous location of all fluid particles that lay upon some given line at an earlier time, we see that $\Delta(s)$ is the normal spacing between two streak lines or two time lines. For example, $\Delta(s)$ is the spacing between two time lines in Fig. 6.4(a). In Fig. 6.4(b), though, $\Delta(s)$ is the spacing between two streak lines along AA' and BB' and between two time lines along A'B'. If the flow is steady, then the streak lines along AA' and BB' coincide with streamlines and conservation of mass requires that

$$\frac{\Delta(\ell)}{\Delta(s)} = \frac{u(s)}{u(\ell)} \tag{6.91}$$

in which $u = |\bar{u}|$. In this case we would also set $\theta(s) = \pi/2$ along AA' and BB' but not, necessarily, along A'B'.

After $\delta(\ell)$ has been calculated from Eq. 6.90, then the concentration distribution within the boundary layer can be calculated with

$$\frac{c}{c_0} = \frac{1}{2}\,\text{erfc}\left(3.29\,\frac{n}{\delta}\right) \tag{6.92}$$

in which n = arc length normal to the boundary of discontinuity in c in Fig. 6.3. The coordinate n is taken as positive in the outward direction, so that Eq. 6.92 gives

$$\underset{n\to\delta/2}{\text{Limit}}\ \frac{c}{c_0} = \frac{1}{2}\,\text{erfc}(1.65) = 0.01 \tag{6.93}$$

$$\underset{n\to-\delta/2}{\text{Limit}}\ \frac{c}{c_0} = \frac{1}{2}\,\text{erfc}(-1.65) = 0.99 \tag{6.94}$$

Thus, the outer and inner edges of the concentration boundary layer have been defined as the points where c is 1% and 99%, respectively, of c_0.

Hunt (1978b) has applied Eqs. 6.90 and 6.92 to unsteady flow from a well in a homogeneous, isotropic aquifer and to steady flow in a heterogeneous, isotropic aquifer. As a first application, though, we will consider the relatively simple case of steady, uniform flow shown in Fig. 6.5(a). At $t = 0$ we will set $c = c_0$ along AB and maintain this value of c along AB for all $t > 0$. The solution for zero dispersion is shown in Fig. 6.5(a). Since the lines AA' and BB' coincide with the uniform flow streamlines, Eq. 6.91 shows that $\Delta(\ell)/\Delta(s) = 1$ and Eq. 6.90 gives, after setting $\theta = \pi/2$,

$$\delta = 6.58\sqrt{\alpha_2 \ell} \tag{6.95}$$

The distance ℓ is measured from point A to the point at which δ is calculated along AA' and from point B to the point at which δ is calculated

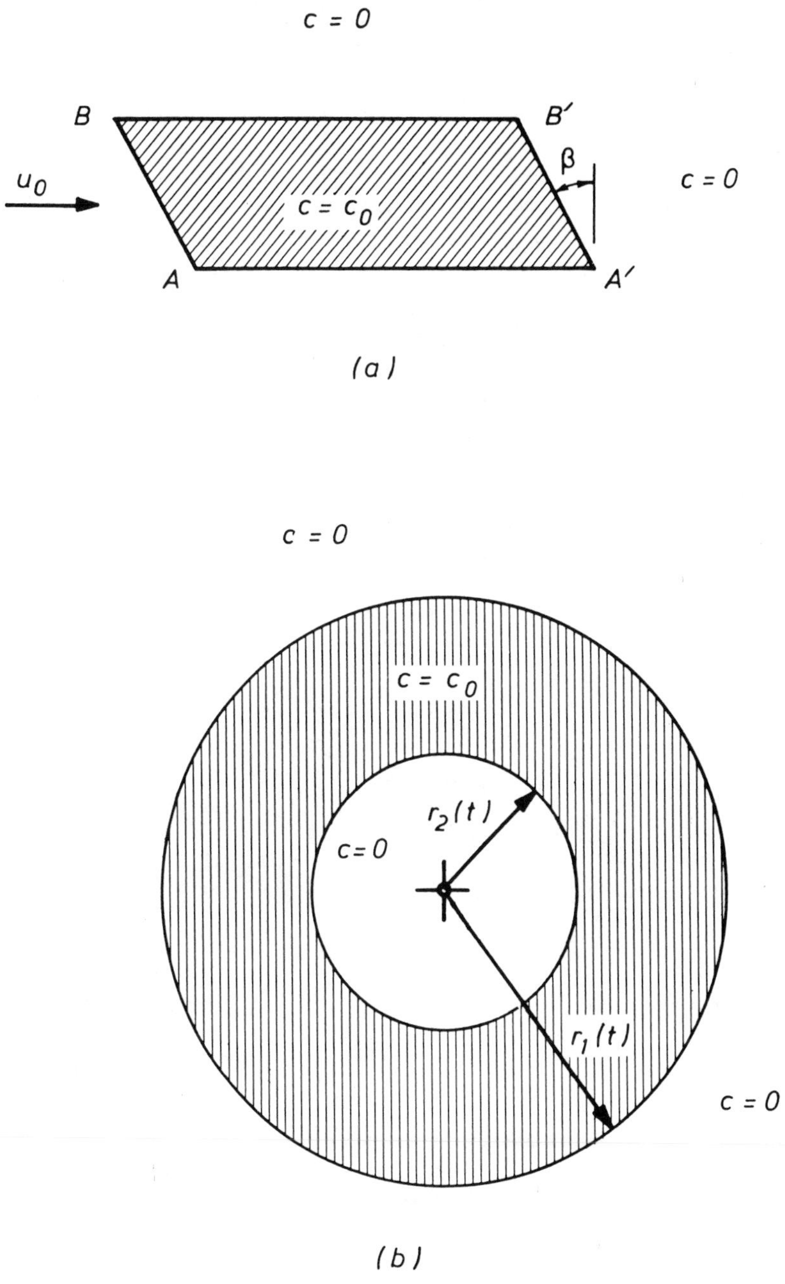

Fig. 6.5 - Applications of Eqs. 6.90 and 6.91 in (a) uniform flow and (b) radial flow from a well.

along BB'. Since $\Delta(\ell)/\Delta(s) = 1$ and $\theta = \beta$ along B'A', the boundary layer thickness along A'B' is given by

$$\delta = 6.58\sqrt{[\alpha_2 + (\alpha_1 - \alpha_2)\cos^2\beta]\ell} \tag{6.96}$$

in which $\ell = u_0 t/\sigma$. Since the solution is a perturbation solution that assumes δ to be relatively small, one would normally expect the solution to become invalid after the inner edge of the two boundary layers along AA' and BB' meet. This, however, is the point at which numerical solution techniques become feasible so that the singular perturbation solution can serve as an aid to obtaining accurate numerical solutions.

A second example in Fig. 6.5(b) is concerned with steady flow from a well. At $t = 0$, the concentration of well water becomes $c = c_0$, and at $t = \tau$ the injection of pollutant ceases ($c = 0$ at $r = 0$ for $t > \tau$). The radial velocity is given by

$$u = \frac{Q}{2\pi B r} \tag{6.97}$$

in which B = aquifer thickness, Q = well flow rate (which remains unchanged for all t) and r = radial coordinate. Thus, inserting Eq. 6.97 into Eq. 6.89 and integrating gives

$$r_1(t) = \sqrt{\frac{Q}{\pi B\sigma} t} \quad \text{for } t \geqq 0 \tag{6.98}$$

$$r_2(t) = \sqrt{\frac{Q}{\pi B\sigma}(t - \tau)} \quad \text{for } t \geqq \tau \tag{6.99}$$

Since a time line just outside the time line $r = r_1(t)$ has the equation $r = r_1(t + t')$, in which t' is small,

$$\Delta(r_1) = r_1(t + t') - r_1(t) \simeq \frac{t'}{2}\sqrt{\frac{Q}{\pi B\sigma t}} = \frac{t'}{2}\,\frac{Q}{\pi B\sigma r_1} \tag{6.100}$$

Thus, $\Delta(r_1)/\Delta(r)$ is given by

$$\frac{\Delta(r_1)}{\Delta(r)} = \frac{r}{r_1} \tag{6.101}$$

and substituting Eq. 6.101 and $\theta = 0$ into Eq. 6.90 gives

$$\delta(r_1) = 6.58\sqrt{\alpha_1\int_0^{r_1}\left(\frac{r}{r_1}\right)^2 dr} = 6.58\sqrt{\frac{\alpha_1 r_1(t)}{3}} \tag{6.102}$$

A similar calculation along the boundary $r = r_2(t)$ gives

$$\delta(r_2) = 6.58\sqrt{\frac{\alpha_1 r_2(t)}{3}} \tag{6.103}$$

Once again, this solution becomes invalid after the inner edges of the two boundary layers meet in the crosshatched region in Fig. 6.5(b).

REFERENCES

Abramowitz, M. and Stegun, I.A. (Editors) 1970. Handbook of Mathematical Functions, ninth printing, U.S. Bureau of Standards, Applied Mathematics Series 55, Washington, Ch.7.

Bear, J. 1972. Dynamics of Fluids in Porous Media, American Elsevier, New York, p. 607.

Carrier, G.F., Krook, M. and Pearson, C.E. 1966. Functions of a Complex Variable, McGraw-Hill Book Co., New York, Ch.6.

Case, C.M. and Addiego, J.C. 1977. "Note on a Series Representation of the Leaky Aquifer Well Function," Journal of Hydrology (The Netherlands), Vol. 32, pp. 393-397.

Gradshteyn, I.S. and Ryzhik, I.M. 1965. Tables of Integrals, Series, and Products, Fourth Edition, Academic Press, New York, pp. 340, 480.

Hildebrand, F.B. 1962. Advanced Calculus for Applications, Prentice-Hall, Inc., Englewood Cliffs, pp. 304-305.

Hunt, B. 1977. "Calculation of the Leaky Aquifer Function," Journal of Hydrology (The Netherlands), Vol. 33, pp. 179-183.

Hunt, B. 1978a. "Dispersive Sources in Uniform Ground-Water Flow," ASCE Journal of the Hydraulics Division, Vol. 104, No. HY1, January, pp. 75-85.

Hunt, B. 1978b. "Dispersion Calculations in Nonuniform Seepage," Journal of Hydrology (The Netherlands), Vol. 36, pp. 261-277.

Wilson, J.L. and Miller, P.J. 1979. Discussion of their paper "Two-Dimensional Plume in Uniform Ground-Water Flow," ASCE Journal of the Hydraulics Division, Vol. 105, No. HY12, December, pp. 1567-1570.

PROBLEMS

1. The method used in the text to write Eq. 6.16 in different orthogonal coordinate systems depends upon being able to write expressions for the gradient and divergence in these coordinate systems. Verify Eq. 6.25 by substituting

$$\nabla c = \hat{e}_r F_r + \hat{e}_\theta F_\theta + \hat{e}_z F_z$$

$$d\bar{r} = \hat{e}_r dr + \hat{e}_\theta r d\theta + \hat{e}_z dz$$

into the identity

$$d\bar{r}.\bar{\nabla}c = dc = \frac{\partial c}{\partial r}\,dr + \frac{\partial c}{\partial \theta}\,d\theta + \frac{\partial c}{\partial z}\,dz$$

to obtain

$$F_r = \frac{\partial c}{\partial r},\quad F_\theta = \frac{1}{r}\frac{\partial c}{\partial \theta},\quad F_z = \frac{\partial c}{\partial z}$$

2. Use the method explained in the previous problem to verify Eq. 6.30. Note that Fig. 6.2b gives

$$d\bar{r} = \hat{e}_r dr + r d\phi \hat{e}_\phi + r d\theta \sin\phi \hat{e}_\theta$$

3. The unit base vectors $\hat{e}_r$, $\hat{e}_\theta$ and $\hat{e}_z$ all have a constant length of unity in Fig. 6.2a. However, $\hat{e}_r$ and $\hat{e}_\theta$ are functions of θ because they change their directions when the θ coordinate of a point changes.

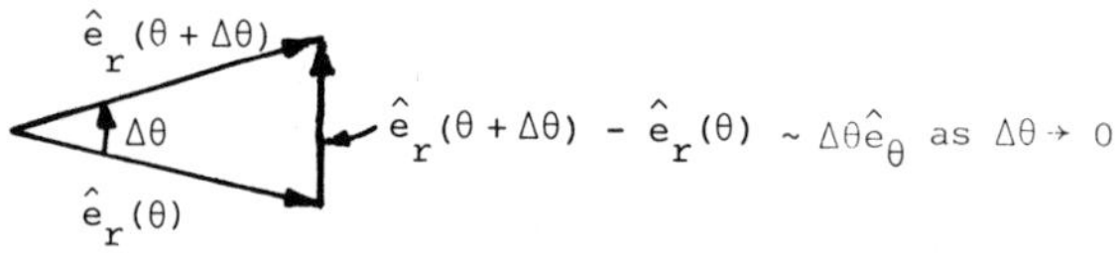

The geometry of the above sketch shows that

$$\frac{d\hat{e}_r(\theta)}{d\theta} = \underset{\Delta\theta\to 0}{\text{Limit}}\ \frac{\hat{e}_r(\theta + \Delta\theta) - \hat{e}_r(\theta)}{\Delta\theta} = \hat{e}_\theta$$

In the same way, the geometry of the following sketch shows that

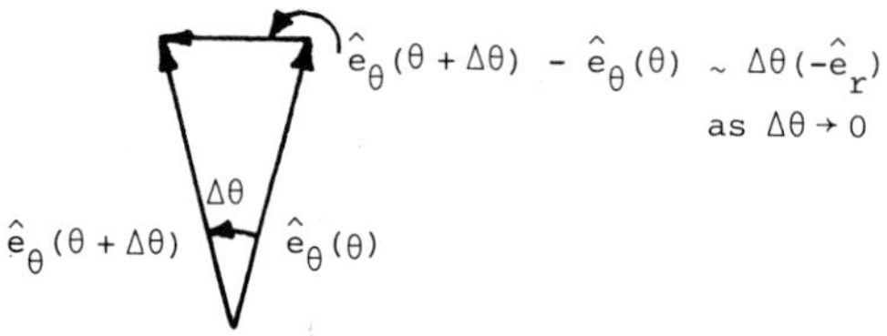

$$\frac{d\hat{e}_\theta(\theta)}{d\theta} = \underset{\Delta\theta\to 0}{\text{Limit}}\ \frac{\hat{e}_\theta(\theta + \Delta\theta) - \hat{e}_\theta(\theta)}{\Delta\theta} = -\hat{e}_r$$

Use these two expressions for the derivatives of $\hat{e}_r$ and $\hat{e}_\theta$ to derive Eq. 6.26 by expanding the following expression for the divergence of $\bar{F}$ in cylindrical coordinates:

$$\bar{\nabla}.\bar{F} = \left(\hat{e}_r \frac{\partial}{\partial r} + \hat{e}_\theta \frac{1}{r}\frac{\partial}{\partial \theta} + \hat{e}_z \frac{\partial}{\partial z}\right)\cdot\left(\hat{e}_r F_r + \hat{e}_\theta F_\theta + \hat{e}_z F_z\right)$$

4. Use the geometry of Fig. 6.2b to show that

$$\frac{\partial \hat{e}_r(\phi,\theta)}{\partial \phi} = \hat{e}_\phi,\qquad \frac{\partial \hat{e}_r(\phi,\theta)}{\partial \theta} = \sin\phi\ \hat{e}_\theta$$

$$\frac{\partial \hat{e}_\phi(\phi,\theta)}{\partial \phi} = -\hat{e}_r,\qquad \frac{\partial \hat{e}_\phi(\phi,\theta)}{\partial \theta} = \cos\phi\ \hat{e}_\theta$$

$$\frac{d\hat{e}_\theta(\theta)}{d\theta} = -(\hat{e}_r \sin\phi + \hat{e}_\phi \cos\phi)$$

Hence, use these expressions to verify Eq. 6.31 by expanding the following expression for the divergence of $\overline{F}$ in spherical coordinates:

$$\overline{\nabla}.\overline{F} = \left(\hat{e}_r \frac{\partial}{\partial r} + \hat{e}_\phi \frac{1}{r}\frac{\partial}{\partial \phi} + \hat{e}_\theta \frac{1}{r\sin\phi}\frac{\partial}{\partial\theta}\right).\left(\hat{e}_r F_r + \hat{e}_\phi F_\phi + \hat{e}_\theta F_\theta\right)$$

5. Use the Fourier integral theorem, Eqs. 6.37 - 6.38, to derive Eq. 6.51 as a solution to the problem

$$D_1 \frac{\partial^2 c}{\partial x^2} = \frac{u_x}{\sigma}\frac{\partial c}{\partial x} + \frac{\partial c}{\partial t} + \lambda c$$

$$c(x, 0) = \frac{M_1}{\sigma}\delta(x)$$

6. The integration in Eq. 6.52 can be carried out by setting $\xi = (t - \tau)^{-\frac{1}{2}}$, $2d\xi = (t - \tau)^{-3/2}\, d\tau$ to obtain

$$c(x, y, z, t) = \frac{\dot{M}\exp\left(\frac{xu_x}{2\sigma D_1}\right)}{4\sigma\sqrt{\pi^3 D_1 D_2{}^2}} \int_{1/\sqrt{t}}^{\infty} \exp\left[-\left(\frac{R^2}{4D_1}\right)\xi^2 - \frac{a^2}{\xi^2}\right] d\xi$$

The integral can be calculated by using an indefinite integral given by Abramowitz and Stegun (1970). This indefinite integral can be calculated by differentiating

$$\operatorname{erf}\left(ax + \frac{b}{x}\right) = \frac{2}{\sqrt{\pi}} \int_0^{ax + \frac{b}{x}} \exp(-\xi^2)\, d\xi$$

with respect to x to obtain

$$\frac{d}{dx}\operatorname{erf}\left(ax + \frac{b}{x}\right) = \frac{2}{\sqrt{\pi}}\exp\left[-\left(ax + \frac{b}{x}\right)^2\right]\left(a - \frac{b}{x^2}\right)$$

$$= \frac{2}{\sqrt{\pi}}\exp(-2ab)\exp\left(-a^2x^2 - \frac{b^2}{x^2}\right)\left(a - \frac{b}{x^2}\right)$$

Replace b with -b to obtain

$$\frac{d}{dx}\operatorname{erf}\left(ax - \frac{b}{x}\right) = \frac{2}{\sqrt{\pi}}\exp(2ab)\exp\left(-a^2x^2 - \frac{b^2}{x^2}\right)\left(a + \frac{b}{x^2}\right)$$

Thus, multiply the first equation by exp(2ab) and the last equation by exp(-2ab) and add to obtain

$$\exp(2ab)\frac{d}{dx}\operatorname{erf}\left(ax + \frac{b}{x}\right) + \exp(-2ab)\frac{d}{dx}\operatorname{erf}\left(ax - \frac{b}{x}\right)$$

$$= \frac{4a}{\sqrt{\pi}}\exp\left(-a^2x^2 - \frac{b^2}{x^2}\right)$$

Finally, taking the indefinite integral of both sides gives the result

$$\int \exp\left(-a^2x^2 - \frac{b^2}{x^2}\right) dx = \frac{\sqrt{\pi}}{4a}\left[\exp(2ab)\ \mathrm{erf}\left(ax + \frac{b}{x}\right) + \exp(-2ab)\ \mathrm{erf}\left(ax - \frac{b}{x}\right)\right]$$

This indefinite integral, together with the identities

$$\mathrm{erf}(\infty) = -\mathrm{erf}(-\infty) = 1$$

$$\mathrm{erfc}(x) = 1 - \mathrm{erfc}(x)$$

can be used to calculate the result given in Eq. 6.53. Use this same indefinite integral to derive Eq. 6.58.

7. The reference by Hantush (1964) in chapter V gives

$$W(0.1,\ 0.001) = 1.8229$$

This result was calculated by using numerical integration. Use Eq. 6.73 to verify the result. Remember that the calculation of $E_n(x)$ can be carried out with Eq. 5.30 and Table 2 of Chapter 5.

8. The reference by Hantush (1964) in chapter V gives

$$W(2.0,\ 4.0) = 0.0112$$

Use both Eq. 6.74 and Eq. 6.80 to verify the result.

9. The error function is defined as

$$\mathrm{erf}(z) = \frac{2}{\sqrt{\pi}} \int_0^z \exp(-x^2)\ dx$$

Use the series expansion

$$\exp(z) = \sum_{n=0}^{\infty} \frac{z^n}{n!}, \quad (0 \leq |z| < \infty)$$

to obtain

$$\mathrm{erf}(z) = \frac{2}{\sqrt{\pi}} \sum_{n=0}^{\infty} \frac{(-1)^n\ z^{2n+1}}{n!\ (2n+1)}, \quad (0 \leq |z| < \infty)$$

Note that erf(-z) = -erf(z).

10. The complementary error function is defined as

$$\mathrm{erfc}(z) = \frac{2}{\sqrt{\pi}} \int_z^{\infty} \exp(-y^2)\ dy$$

[Note that this definition and the definition for erf(z) in problem 9 lead to the identity erf(z) + erfc(z) = 1.] Use the substitution

$$y = z\sqrt{1 + x^2}$$

to obtain an integral similar in form to Eq. 6.77. Then use Eq. 6.79 and the definition of the gamma function

$$\int_0^\infty x^n \exp(-x)\, dx = \Gamma(n+1) = n!$$

to obtain the asymptotic expansion

$$\mathrm{erfc}(z) \sim \frac{\exp(-z^2)}{z\sqrt{\pi}} \left[1 + \sum_{n=1}^{\infty} (-1)^n \frac{1.3.5.\ldots.(2n-1)}{(2z^2)^n} \right]$$

11. The exponential integral is defined as

$$E_n(z) = \int_1^\infty \exp(-zt) \frac{dt}{t^n}, \; [n \text{ an integer, } \mathrm{Re}\{z\} > 0]$$

a. Set $u = t^{-n}$, $dv = \exp(-zt)dt$ and integrate once, by parts, to obtain the recurrence formula

$$E_{n+1}(z) = \frac{1}{n} [\exp(-z) - zE_n(z)] \text{ for } n = 1, 2, 3, \ldots$$

b. The usual form for $E_1(z)$ is obtained by setting $x = zt$ to obtain

$$E_1(z) = \int_z^\infty \exp(-x) \frac{dx}{x}$$

Differentiating this gives

$$\frac{dE_1(z)}{dz} = -\frac{\exp(-z)}{z}$$

Use a Taylor's series expansion of $\exp(-z)$ about $z = 0$ and indefinite integration to show that

$$E_1(z) = C - \ln(z) - \sum_{n=1}^{\infty} \frac{(-1)^n z^n}{n!n}$$

Note that this series expansion is identical with the series expansion in Table 2 of Chapter 5 if the integration constant is taken as

$$C = \ln(0.56145948) = -\gamma$$

in which γ = Euler's constant = 0.5772····.

12. The substitution $x = z(t+1)$ in the integral

$$E_1(z) = \int_z^\infty \exp(-x) \frac{dx}{x}$$

gives the result

$$E_1(z) = \exp(-z) \int_0^\infty \exp(-zt) \frac{dt}{1+t}$$

The function exp(-zt) has no relative maximum upon the integration interval. It does, however, have an absolute maximum on the integration interval at $t = 0$, and the exponential function decays very rapidly for $t > 0$ when $\text{Re}\{z\}$ is a large, positive number. Hence, use the infinite series

$$\frac{1}{1+t} = \sum_{n=0}^{\infty} (-1)^n t^n, \quad (0 \leq t < 1)$$

and the definition of the gamma function in problem 10 to obtain the asymptotic series

$$E_1(z) \sim \frac{\exp(-z)}{z} \sum_{n=0}^{\infty} \frac{(-1)^n n!}{z^n}$$

The first four terms of this expansion are written out in Table 2 of Chapter 5.

13. In Fig. 6.5(a), set the length of AB equal to 100 meters, $\beta = 0$, $\alpha_1 = 1$ meter and $\alpha_2 = 0.1$ meter. Calculate the distance downstream to the point where the inner edges of the boundary layers along AA' and BB' meet. Then calculate the boundary layer thickness, δ, along A'B' at this point.

14. A pollutant is distributed uniformly and instantaneously over a circular region of radius 50 meters. Assume that $\alpha_1 = 5$ meters and $\alpha_2 = 0.5$ meter. Plot, to scale, the inner and outer edges of the pollutant boundary layer after the center of the polluted region has moved a distance of 20 meters downstream. Then calculate the distance downstream to the point at which the inner edges of the boundary layer will first meet. What is this distance if $\alpha_1 = 10$ meters and $\alpha_2 = 1$ meter?

7 The Exact Solution of Steady-Flow Problems

32. Problem Formulations

This chapter is concerned with the mathematical solution of problems governed by the Laplace equation in two dimensions. In particular, we will consider steady-flow problems in homogeneous, nonleaky aquifers. This means that the number and type of problems treated in this chapter will be considerably more restricted than those treated in Chapter IV. On the other hand, the convenience and efficiency of closed-form, mathematical solutions almost always makes them a more desirable alternative than the use of either numerical or experimental solutions. The only catch, of course, is that mathematical solution techniques require greater amounts of mathematical skill and understanding.

For two-dimensional flows in a vertical, (x, y) plane, we will set

$$\bar{u} = u\hat{i} + v\hat{j} \tag{7.1}$$

and

$$\phi = -Kh, \quad \left(h = \frac{P}{\rho g} - \hat{g}.\bar{r}\right) \tag{7.2}$$

in which K is a scalar constant. Thus, Darcy's law, Eq. 2.11, and the continuity equation, Eq. 2.22, take the form

$$(u,\ v) = \left(\frac{\partial\phi}{\partial x}, \frac{\partial\phi}{\partial y}\right) \tag{7.3}$$

$$\frac{\partial u}{\partial x} + \frac{\partial v}{\partial y} = 0 \tag{7.4}$$

Eliminating u and v between Eqs. 7.3 and 7.4 shows that ϕ is a harmonic function (i.e. - that ϕ is a solution of the Laplace equation).

$$\frac{\partial^2\phi}{\partial x^2} + \frac{\partial^2\phi}{\partial y^2} = 0 \tag{7.5}$$

Equation 7.4 can also be satisfied by defining a stream function, ψ, in the following way:

$$(u,\ v) = \left(\frac{\partial\psi}{\partial y}, -\frac{\partial\psi}{\partial x}\right) \tag{7.6}$$

Substituting Eq. 7.6 into Eq. 7.4 shows that the continuity equation is satisfied by any and all choices for ψ that have continuous first and second derivatives.

$$\frac{\partial u}{\partial x} + \frac{\partial v}{\partial y} = \frac{\partial^2\psi}{\partial x\partial y} - \frac{\partial^2\psi}{\partial y\partial x} = 0 \tag{7.7}$$

This does not mean, though, that ψ can be chosen in an arbitrary manner since Eqs. 7.3 and 7.6 require

$$\frac{\partial\phi}{\partial x} = \frac{\partial\psi}{\partial y} \tag{7.8}$$

$$\frac{\partial\phi}{\partial y} = -\frac{\partial\psi}{\partial x} \tag{7.9}$$

Thus, differentiating Eq. 7.8 with respect to y, Eq. 7.9 with respect to x and subtracting to eliminate ϕ shows that ψ is also a harmonic function.

$$\frac{\partial^2\psi}{\partial x^2} + \frac{\partial^2\psi}{\partial y^2} = 0 \tag{7.10}$$

It is perfectly possible to reverse the signs in Eq. 7.6. If this is done, then one should also reverse the sign in Eq. 7.2 and place a minus sign in front of each term on the right side of Eq. 7.3 so that Eqs. 7.8 - 7.9 remain unaltered.

It was shown in Chapter I that $\overline{\nabla}\phi$ and $\overline{\nabla}\psi$ are vectors that are orthogonal to curves of constant ϕ and ψ, respectively. Computation of the dot product $\overline{\nabla}\phi.\overline{\nabla}\psi$ gives

$$\overline{\nabla}\phi.\overline{\nabla}\psi = \frac{\partial\phi}{\partial x}\frac{\partial\psi}{\partial y} + \frac{\partial\phi}{\partial y}\frac{\partial\psi}{\partial y} \tag{7.11}$$

The use of Eqs. 7.8 - 7.9 to eliminate either ϕ or ψ in Eq. 7.11 shows that

$$\overline{\nabla}\phi.\overline{\nabla}\psi = 0 \tag{7.12}$$

Thus, curves of constant ϕ and ψ are orthogonal to each other at all points where $|\overline{\nabla}\phi|$ and $|\overline{\nabla}\psi|$ are finite and different from zero. Since Eq. 7.3 shows that curves of constant ϕ are orthogonal to $\overline{u}$, Eq. 7.12 shows that curves of constant ψ are tangent to $\overline{u}$. In other words, curves of constant ψ coincide with streamlines, which is the reason for calling ψ a stream function.

The tangential and normal derivatives of ϕ and ψ along a curve are related. For example, Fig. 7.1 shows the unit tangent and normal to a curve whose equation is given parametrically as a function of arc length. The tangential derivative of ϕ is given by

$$\frac{d\phi}{ds} = \frac{\partial\phi}{\partial x}\frac{dx(s)}{ds} + \frac{\partial\phi}{\partial y}\frac{dy(s)}{ds} \tag{7.13}$$

However, since the cross product of $\hat{k}$ and $\hat{e}_t$ yields $\hat{e}_n$, the direction cosines of $\hat{e}_t$ and $\hat{e}_n$ are related:

$$\frac{dx}{ds} = \frac{dy}{dn} \tag{7.14}$$

$$\frac{dy}{ds} = -\frac{dx}{dn} \tag{7.15}$$

Thus, the use of Eqs. 7.8 - 7.9 and 7.14 - 7.15 in Eq. 7.13 gives

$$\frac{d\phi}{ds} = \frac{\partial\psi}{\partial y}\frac{dy}{dn} + \frac{\partial\psi}{\partial x}\frac{dx}{dn} = \frac{d\psi}{dn} \tag{7.16}$$

Likewise, calculation of the tangential derivative of ψ gives

$$\frac{d\psi}{ds} = -\frac{d\phi}{dn} \tag{7.17}$$

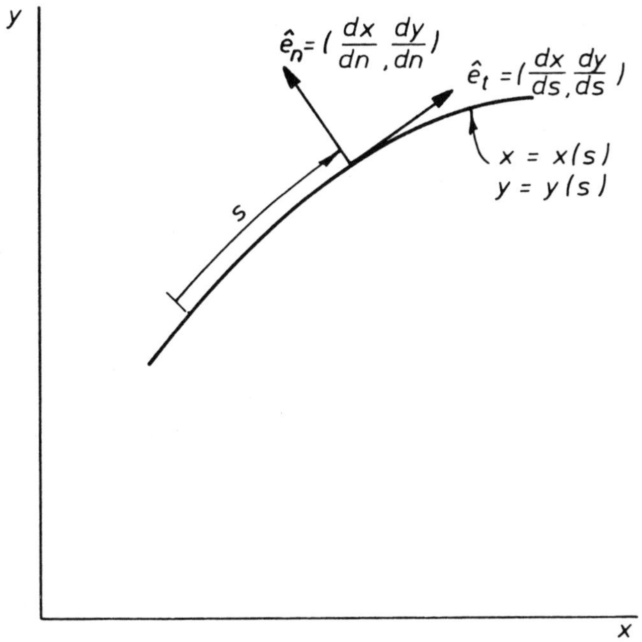

Fig. 7.1 - The unit tangent and normal to a curve whose equation is given parametrically as a function of arc length.

The signs are reversed in Eqs. 7.16 - 7.17 if the direction of $\hat{e}_n$ is reversed in Fig. 7.1.

Equations 7.16 and 7.17 are really the result of writing Eqs. 7.8 - 7.9 in curvilinear coordinates. If s is taken as arc length along a line of constant ψ, then Eq. 7.17 provides a second proof of the result that lines of constant ψ coincide with streamlines (since Eq. 7.17 shows that $d\phi/dn = \bar{u}.\hat{e}_n = 0$ along lines of constant ψ). Furthermore, integrating Eq. 7.16 along a line normal to the streamlines gives

$$\psi_2 - \psi_1 = \int_1^2 \frac{d\psi}{dn}\, dn = \int_1^2 \frac{d\phi}{ds}\, dn = \int_1^2 \bar{u}.\hat{e}_t dn \tag{7.18}$$

Thus, the constant values of ψ upon each streamline are determined from the requirement

$$\psi_2 - \psi_1 = \pm q \tag{7.19}$$

in which q = two-dimensional flow rate between streamlines 1 and 2. The sign on the right side of Eq. 7.19 is determined by the relative directions of $\bar{u}$ and $\hat{e}_t$ in the right side of Eq. 7.18. Since unique flow rates can be expected to exist for any well-posed, physical problem, Eq. 7.19 shows that $\psi(x, y)$ is determined uniquely for any problem to within one, arbitrary, additive constant. Sometimes we will fix the magnitude of this additive constant for a given problem by requiring that $\psi = 0$ upon one particular streamline.

For steady, two-dimensional flows in horizontal, unconfined aquifers, we will set

$$\bar{q} = q_x\hat{i} + q_y\hat{j} \tag{7.20}$$

$$\phi = -\frac{1}{2} Kh^2 \tag{7.21}$$

in which $\bar{q} = \bar{u}h$ = flux vector per unit arc length in the horizontal (x, y) plane, h = free surface elevation above the horizontal aquifer bottom and K is a scalar constant. Thus, Eqs. 2.37 - 2.38 and 2.41 reduce, for steady flow with no vertical inflow or outflow, to

$$(q_x, q_y) = \left(\frac{\partial\phi}{\partial x}, \frac{\partial\phi}{\partial y}\right) \tag{7.22}$$

$$\frac{\partial q_x}{\partial x} + \frac{\partial q_y}{\partial y} = 0 \tag{7.23}$$

Equation 7.23 implies the existence of a stream function, ψ, given by

$$(q_x, q_y) = \left(\frac{\partial\psi}{\partial y}, -\frac{\partial\psi}{\partial x}\right) \tag{7.24}$$

Since ϕ and ψ are seen from Eqs. 7.22 and 7.24 to satisfy Eqs. 7.8 - 7.9, we can conclude that lines of constant ϕ and ψ are orthogonal, that lines of constant ψ coincide with streamlines and that numerical values of ψ upon each streamline are fixed by the requirement

$$\psi_2 - \psi_1 = \pm Q \tag{7.25}$$

in which Q = three dimensional flow rate between the streamlines 1 and 2. The curves of constant ϕ and ψ are actually vertical surfaces that appear as lines in the horizontal, (x, y) plane.

An analogous set of equations can be worked out for steady, two-dimensional flows in horizontal, confined aquifers. The result is

$$\bar{q} = q_x\hat{i} + q_y\hat{j} \tag{7.26}$$

$$\phi = - Th, \quad (T = KB = \text{constant}) \tag{7.27}$$

$$\frac{\partial q_x}{\partial x} + \frac{\partial q_y}{\partial y} = 0 \tag{7.28}$$

$$(q_x, q_y) = \left(\frac{\partial\psi}{\partial y}, - \frac{\partial\psi}{\partial x}\right) = \left(\frac{\partial\phi}{\partial x}, \frac{\partial\phi}{\partial y}\right) \tag{7.29}$$

$$\psi_2 - \psi_1 = \pm Q \tag{7.30}$$

Equations 7.26 - 7.30 can also be used to describe unconfined flow problems when changes in h are small compared with the saturated aquifer thickness, as explained in section 10. In either case, it is always possible to interpret h in Eq. 7.27 as the change in h from some previous, steady-flow distribution of h. Then the drawdown is given by $|h|$.

Sea-water intrusion problems, as discussed in section 12, can be described by Eqs. 7.26 and 7.28 - 7.30 when Eq. 7.27 is replaced with

$$\begin{aligned} \phi &= - \tfrac{1}{2} K[(h + D)^2 - (1 + \varepsilon)D^2] \text{ for x, y in region I} \\ &= - \tfrac{1}{2} K\left(1 + \frac{1}{\varepsilon}\right) h^2 \text{ for x, y in region II} \end{aligned} \tag{7.31}$$

in which region II contains the salt water interface. Here again, it has been assumed that the flow is steady, that K is a scalar constant and that vertical inflow or outflow is zero. Wells can be accounted for in any of the problems treated in this chapter by requiring ϕ to have a logarithmic singularity at each well rather than by adding a term $Q\delta(x - x_0)\ \delta(y - y_0)$ to the continuity equation.

The effects of anisotropy can be included if it is assumed that the aquifer is homogeneous (i.e. - if the principal values and directions of the permeability tensor are everywhere constant). For example, if the x and y axes point in the principal directions, Eq. 2.12 becomes

$$\bar{u} = - K_x \frac{\partial h}{\partial x}\hat{i} - K_y \frac{\partial h}{\partial y}\hat{j} \tag{7.32}$$

and the continuity equation remains identical with Eq. 7.4. Using the coordinate transformations

$$X = x\sqrt{K_y/K_x} \tag{7.33}$$

in Eqs. 7.4 and 7.32, together with the equation

$$U = u\sqrt{K_y/K_x} \tag{7.34}$$

which follows by differentiating both sides of Eq. 7.33 with respect to time, leads to the result

$$\phi = -K_y h \tag{7.35}$$

$$(U, v) = \left(\frac{\partial\phi}{\partial X}, \frac{\partial\phi}{\partial y}\right) \tag{7.36}$$

$$\frac{\partial U}{\partial X} + \frac{\partial v}{\partial y} = 0 \tag{7.37}$$

$$(U, v) = \left(\frac{\partial\psi}{\partial y}, -\frac{\partial\psi}{\partial X}\right) \tag{7.38}$$

It follows from Eqs. 7.36 and 7.38 that the curves of constant ϕ and ψ are orthogonal in the (X, y) plane but not in the physical (x, y) plane. The physical significance of curves of constant ψ can be examined by using Eqs. 7.33, 7.34 and 7.38 to obtain expressions for u and v in the (x, y) plane.

$$u = \sqrt{K_x/K_y}\; U = \sqrt{K_x/K_y}\,\frac{\partial\psi}{\partial y} \tag{7.39}$$

$$v = -\frac{\partial\psi}{\partial X} = -\sqrt{K_x/K_y}\,\frac{\partial\psi}{\partial x} \tag{7.40}$$

Thus, if s and n are arc length along and normal to any curve in the (x, y) plane, Eqs. 7.14, 7.15, 7.39 and 7.40 give

$$\bar{u}\cdot\hat{e}_n = -\sqrt{K_x/K_y}\,\frac{d\psi}{ds} \tag{7.41}$$

Since $\bar{u}\cdot\hat{e}_n = 0$ along a streamline, Eq. 7.41 shows that curves of constant ψ coincide with streamlines. If we choose s to be arc length along any curve joining two lines of constant ψ (streamlines), then integrating Eq. 7.41 shows that

$$\psi_2 - \psi_1 = \pm q\,\sqrt{K_y/K_x} \tag{7.42}$$

The results in Eqs. 7.32 - 7.42 become identical with the results given by Eqs. 7.1 - 7.19 when $K_x = K_y = K$.

The similarities between Eqs. 7.32 - 7.42 and 7.1 - 7.19 suggest that the following approach can be used to find solutions for homogeneous, anisotropic aquifers:

1. Use Eq. 7.33 to transform the aquifer geometry in the physical (x, y) plane into a corresponding geometry in the (X, y) plane. Note that straight lines will transform into straight lines under this transformation.

2. Solve the problem for a homogeneous, isotropic aquifer with a permeability of K in the (X, y) plane.

3. Use Eq. 7.33 to rewrite the solution in terms of x and y, replace K with K_y and replace q with $q\sqrt{K_y/K_x}$ to obtain the solution for the anisotropic aquifer in the (x, y) plane.

Since K only appears in the (X, y) plane solution in the dimensionless ratio

$$\frac{q}{HK}$$

in which H = a characteristic value of h, this procedure leads to a solution rewritten in terms of the dimensionless ratio

$$\frac{q}{H\sqrt{K_x K_y}}$$

Hence, it is sometimes stated in texts that the anisotropic problem has an equivalent permeability of

$$K_{eq} = \sqrt{K_x K_y} \qquad (7.43)$$

33. Analytic Functions of a Complex Variable

We will let z represent the complex variable

$$z = x + iy, \quad (i = \sqrt{-1}) \qquad (7.44)$$

A number, z_1, is located in a complex z plane by plotting its real part, x_1, along the abscissa and its imaginary part, y_1, along the ordinate, as shown in Fig. 7.2. Polar coordinates can also be used to write $x_1 = r\cos\theta$ and $y_1 = r\sin\theta$ in Eq. 7.44:

$$z_1 = r(\cos\theta + i\sin\theta) \qquad (7.45)$$

An exponential function can be expanded in a Taylor's series in the form

$$e^x = \sum_{n=0}^{\infty} \frac{x^n}{n!}, \quad (0 \leqq x < \infty) \qquad (7.46)$$

Thus, setting x = iθ in Eq. 7.46 and separating real and imaginary parts gives

$$e^{i\theta} = \left(1 - \frac{\theta^2}{2!} + \frac{\theta^4}{4!} - \frac{\theta^6}{6!} + \ldots\right) + i\left(\theta - \frac{\theta^3}{3!} + \frac{\theta^5}{5!} - \frac{\theta^7}{7!} + \ldots\right) \qquad (7.47)$$

The infinite series expansions on the right side of Eq. 7.47 converge, for all finite values of θ, to cos θ and sin θ.

$$e^{i\theta} = \cos\theta + i\sin\theta \qquad (7.48)$$

Thus, Eq. 7.45 can also be written in the more compact, exponential form

$$z_1 = re^{i\theta}, \quad \left[r = \sqrt{x_1^2 + y_1^2}, \ \theta = \tan^{-1}\left(\frac{y_1}{x_1}\right)\right] \qquad (7.49)$$

Since two complex numbers are added or subtracted by adding or subtracting their real and imaginary parts separately, it is possible to use a graphical representation for the sum or difference of two complex numbers, as shown in Fig. 7.3. These sums and differences can also be written in exponential forms.

$$z_1 + z_2 = re^{i\theta}, \quad \left[r = \sqrt{(x_1 + x_2)^2 + (y_1 + y_2)^2}, \; \theta = \tan^{-1}\left(\frac{y_1 + y_2}{x_1 + x_2}\right)\right] \tag{7.50}$$

$$z_2 - z_1 = re^{i\theta}, \quad \left[r = \sqrt{(x_2 - x_1)^2 + (y_2 - y_1)^2}, \; \theta = \tan^{-1}\left(\frac{y_2 - y_1}{x_2 - x_1}\right)\right] \tag{7.51}$$

A complex function, $w(z)$, can be formed by letting z take on any of the values in a given region or domain in the z plane. As an example, let the domain in the z plane be the interior of the circle $|z| = r = 2$, as shown in Fig. 7.4. Then

$$w(z) = z - 1 = re^{i\theta}, \quad \left[r = \sqrt{(x-1)^2 + y^2}, \; \theta = \tan^{-1}\left(\frac{y}{x-1}\right)\right] \tag{7.52}$$

is a function of z since it gives a value of w for every value of z within the domain. The value for θ in Eq. 7.52 is not unique since any even multiple of 2π may be added or subtracted from θ without changing the value of $\sin\theta$, $\cos\theta$ or $\tan\theta$. However, all of these values of θ yield the same real and imaginary parts for $w(z)$, which means that $w(z)$ is single-valued in its given domain.

A second example of a complex function is given by

$$w(z) = \sqrt{z-1} = \sqrt{r}\, e^{i\theta/2} \tag{7.53}$$

This time we will choose for a domain the interior of the circle $|z| = 2$ except that we will exclude the segment along the x axis that joins the points $z = 1$ and $z = 2$, as shown in Fig. 7.5. Values for θ will be made unique by limiting θ to the range $0 < \theta < 2\pi$. The limits on the range for θ could be changed by adding or subtracting any even multiple of 2π. However, we could not choose $-\pi < \theta < \pi$ for this example since $w(z)$ would not vary in a continuous way for all positions of z in the given domain. Letting $y \to 0$ through positive values for $1 < x < 2$ gives

$$w_+(x) = \lim_{\theta \to 0} w(z) = \sqrt{r} \tag{7.54}$$

and letting $y \to 0$ through negative values for $1 < x < 2$ gives

$$w_-(x) = \lim_{\theta \to 2\pi} w(z) = \sqrt{r}\, e^{i\pi} = -\sqrt{r} \tag{7.55}$$

Thus, $w(z)$ is not single-valued along the segment of the real axis joining $z = 1$ and $z = 2$, which is the reason for excluding this line segment from the domain of $w(z)$. The path surrounding this line segment is called a branch cut, and $z = 1$ is called a branch point. In applied problems, this branch cut often

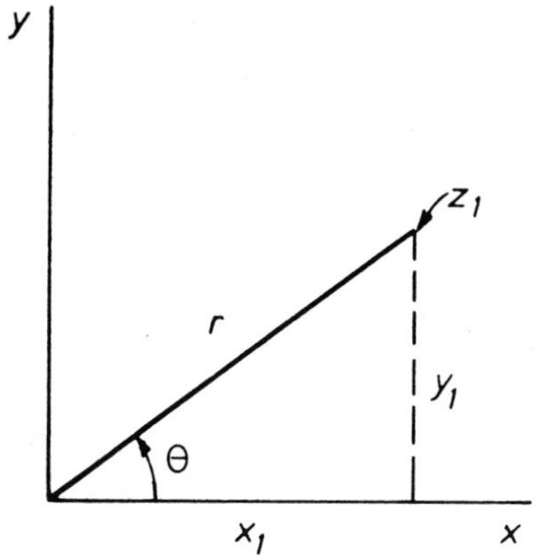

Fig. 7.2 - Plotting a complex number z_1.

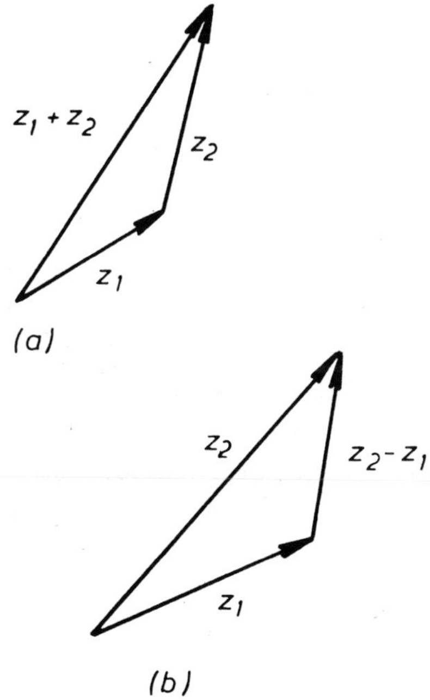

Fig. 7.3 - Graphical representation of (a) the sum of z_1 and z_2 and (b) the difference of z_2 and z_1.

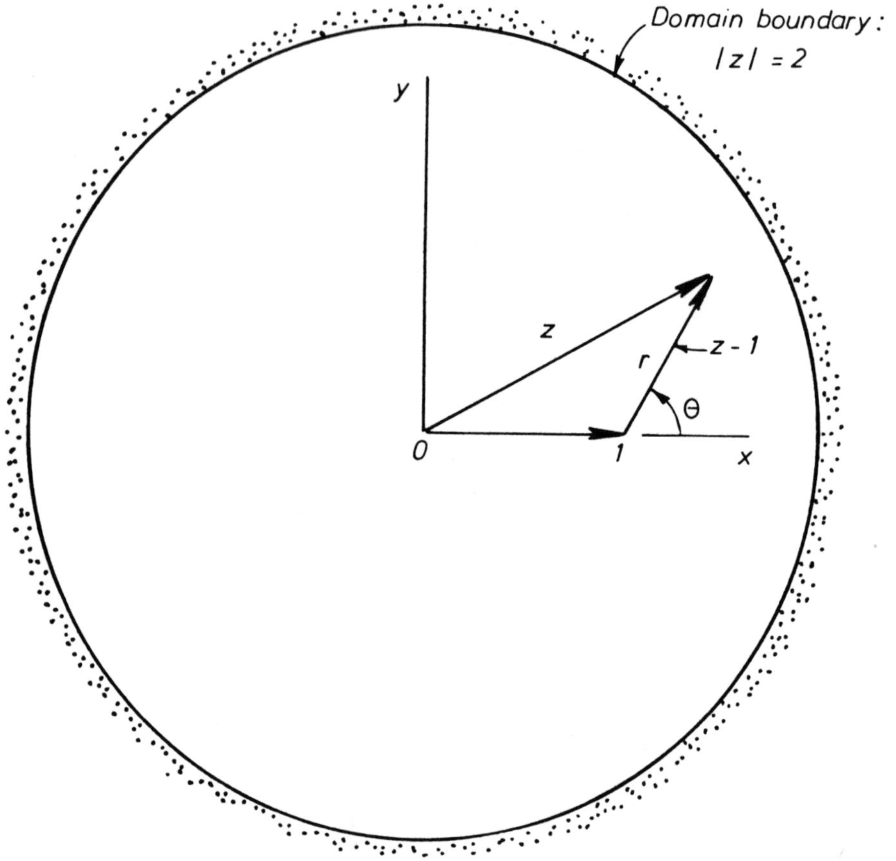

Fig. 7.4 - A domain and graphical representation for the function w(z) = z-1.

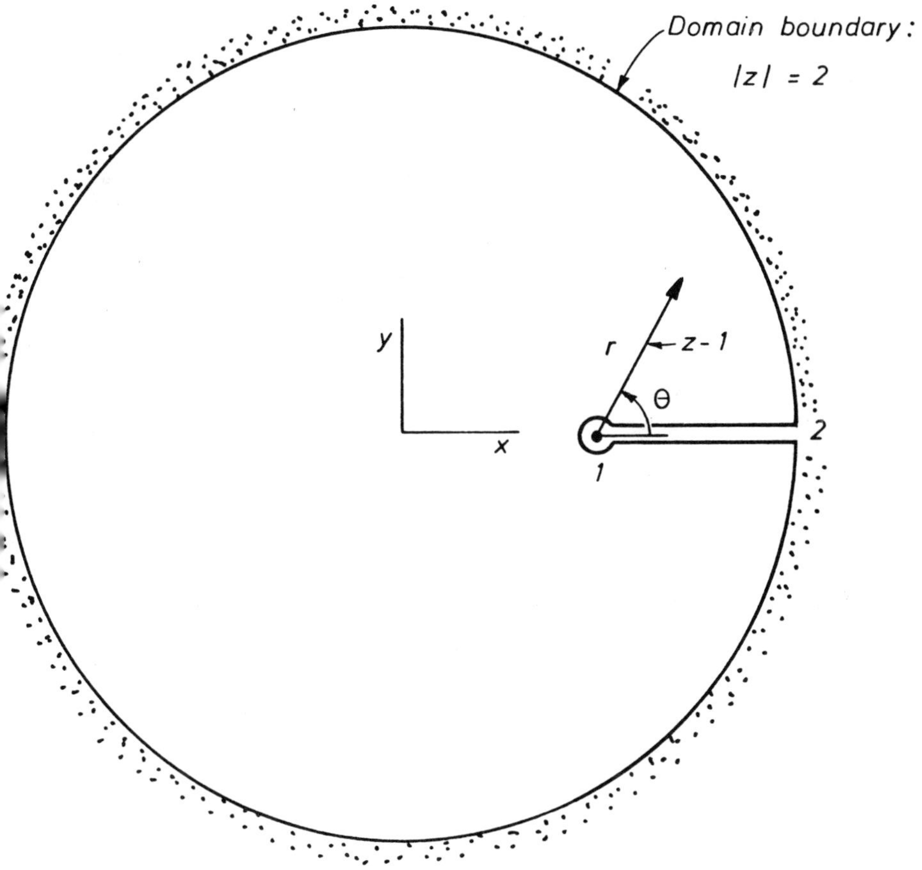

Fig. 7.5 - A domain for the function $\sqrt{z-1}$.

coincides with a physical boundary. In fact, for this particular example we could choose the branch cut to be any straight or curved line joining the branch point at $z = 1$ to any point on $|z| = 2$. This assumes, of course, that a range is also chosen for θ that allows $w(z)$ to vary continuously for all positions of z within the resulting domain. In other words, a branch cut is a geometrical construction that keeps a function single-valued within its given domain.

It is always possible to substitute Eq. 7.44 into a function $w(z)$ and then separate its real and imaginary parts. Traditionally, we show this by writing

$$w(z) = \phi(x, y) + i\psi(x, y) \tag{7.56}$$

in which ϕ and ψ are the real and imaginary parts of w. The derivative of w is defined as

$$\frac{dw(z_1)}{dz} = \underset{z_2 \to z_1}{\text{Limit}} \; \frac{w(z_2) - w(z_1)}{z_2 - z_1} \tag{7.57}$$

Equation 7.57 is a natural extension of the definition for the derivative of a function in real variable theory. It is, however, more general in the sense that z_2 and z_1 have real and imaginary parts so that z_2 can approach z_1 along any one of an infinite number of paths joining the two points. For example, if $z_2 \to z_1$ along the path $y = c$, then $z_1 = x_1 + ic$, $z_2 = x_2 + ic$ and

$$\frac{dw}{dz} = \frac{\partial w}{\partial x} = \frac{\partial \phi}{\partial x} + i\,\frac{\partial \psi}{\partial x} \tag{7.58}$$

Likewise, if $z_2 \to z_1$ along the path $x = c$, then $z_1 = c + iy_1$, $z_2 = c + iy_2$ and

$$\frac{dw}{dz} = \frac{1}{i}\frac{\partial w}{\partial y} = \frac{1}{i}\left(\frac{\partial \phi}{\partial y} + i\,\frac{\partial \psi}{\partial y}\right) \tag{7.59}$$

However, if dw/dz is single-valued at z_1, then the right sides of Eqs. 7.58 and 7.59 must be equal. Since two complex numbers can be equal only if their real and imaginery parts are equal, this gives

$$\frac{\partial \phi}{\partial x} = \frac{\partial \psi}{\partial y} \tag{7.60}$$

$$\frac{\partial \phi}{\partial y} = -\frac{\partial \psi}{\partial x} \tag{7.61}$$

Equations 7.60 and 7.61 are known as the Cauchy-Riemann conditions. If a function, $w(z)$, and its derivative, $dw(z)/dz$, are finite and single-valued at all points within a given domain, then $w(z)$ is said to be analytic, regular or holomorphic within that domain. Thus, Eqs. 7.60 - 7.61 are necessary conditions for $w(z)$ to be analytic within a given domain.

Differentiating Eq. 7.60 with respect to x, Eq. 7.61 with respect to y and adding gives

$$\frac{\partial^2 \phi}{\partial x^2} + \frac{\partial^2 \phi}{\partial y^2} = 0 \tag{7.62}$$

Differentiating Eq. 7.60 with respect to y, Eq. 7.61 with respect to x and subtracting gives

$$\frac{\partial^2\psi}{\partial x^2} + \frac{\partial^2\psi}{\partial y^2} = 0 \qquad (7.63)$$

Equations 7.60 - 7.63 are identical with Eqs. 7.5 and 7.8 - 7.10 of the previous section. Hence, we can call w(z) a complex potential function and state that its real and imaginary parts will be a velocity potential and stream function for some groundwater problem in any domain in which w(z) is analytic. Furthermore, section 32 and Eqs. 7.58 - 7.61 show that the derivative of w(z) gives

$$\frac{dw(z)}{dz} = u - iv \quad \text{when the (x, y) plane is vertical}$$

$$= q_x - iq_y \quad \text{when the (x, y) plane is horizontal} \qquad (7.64)$$

Equation 7.64 states that dw(z)/dz is the complex conjugate of either the flux velocity vector, $\bar{u}$, or the flux vector, $\bar{q}$. Equation 7.64 can also be written in exponential form

$$\frac{dw(z)}{dz} = Ve^{-i\theta}, \quad \left[V = \sqrt{u^2 + v^2}, \quad \theta = \tan^{-1}\left(\frac{v}{u}\right)\right]$$

$$= qe^{-i\theta}, \quad \left[q = \sqrt{q_x^2 + q_y^2}, \quad \theta = \tan^{-1}\left(\frac{q_y}{q_x}\right)\right] \qquad (7.65)$$

34. Some Basic Potential Functions

The complex potential for uniform flow is easily calculated from Eq. 7.65. If the streamlines make a constant angle, θ_0, with the x axis, and if the velocity magnitude has the constant value of V_0, then integration of Eq. 7.65 gives

$$w(z) = zV_0e^{-i\theta_0} \qquad (7.66)$$

A complex integration constant can also be added to the right side of Eq. 7.66 to cause ϕ and ψ to have specified values upon one equipotential line and one streamline. In horizontal flows, V_0 in Eq. 7.66 is replaced with q_0.

Since flow from a source has radial streamlines (θ = constant), the complex potential for a source at the point $z = z_0$ must have the form

$$w(z) = A\ln(z - z_0) = A[\ln(r) + i\theta] \qquad (7.67)$$

in which A is a real constant. Taking the imaginary part of Eq. 7.67 gives

$$\psi = A\theta \qquad (7.68)$$

Thus, Eq. 7.67 is multiple-valued, and we will choose a branch cut that coincides with any radial streamline that starts at r = 0 and ends at r = ∞. Choosing any range for θ that allows ψ to vary in a continuous way in the region exterior to the branch cut, we then find a discontinuity in ψ along the branch cut of

$$\psi_+ - \psi_- = 2\pi A \tag{7.69}$$

Thus, Eqs. 7.19 and 7.69 give

$$A = \frac{q}{2\pi} \tag{7.70}$$

$$w(z) = \frac{q}{2\pi} \ln(z - z_0) \tag{7.71}$$

in which q = two-dimensional flow rate emitted by the source in the vertical, (x, y) plane. Equations 7.25, 7.30 and 7.69 show that Eq. 7.67 takes the following form for horizontal flows:

$$w(z) = \frac{Q}{2\pi} \ln(z - z_0) \tag{7.72}$$

Calculation of the complex velocity readily shows that the velocity is radially outward from z_0, which means that the correct sign was used in calculating A. Flow toward a sink is modeled simply by replacing q and Q with -q and -Q in Eqs. 7.71 and 7.72, respectively.

35. Superposition of Solutions

If $w_1(z)$ and $w_2(z)$ are each analytic in a given domain, then

$$w(z) = w_1(z) + w_2(z) \tag{7.73}$$

is also analytic in the same domain. Differentiating Eq. 7.73 and using Eq. 7.64 leads to the result

$$u - iv = (u_1 + u_2) - i(v_1 + v_2) \tag{7.74}$$

Thus, adding two complex potentials gives a third potential with a velocity field equal to the vector sum of the velocity fields generated separately by $w_1(z)$ and $w_2(z)$. The extension of this result to include the sum or difference of any number of complex potential functions is straightforward.

As an example of an application of the superposition principle, consider the problem of a well next to a river in an unconfined aquifer. The well is located at $z = z_0$, and the river edge coincides with the x axis, as shown in Fig. 7.6. Near $z = z_0$, the complex potential must have the behaviour

$$w(z) \sim -\frac{Q}{2\pi} \ln(z - z_0) \tag{7.75}$$

The river must be a constant potential boundary with

$$\phi(x, 0) = \mathrm{Re}\{w(x)\} = -\tfrac{1}{2}KH^2 \tag{7.76}$$

in which H = elevation of the water surface in the river above the bottom boundary of the horizontal aquifer. The image point of $z_0 = x_0 + iy_0$ is the complex conjugate point, $\bar{z}_0 = x_0 - iy_0$. Thus, the solution to the problem is

$$w(z) = -\frac{Q}{2\pi}\ln(z - z_0) + \frac{Q}{2\pi}\ln(z - \bar{z}_0) - \tfrac{1}{2}KH^2 \qquad (7.77)$$

A proof that Eq. 7.77 solves the problem is obtained easily by noting that $w(z)$ is analytic everywhere in the upper half z plane except at the branch point z_0 and along a branch cut that leaves z_0 and finishes at a point either along the river or at infinity, that $w(z)$ has the correct logarithmic singularity at z_0 and that the real part of $w(z)$ satisfies Eq. 7.76 when z becomes a point on the x axis ($|z - z_0| = |z - \bar{z}_0|$ when $z \to x + i0$). If $z = z_0 + \Delta$ is a point on the well perimeter and $h = h_0$ within the well, then taking the real part of Eq. 7.77 after setting $\phi = -\tfrac{1}{2}Kh_0^2$ and $z = z_0 + \Delta$ gives

$$-\tfrac{1}{2}Kh_0^2 = -\frac{Q}{2\pi}\ln|\Delta| + \frac{Q}{2\pi}|z_0 - \bar{z}_0 + \Delta| - \tfrac{1}{2}KH^2 \qquad (7.78)$$

Thus, if $|\Delta| = \delta$ = well radius and $\delta/y_0 << 1$ so that $|z_0 - \bar{z}_0 + \Delta|$ $= |2iy_0 + \Delta| \simeq 2y_0$, Eq. 7.78 gives the result

$$\frac{Q}{K(H^2 - h_0^2)} = \frac{\pi}{\ln\left(\dfrac{2y_0}{\delta}\right)} \qquad (7.79)$$

A well-interferance problem occurs when two wells, at $z = z_0$ and $z = z_1$, each pump a flow rate of Q' from the aquifer in Fig. 7.6. In this case, a procedure very similar to the one used to obtain Eq. 7.79 gives

$$w(z) = -\frac{Q'}{2\pi}\ln(z - z_0) + \frac{Q'}{2\pi}\ln(z - \bar{z}_0) - \frac{Q'}{2\pi}\ln(z - z_1) + \frac{Q'}{2\pi}\ln(z - \bar{z}_1) - \tfrac{1}{2}KH^2 \qquad (7.80)$$

$$-\tfrac{1}{2}Kh_0^2 = -\frac{Q'}{2\pi}\ln|\Delta| + \frac{Q'}{\pi}\ln|z_0 - \bar{z}_0 + \Delta| - \frac{Q'}{2\pi}\ln|z_0 - z_1 + \Delta| + \frac{Q'}{2\pi}\ln|z_0 - \bar{z}_1 + \Delta| - \tfrac{1}{2}KH^2 \qquad (7.81)$$

$$\frac{Q'}{K(H^2 - h_0^2)} = \frac{\pi}{\ln\left[\dfrac{2y_0}{\delta}\sqrt{\dfrac{(x_0 - x_1)^2 + (y_0 + y_1)^2}{(x_0 - x_1)^2 + (y_0 - y_1)^2}}\right]} \qquad (7.82)$$

Equation 7.82 shows that $Q' < Q$. However, a spacing between the two wells can be calculated for which Q and Q' are nearly the same by setting

$$\frac{Q - Q'}{Q} = 1\% = \frac{1}{100} \qquad (7.83)$$

Thus, Eqs. 7.79, 7.82 and 7.83 allow the well spacing at which interference becomes negligible to be calculated from

$$\frac{2y_0}{\delta} = \left[\frac{(x_0 - x_1)^2 + (y_0 + y_1)^2}{(x_0 - x_1)^2 + (y_0 - y_1)^2}\right]^{49.5} \qquad (7.84)$$

It is useful to note that Eq. 7.84 is independent of K, Q, Q', H and h_0.

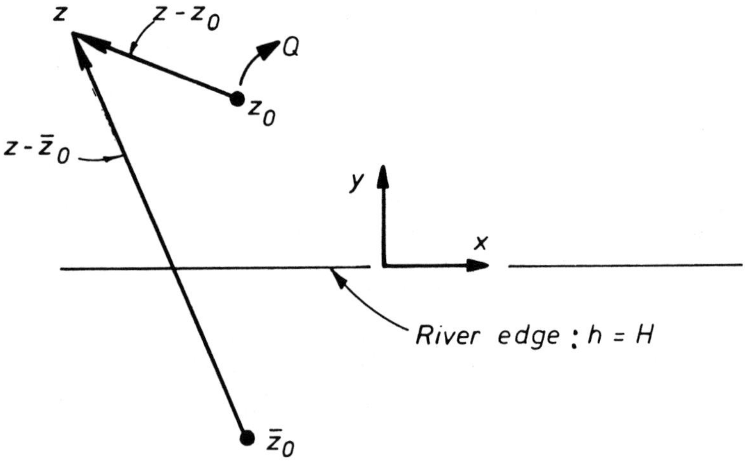

Fig. 7.6 - A well located next to a river.

Superposition and the method of images can be used to model flow to a well in a corner. For example, Fig. 7.7(a) shows flow to a well in which the positive real axis is an equipotential boundary and the positive imaginary axis is a streamline. Figure 7.7(b) has the same geometry except that the positive real and imaginary axes are both held at the same constant potential. The flows in Figs. 7.7 (c) and (d) both have the entire real axis as a constant potential boundary. Since Fig. 7.7(a) has zero horizontal velocity components along the positive imaginary axis, adding the potentials for (c) and (d) will give the correct flow pattern for (a). Likewise, since Fig. 7.7(b) has zero vertical velocity components along the positive imaginary axis, subtracting the potential for (d) from the potential for (c) will give the flow pattern for (b). In either case, it may also be necessary to add a complex constant to the solution so that ϕ and ψ attain their specified, constant values along the real and imaginary axes. Obviously, this same method can be used to obtain solutions for any number of wells in Fig. 7.7(a) or (b).

For another example of superposition, consider the flow to a well along a seacoast, as shown in Fig. 7.8. The complex potential for this flow is

$$w(z) = iq_0 z - \frac{Q}{2\pi}\ln(z - z_0) + \frac{Q}{2\pi}\ln(z - \bar{z}_0) \qquad (7.85)$$

The critical condition discussed in section 12 of chapter III occurs when the toe of the salt water wedge intersects the bottom aquifer boundary at a point that lies directly beneath a saddle point of the free surface. At this saddle point

$$\frac{dw(z)}{dz} = 0 = iq_0 - \frac{Q}{2\pi}\left(\frac{1}{z - z_0} - \frac{1}{z - \bar{z}_0}\right) \qquad (7.86)$$

The Ghyben-Herzberg approximation, Eq. 3.37, also requires, at the saddle point, that

$$\phi = -\tfrac{1}{2}K\varepsilon(1 + \varepsilon)D^2 = -q_0 y + \frac{Q}{2\pi}\ln\left|\frac{z - \bar{z}_0}{z - z_0}\right| \qquad (7.87)$$

Equations 7.86 - 7.87 contain three real equations with three unknowns: Q = critical flow rate and (x, y) = coordinates of the saddle point on the free surface. Setting $z_0 = iy_0$ in Eqs. 7.86 - 7.87 leads to the result

$$x = 0 \qquad (7.88)$$

$$\frac{KD^2}{q_0 y_0}\varepsilon(1 + \varepsilon) = 2\frac{y}{y_0} - \left(1 - \frac{y^2}{y_0^2}\right)\ln\left|\frac{1 + y/y_0}{1 - y/y_0}\right| \qquad (7.89)$$

$$\frac{Q}{q_0 y_0} = \pi\left(1 - \frac{y^2}{y_0^2}\right) \qquad (7.90)$$

Equation 7.88 shows that the saddle point lies along the y axis, Eq. 7.89 is

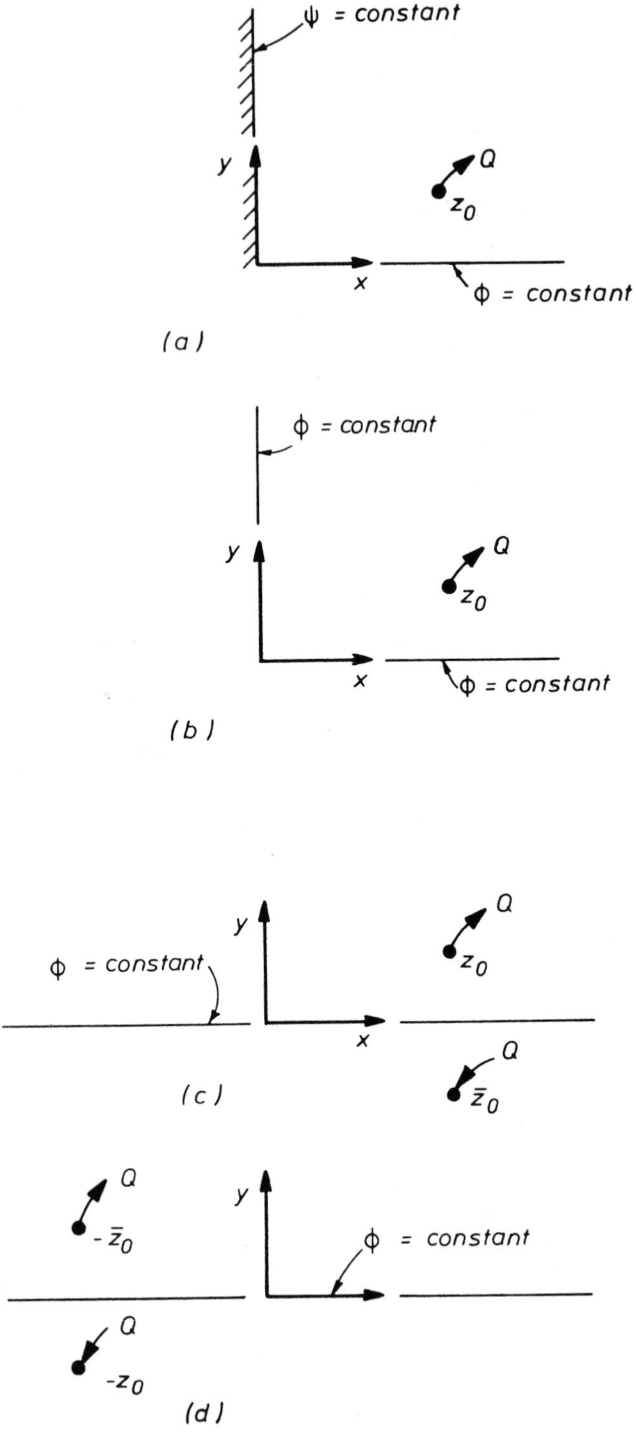

Fig. 7.7 - Flow to a well in a corner.

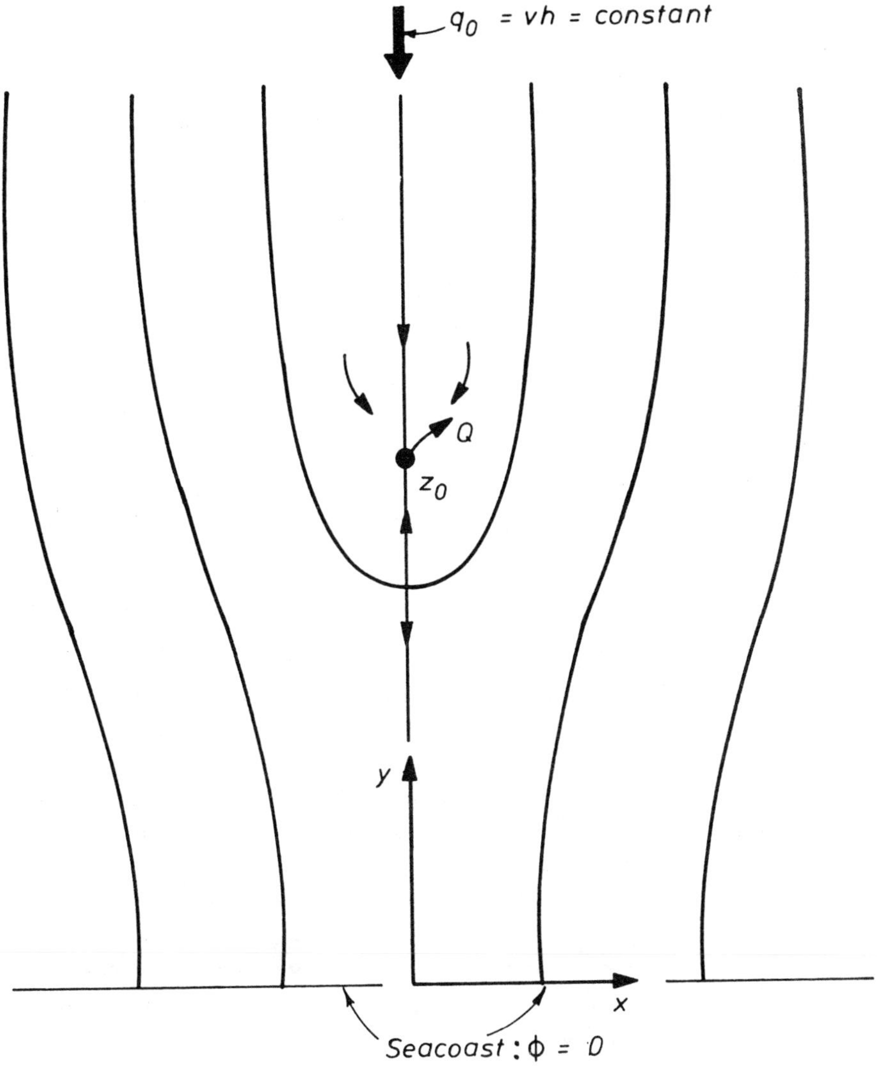

Fig. 7.8 - Flow to a well along a seacoast.

a transcendental equation that can be solved for the y coordinate of the saddle point and Eq. 7.90 allows the critical flow rate, Q, to be calculated after y has been found from Eq. 7.89. The solution to this problem was calculated first by Strack (1976).

36. Analytic Coordinate Transformations

Let t represent the complex variable

$$t = \xi + i\eta \qquad (7.91)$$

in which ξ and η are real variables. Then separating real and imaginary parts in the function

$$z = z(t) \qquad (7.92)$$

gives the two real equations

$$x = x(\xi, \eta) \qquad (7.93)$$

$$y = y(\xi, \eta) \qquad (7.94)$$

Either Eq. 7.92 or Eqs. 7.93 - 7.94 can be looked upon as giving a coordinate transformation between the z and t planes, as shown in Fig. 7.9. Under this transformation, a given domain, D_z, and boundary, Γ_z, in the z plane will be transformed or mapped into some domain, D_t, and boundary, Γ_t, in the t plane. Since an analytic function is single-valued, the transformation between D_z and D_t will be one to one at all points where z(t) is analytic.

The use of Eq. 7.92 in the complex potential function, w(z), gives

$$w = w[z(t)] = \phi(\xi, \eta) + i\psi(\xi, \eta) \qquad (7.95)$$

Thus, w is an analytic function of t, and the real and imaginary parts of w are harmonic functions of ξ and η in the t plane. This suggests that a transformation like Eq. 7.92 might be used to map a relatively complicated solution domain in the z plane into a simpler geometry in the t plane. Then a transformation of the boundary conditions for w into the t plane allows w to be calculated as an analytic function of t in the t plane. Finally, Eq. 7.92 and w = w(t) can be used to calculate the solution in the z plane by eliminating the parameter t. This procedure becomes particularly simple when either ϕ or ψ are prescribed constants along all portions of the solution domain boundary, since these constants have the same magnitude along corresponding portions of the boundary in both the z and t planes.

A simple example of this procedure is shown in Fig. 7.10, where the wedge-shaped region in the z plane is mapped into the upper half t plane by using the transformation

$$t = z^{\pi/\alpha} \qquad (7.96)$$

A well at z_0 is mapped into the point

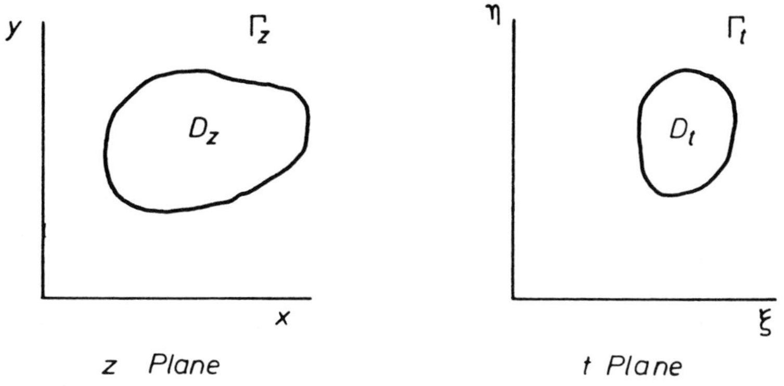

Fig. 7.9 - Use of the analytic function $z = z(t)$ to map a domain in the z plane into a domain in the t plane.

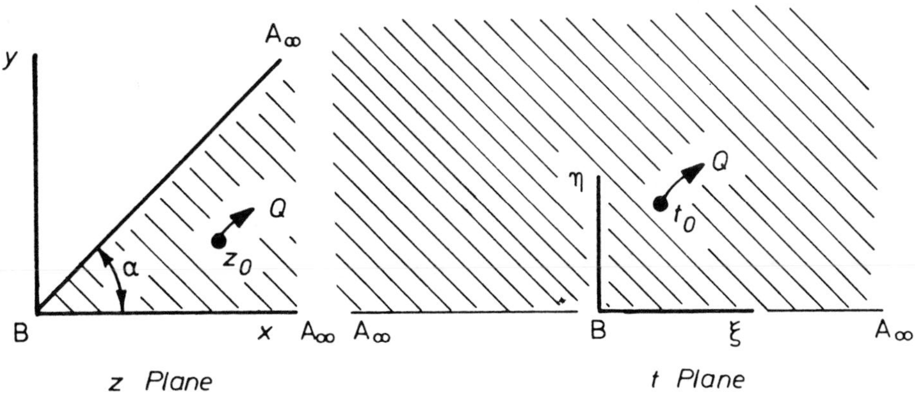

Fig. 7.10 - Flow to a well in a wedge-shaped region.

$$t_0 = z_0^{\pi/\alpha} \tag{7.97}$$

If the solution domain in the physical z plane is the plan view of an unconfined aquifer, and if the boundary ABA is a river or reservoir boundary, then ϕ has the same constant value along ABA in both planes.

$$\phi = -\tfrac{1}{2}KH^2 \text{ along ABA} \tag{7.98}$$

in which H = constant elevation of the reservoir free surface above the bottom aquifer boundary. The constant values of ψ on the streamlines to the well will be identical in both planes, so that the well abstracts the same flow, Q, in both planes. Thus, the solution to the problem in the t plane is

$$w = -\frac{Q}{2\pi}\ln(t - t_0) + \frac{Q}{2\pi}\ln(t - \bar{t}_0) - \tfrac{1}{2}KH^2 \tag{7.99}$$

and eliminating the parameter t between Eqs. 7.96 and 7.99 gives the solution in the z plane.

$$w = \frac{Q}{2\pi}\ln\left\{\frac{z^{\pi/\alpha} - \bar{z}_0^{\pi/\alpha}}{z^{\pi/\alpha} - z_0^{\pi/\alpha}}\right\} - \tfrac{1}{2}KH^2 \tag{7.100}$$

The relationship between the flow rate, Q, and piezometric head in the well, h_0, can be found by setting $\text{Re}\{w\} = \phi^{\cdot} = -Kh_0^2/2$ and $z = z_0 + \Delta$ in which $|\Delta| = \delta$ = well radius. Then retaining only first-order terms in an expansion for small Δ and taking the real part of both sides of the equation leads to the result

$$\frac{\pi K(H^2 - h_0^2)}{Q} = \ln\frac{|z_0^{\pi/\alpha} - \bar{z}_0^{\pi/\alpha}|}{\frac{\pi}{\alpha}|\Delta|\,|z_0|^{\pi/\alpha - 1}} = \ln\left[\frac{2\alpha}{\pi}\,\frac{r}{\delta}\,\sin\left(\pi\,\frac{\theta}{\alpha}\right)\right] \tag{7.101}$$

in which $\delta = |\Delta|$ = well radius, $r = |z_0| = \sqrt{x_0^2 + y_0^2}$ and $\theta = \text{Arg}\{z_0\} = \tan^{-1}(y_0/x_0)$. The correct branch for θ is the one for which $0 < \theta < \alpha$.

The mapping, Eq. 7.96, used in Fig. 7.10 is easily verified by setting $z = re^{i\theta}$.

$$t = z^{\pi/\alpha} = r^{\pi/\alpha}e^{i\pi\theta/\alpha} \tag{7.102}$$

Thus, $\theta = 0$ maps into the positive real t axis, $\theta = \alpha$ maps into the negative real t axis and the arc r = constant for $0 \leqq \theta \leqq \alpha$ maps into the half circle $|t|$ = constant with $0 \leqq \text{Arg}\{t\} \leqq \pi$, as shown in Fig. 7.11. Also shown in Fig. 7.11 are three closely spaced points, labeled 1, 2 and 3, in the z plane and the images of these points in the t plane. Thus, under the transformation $z = z(t)$,

$$\frac{z_3 - z_1}{z_2 - z_1} = \frac{z(t_3) - z(t_1)}{z(t_2) - z(t_1)} \simeq \frac{z'(t_1)}{z'(t_1)}\,\frac{t_3 - t_1}{t_2 - t_1} \tag{7.103}$$

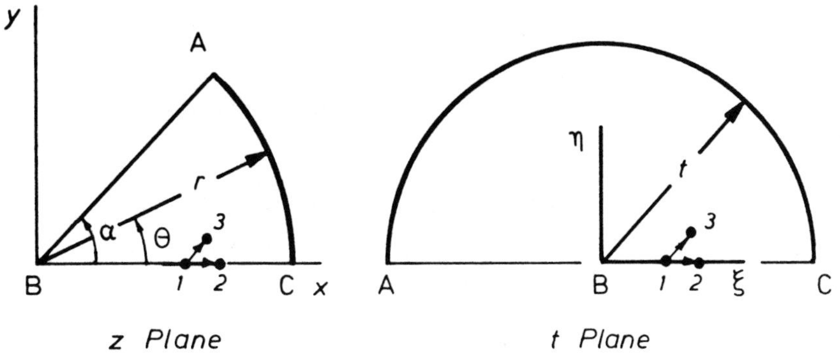

Fig. 7.11 - The mapping used in Fig. 7.10.

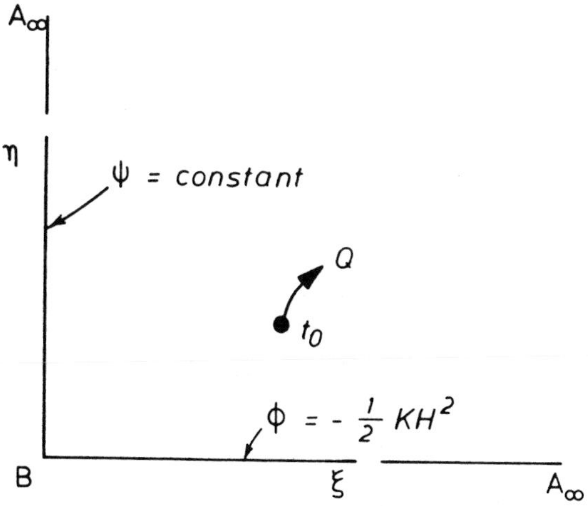

Fig. 7.12 - Mapping the solution domain in Fig. 7.10 into the first quadrant of the t plane.

in which only the first-order terms in the Taylor's series expansions of the numerator and denominator have been retained. But the ratio of derivatives is unity if z(t) is analytic at t_1. Thus, taking the argument of both sides gives (this can be carried out formally by taking the logarithm of both sides of Eq. 7.103 and equating imaginary parts on both sides of the equation)

$$\text{Arg}\{z_3 - z_1\} - \text{Arg}\{z_2 - z_1\} = \text{Arg}\{t_3 - t_1\} - \text{Arg}\{t_2 - t_1\} \quad (7.104)$$

In words, Eq. 7.104 states that the angle between the infinitesimal line segments joining points 1 and 3 and 1 and 2 is preserved at all points where the mapping transformation is analytic. This is the reason for calling an analytic mapping transformation conformal, and it also leads to the important fact that the interior of the domain lies on the same side of an observer in both planes if he traverses the boundary points in the same order in both planes. In other words, traversing the boundary from A to B to C to A in Fig. 7.11 shows that the solution domain lies on the left side of the observer in both planes. This extremely useful device requires that the third coordinate axis in both planes point in the same direction. It is also worth pointing out that all points on the circular arc AC in Fig. 7.11 become one and the same point as the radius of the arc becomes infinite, which is the reason for replacing point C in Fig. 7.11 with point A in Fig. 7.10.

The wedge-shaped solution domain in Fig. 7.10 is mapped into the first quadrant in the t plane with the transformation

$$t = z^{\pi/(2\alpha)} \quad (7.105)$$

Thus, if the boundary $\theta = \alpha$ is a streamline and $\theta = 0$ is an equipotential line, the methods of images discussed in the previous section can be used to write the solution

$$w = -\frac{Q}{2\pi}\ln(t - t_0) + \frac{Q}{2\pi}\ln(t - \bar{t}_0) - \frac{Q}{2\pi}\ln(t + \bar{t}_0) + \frac{Q}{2\pi}\ln(t + t_0) - \tfrac{1}{2}KH^2 \quad (7.106)$$

The dimensionless flow rate is computed by setting $\phi = \text{Re}\{w\} = -Kh_0^2/2$ and $z = z_0 + \Delta$ in which $|\Delta| = \delta$ = well radius. Then retaining only the first-order terms in the expansion for small Δ gives the result

$$\frac{\pi K(H^2 - h_0^2)}{Q} = \ln\left[\frac{4\alpha}{\pi}\frac{r}{\delta}\tan\left(\frac{\pi\theta}{2\alpha}\right)\right] \quad (7.107)$$

in which $r = |z_0| = \sqrt{x_0^2 + y_0^2}$ and $\theta = \text{Arg}\{z_0\} = \tan^{-1}(y_0/x_0)$ with $0 < \theta < \alpha$.

37. The Schwarz-Christoffel Transformation

The use of analytic tranformations to solve harmonic boundary-value problems depends upon finding a transformation that maps the solution domain into a simpler solution domain in the t plane. A number of mapping transformations have been tabulated, for example, by Churchill (1960) and Kober (1957). However, a large number of useful transformations can be calculated directly with the Schwarz-Christoffel transformation, a transformation that maps the interior of regions bounded by straight lines (polygons) into either an upper or lower half plane.

The Schwarz-Christoffel transformation that maps the interior of the quadrilateral shown in Fig. 7.13 to the lower half t plane is

$$\frac{dz}{dt} = C_1(t - t_1)^{-\alpha_1/\pi}(t - t_2)^{-\alpha_2/\pi}(t - t_3)^{-\alpha_3/\pi}(t - t_4)^{-\alpha_4/\pi} \qquad (7.108)$$

in which C_1 = complex constant, t_i = coordinate of the image of vertex i (on the real t axis) and α_i = angle that the unit tangent rotates through when traversing the boundary at vertex i in the z plane. The angle α_i is determined by traversing the boundary so that the polygon interior lies on the left, and the sign of α_i is positive when the unit tangent rotates in the counterclockwise direction for the (x, y) coordinate axes shown in Fig. 7.13. Thus, a negative value of α_i occurs for a reentrant angle at vertex i, and the sum of all values of α_i must equal 2π. Reversing the order of the points t_i along the real t axis will map the polygon interior to the upper half t plane.

Equation 7.108 may be verified by noting that the entire real t axes is a branch cut for the right side of Eq. 7.108 and that branch points are located at the points $t = t_i$. A branch for $t - t_i$ must be chosen so that the argument, θ_i, of $t - t_i$ varies continuously for every position of point t in the lower half plane, and we will choose $-\pi \leqq \theta_i \leqq 0$. Thus, equating the imaginary part of the logarithm of each side of Eq. 7.108 gives

$$\mathrm{Arg}\{dz\} - \mathrm{Arg}\{dt\} = \mathrm{Arg}\{C_1\} - \frac{\alpha_1}{\pi}\mathrm{Arg}\{t - t_1\} - \frac{\alpha_2}{\pi}\mathrm{Arg}\{t - t_2\}$$
$$- \frac{\alpha_3}{\pi}\mathrm{Arg}\{t - t_3\} - \frac{\alpha_4}{\pi}\mathrm{Arg}\{t - t_4\} \qquad (7.109)$$

Since $\mathrm{Arg}\{dt\} = -\pi$ = constant as point t moves along the real t axis from ∞ to $-\infty$, the argument of dz is seen to remain constant for values of $t = \xi + i0\_$ between each of the vertex image points and to increase by an amount α_i as point t traverses around the branch point at the point $t = t_i$. Hence, Eq. 7.108 maps the real t axis into a polygon with the same vertex angles as the polygon in the z plane.

Taking the indefinite integral of Eq. 7.108 introduces another complex constant, C_2, into the solution for z, which gives a total of two complex constants (C_1 and C_2) and four real constants (t_1 through t_4) to determine. These constants fix the position of the vertices in the z plane. Since the vertex angles have been fixed with the values of α_i, the position of three vertices must be fixed in the z plane in order to construct a unique polygon. This means that requiring all vertices to map into the correct image points

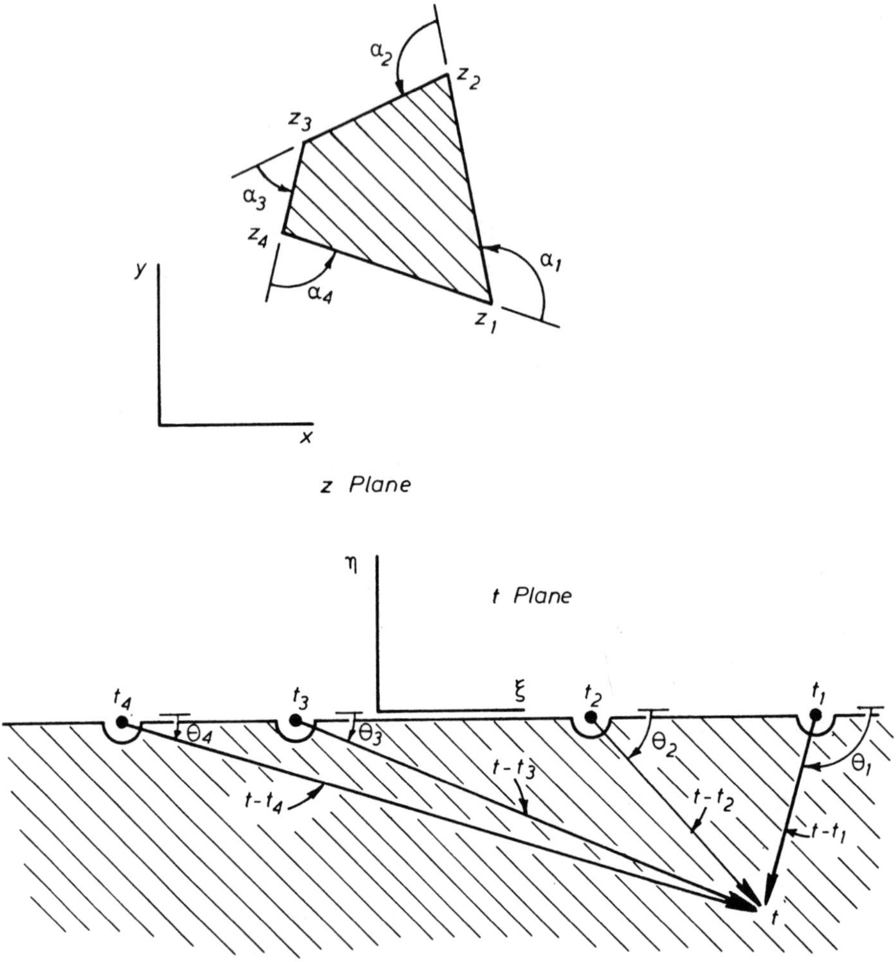

Fig. 7.13 - The Schwarz-Christoffel transformation used to map the interior of a polygon to the lower half t plane.

will give only three independent equations, and three of the six constants can be chosen arbitrarily. Since the constants C_1 and C_2 cannot be chosen arbitrarily if the polygon is to have the correct orientation and position in the z plane, this uniqueness condition reduces to the result that exactly three of the vertex image point coordinates at $t = t_i$ may be chosen arbitrarily. Similar reasoning shows that this same result holds for a polygon with any number of vertices. (i.e. - The location of three vertex image points along the real t axis can be chosen arbitrarily.)

In many applications it is desirable to choose $t_i = \infty$ for one vertex. For example, if the image point for t_1 is chosen at infinity, Eq. 7.108 can be rewritten in the form

$$\frac{dz}{dt} = C_1 t_1^{-\alpha_1/\pi}\left(\frac{t}{t_1} - 1\right)^{-\alpha_1/\pi}(t - t_2)^{-\alpha_2/\pi}(t - t_3)^{-\alpha_3/\pi}(t - t_4)^{-\alpha_4/\pi}$$

Thus, taking the limit $t_1 \to \infty$ gives

$$\frac{dz}{dt} = C_1'(t - t_2)^{-\alpha_2/\pi}(t - t_3)^{-\alpha_3/\pi}(t - t_4)^{-\alpha_4/\pi} \qquad (7.110)$$

in which

$$C_1' = At_1^{-\alpha_1/\pi}(-1)^{-\alpha_1/\pi} \qquad (7.111)$$

In other words, the image of vertex i in the t plane can be chosen to lie at infinity by simply omitting the term $(t - t_i)^{-\alpha_i/\pi}$ on the right side of the Schwarz-Christoffel transformation. Even though this term no longer appears in the transformation, it is important to realize (a) that one of the three arbitrary choices for t_i has been used in the process and (b) that this value of α_i must still be included when summing the values of α_i to check that the result is 2π.

As an example, we will map the semi-infinite strip shown in Fig. 7.14. Traversing the points in the order ABCA shows that the interior of the strip will be mapped to the upper half t plane. Values of α_i are π, $\pi/2$ and $\pi/2$ at points A, B and C, respectively. The three arbitrary choices for t_i have been used to map these vertices to ∞, -1 and 1 along the real t axis. Thus, the transformation is

$$\frac{dz}{dt} = C_1(t + 1)^{-1/2}(t - 1)^{-1/2} \qquad (7.112)$$

Taking the indefinite integral of both sides of Eq. 7.112 gives

$$z = C_1\cosh^{-1}(t) + C_2 \qquad (7.113)$$

It is shown in appendix I how all inverse trigonometric and hyperbolic functions can be rewritten as logarithms. Thus, Eq. 7.113 can be rewritten in the alternative form

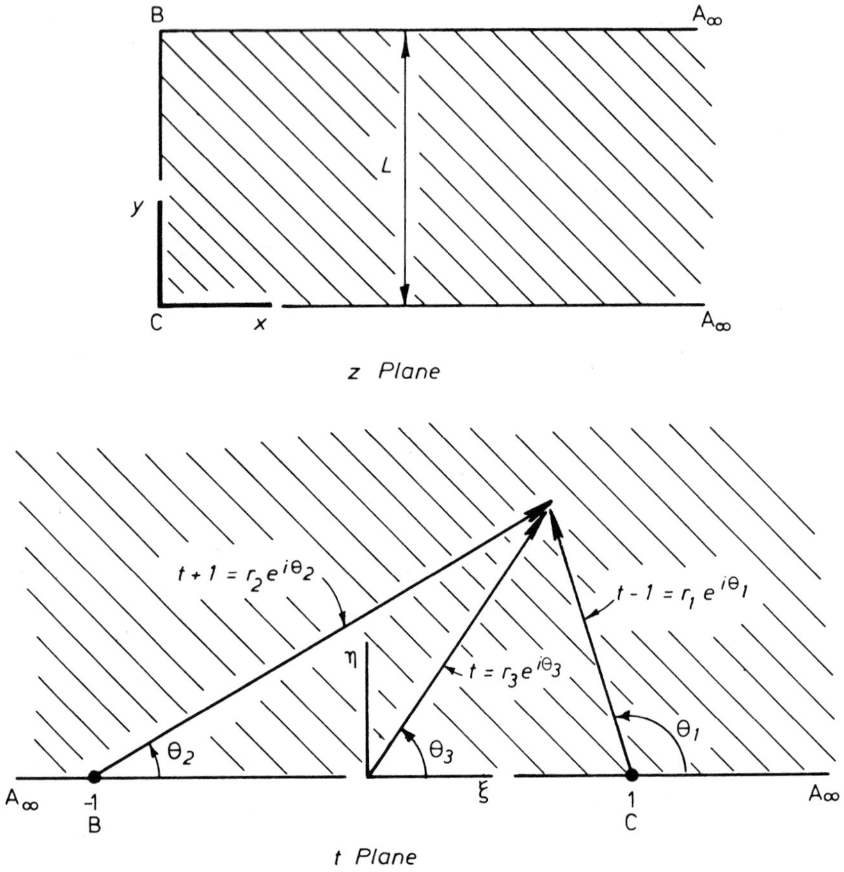

Fig. 7.14 - Use of the Schwarz-Christoffel transformation to map a semi-infinite strip to the upper half t plane.

$$z = C_1 \ln(t + \sqrt{t^2 - 1}) + C_2 \tag{7.114}$$

The constants C_1 and C_2 will be found by requiring points B and C to map into their correct image points.

The logarithm in Eq. 7.114 is multiple-valued, and it is convenient to rewrite it in the form

$$z = C_1 [\ln(r) + i\theta] + C_2 \tag{7.115}$$

in which

$$r = |t + \sqrt{(t - 1)(t + 1)}| \tag{7.116}$$

$$\theta = \text{Arg}\{t + \sqrt{(t - 1)(t + 1)}\} \tag{7.117}$$

An examination of Eq. 7.117 for different positions of point t in the upper half plane shows that we can choose the branch $0 \leqq \theta \leqq \pi$. Setting $t = 1$ for $z = 0$ at point C gives

$$0 = C_1 [0] + C_2 \tag{7.118}$$

and setting $t = -1$ for $z = iL$ at point B gives

$$iL = C_1 [i\pi] + C_2 \tag{7.119}$$

The simultaneous solution of Eqs. 7.118 and 7.119 gives

$$C_1 = \frac{L}{\pi} \tag{7.120}$$

$$C_2 = 0 \tag{7.121}$$

It is interesting to notice that letting $t \to \xi + i0$ for $-1 < \xi < 1$ gives $r = 1$ and

$$\theta = \text{Arg}\{\xi + i\sqrt{1 - \xi^2}\} = \cos^{-1}(\xi) \tag{7.122}$$

in which $0 \leqq \theta \leqq \pi$ for $-1 < \xi < 1$. Thus, Eq. 7.122 and 7.115 and the concept of analytic continuation, as explained by Nehari (1952), can be used to obtain another alternative form for the integral of Eq. 7.112:

$$z = iC_1 \cos^{-1}(t) + C_2 \tag{7.123}$$

Equation 7.123 could have been obtained more directly by rewriting Eq. 7.112 in the form

$$\frac{dz}{dt} = \frac{C_1}{\sqrt{t^2 - 1}} = \frac{-iC_1}{\sqrt{1 - t^2}} \tag{7.124}$$

in which the $-i$ was obtained by using the chosen branch for $\sqrt{t^2 - 1}$. Thus, the substitution $t = \cos\theta$ allows Eq. 7.123 to be obtained as the direct result of integrating Eq. 7.124.

Unwary readers should not be trapped into thinking that mapping the cross-hatched areas in Fig. 7.14 into each other also gives a transformation that maps the exterior of the polygon to the lower half t plane. Nehari (1952) uses the reflection principle to show that the three reflections of the cross-hatched region obtained by replacing the three, straight-line boundaries with mirrors are each mapped into the lower half t plane. In turn, the reflections about the straight-line boundaries of these regions are mapped into the upper half t plane, and the entire process is continued until the entire z plane is covered with semi-infinite strips that are mapped to either the upper or lower half t planes. This is shown with cross hatching in Fig. 7.15. Thus, since one point in the upper half t plane has an infinite number of image points in the z plane, we see why it is necessary to construct a branch cut and choose one branch for our calculations in the t plane.

38. Applications of the Schwarz-Christoffel Transformation

For the first application, assume that the semi-infinite strip in Fig. 7.14 represents the plan view of an unconfined aquifer in a long peninsula. If it is assumed that changes in h are small compared with saturated thicknesses of the horizontal aquifer, then ϕ and ψ can be defined in the way shown in Eqs. 7.26 - 7.30. A well is assumed to be located at $z = z_0$, and we will set $\phi = 0$ along the constant potential boundary ABCA. Thus, our calculations will give the change in h, and the drawdown will be given by $|h|$.

Since the zero drawdown boundary condition and the flow rate to the well remain unaltered in the transformation to the upper half t plane, the problem solution in the t plane is

$$w = -\frac{Q}{2\pi}\ln(t - t_0) + \frac{Q}{2\pi}\ln(t - \bar{t}_0) \tag{7.125}$$

in which z and t are related through either Eq. 7.113, 7.114, 7.115 or 7.123. Eliminating the parameter t between Eqs. 7.113 and 7.125 gives the solution in the z plane.

$$w = \frac{Q}{2\pi}\ln\left[\frac{\cosh\left(\pi\frac{z}{L}\right) - \cosh\left(\pi\frac{\bar{z}_0}{L}\right)}{\cosh\left(\pi\frac{z}{L}\right) - \cosh\left(\pi\frac{z_0}{L}\right)}\right] \tag{7.126}$$

Finally, if the drawdown at the well is H, setting $\phi = TH$ and $z = z_0 + \Delta$, expanding the right side of Eq. 7.126 for small Δ and taking the real part of both sides gives the dimensionless flow rate

$$\frac{2\pi TH}{Q} = \ln\left[\frac{2}{\pi}\frac{L}{\delta}\frac{\sin\left(\pi\frac{y_0}{L}\right)\sinh\left(\pi\frac{x_0}{L}\right)}{\sqrt{\cosh^2\left(\pi\frac{x_0}{L}\right) - \cos^2\left(\pi\frac{y_0}{L}\right)}}\right] \tag{7.127}$$

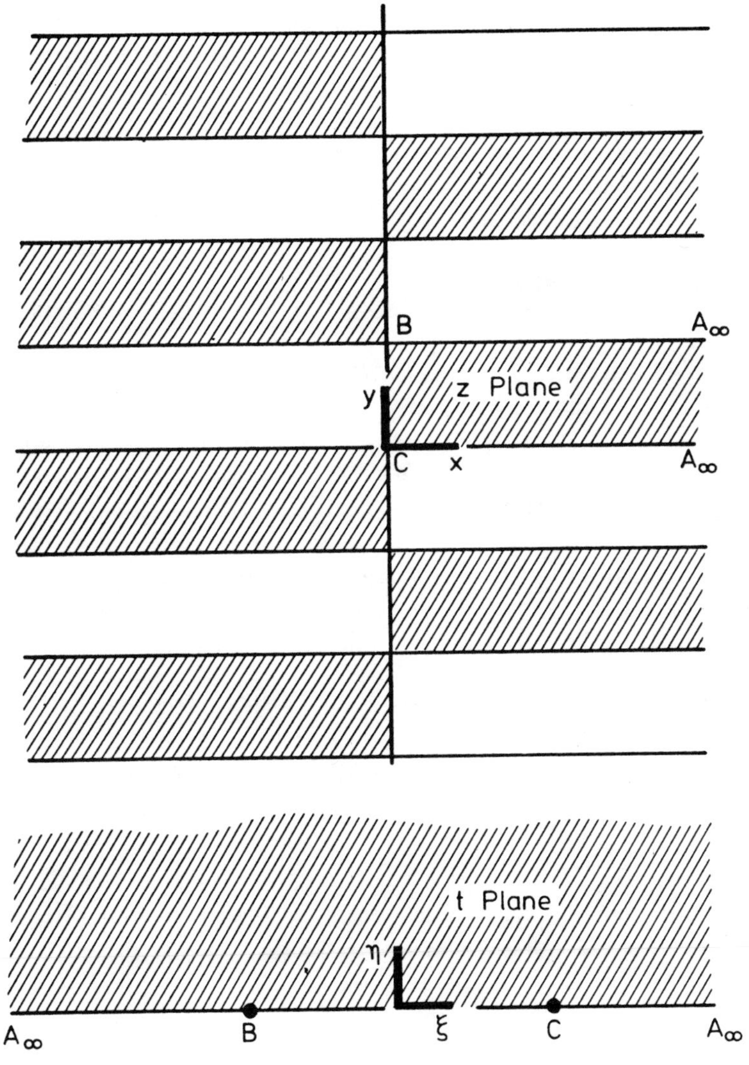

Fig. 7.15 - The multiple-valued nature of the transformation used in Fig. 7.14.

in which L = peninsula width, $\delta = |\Delta|$ = well radius and (x_0, y_0) = well coordinates. The right side of Eq. 7.127 has been simplified by using methods explained in the appendix on trigonometric and hyperbolic functions with complex arguments.

Equation 7.126 can be interpreted as the result of distributing an infinite number of singularities at image points for the boundary geometry in Fig. 7.14. One way to locate these image points is to find the images of t_0 and $\bar{t}_0$ in each of the semi-infinite strips shown in Fig. 7.15. A sink is placed at the image point in each cross-hatched strip and a source at the image point in each of the remaining strips. This result is easily seen when it is remembered that the cross-hatched regions are mapped to the upper half t plane, which contains a sink at the point $t = t_0$, and that the remaining strips are mapped to the lower half t plane, which contains a source at $t = \bar{t}_0$.

For a second application, consider flow to a collection gallery beside a river, as shown in Fig. 7.16. A collection gallery is a ditch or a horizontal, slotted pipe that is buried beneath the free surface of an unconfined water table. The variable that is of most engineering significance is the relationship between the flow, Q, to the gallery, the drawdown, H, at the gallery and the dimensions ℓ and d. A knowledge of this relationship allows an engineer to calculate the position, length and submergence depth that is required for the gallery to deliver a specified flow. A collection gallery is a useful way to obtain water from a shallow aquifer because it creates relatively small drawdowns and velocities when compared with the result of abstracting the same flow rate from a single well.

The solution domain in Fig. 7.16(a) is doubly-connected. However, symmetry allows us to consider flow in the simply-connected, first quadrant of the z plane, as shown in Fig. 7.16(b). This upper right quadrant is mapped to the upper half t plane in Fig. 7.16(c) by integrating

$$\frac{dz}{dt} = C_1 (t - 1)^{-\frac{1}{2}} \qquad (7.128)$$

in which the term for vertex A ($\alpha = 3\pi/2$) has been omitted since it is mapped to infinity in the t plane. Integration of Eq. 7.128 gives

$$z = 2C_1 \sqrt{t - 1} + C_2 \qquad (7.129)$$

Requiring points B and D to map into their correct image points gives $C_1 = (\ell + d)/2$ and $C_2 = 0$ when the branch in the t plane is chosen for which $0 \leqq \mathrm{Arg}\{t - 1\} \leqq \pi$. Thus, Eq. 7.129 becomes

$$z = (\ell + d) \sqrt{t - 1} \qquad (7.130)$$

and requiring point C to map into its image gives the coordinate of point C along the real t axis:

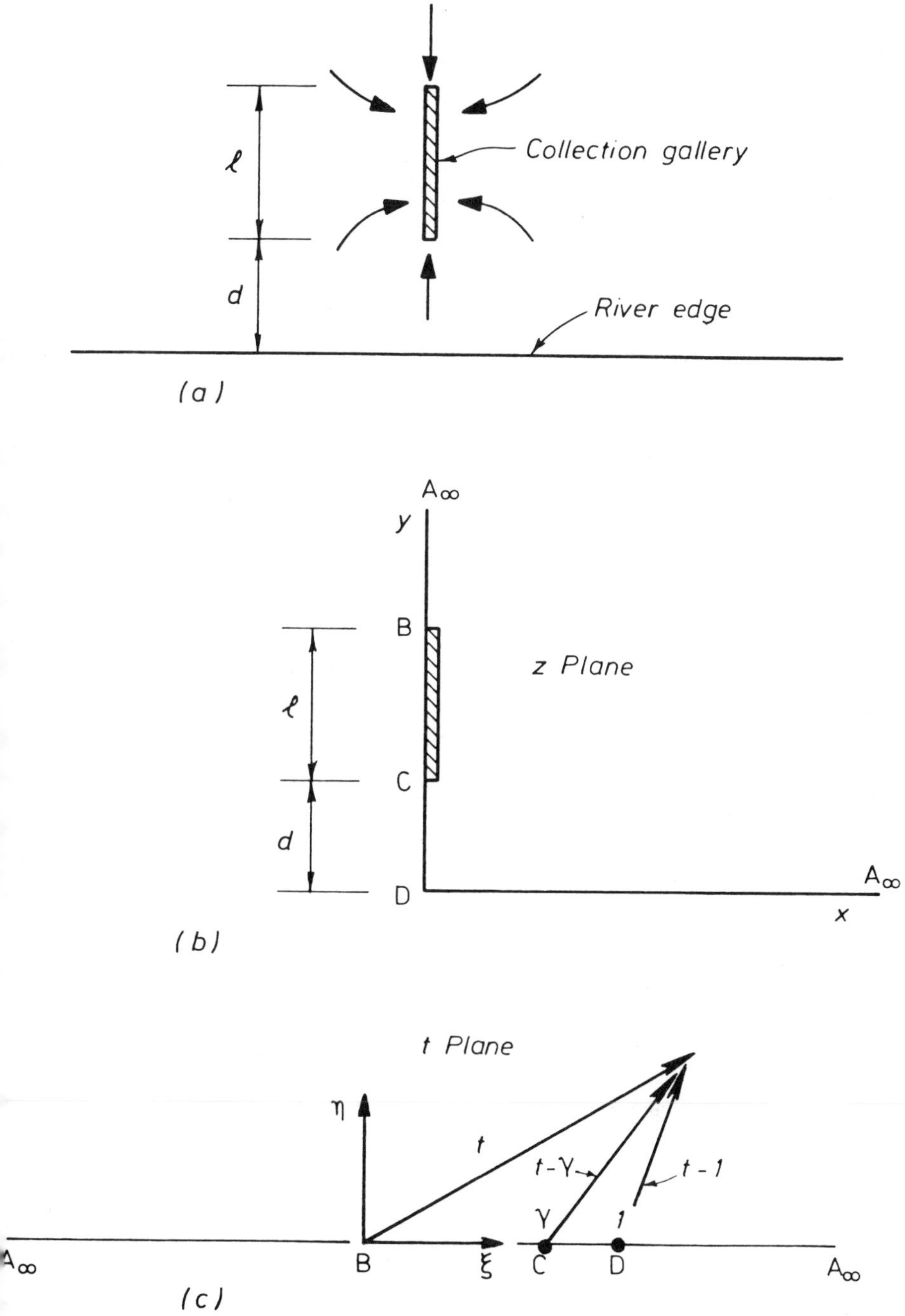

Fig. 7.16 - Flow to a collection gallery near a river.

$$\gamma = 1 - \left(\frac{d}{\ell + d}\right)^2 \qquad (7.131)$$

The transformation given by Eq. 7.130 allows the problem to be reduced to the calculation of a function w(t) which is analytic in the upper half t plane and whose real and imaginary parts satisfy the following boundary conditions:

$$\begin{aligned} \phi &= TH \quad \text{along BC} \\ \phi &= 0 \quad \text{along AD} \\ \psi &= 0 \quad \text{along AB} \\ \psi &= \tfrac{1}{2} Q \quad \text{along CD} \end{aligned} \qquad (7.132\text{ a, b, c, d})$$

Since H = drawdown = $|h|$ along BC, Eqs. 7.132a, b indicate that we are using the formulation given by Eqs. 7.26 - 7.30 to calculate relatively small changes in h from an existing steady-flow distribution of h. Equation 7.132c makes the solution for ψ unique by fixing a numerical value of ψ upon the streamline AB, and Eq. 7.132d fixes the value of ψ along the streamline CD in accordance with Eq. 7.30. The sign on the right side of Eq. 7.132d is fixed by noting that $\partial\psi/\partial y = u < 0$ in the upper right quadrant of the z plane. (i.e. - ψ decreases with y, so choosing $\psi = 0$ on AB means that ψ must be a positive constant on CD)

Equations 7.132 shows that values of ϕ or ψ are constant upon all boundaries of the flow region. This means that the flow region in the w plane is a polygon, and mapping this polygon upon the upper half t plane will give a solution for w(t) that is analytic in the upper half t plane and that satisfies the boundary conditions given by Eqs. 7.132. Polubarinova-Kochina (1962) states that N.N. Pavlovsky used this method in 1922 to obtain solutions for seepage beneath hydraulic structures. The flow region constructed from Eqs. 7.132 is shown in Fig. 7.17 as a rectangle in the first quadrant of the w plane.

Since the vertex at A is mapped to infinity in the t plane, the transformation between the w and t planes is found by integrating

$$\frac{dw}{dt} = C_1 \, t^{-\frac{1}{2}} \, (t - \gamma)^{-\frac{1}{2}} \, (t - 1)^{-\frac{1}{2}} \qquad (7.133)$$

Equation 7.133 can not be integrated in closed form, but noticing that w = TH when t = 0 and taking the definite integral of Eq. 7.133 gives

$$w = C_1 \int_0^t \frac{dt}{\sqrt{t(t-\gamma)(t-1)}} + TH \qquad (7.134)$$

The integral in Eq. 7.134 is a line integral, and, since its integrand is an analytic function of t in the upper half t plane if we choose the arguments of t, t - γ and t - 1 to lie between 0 and π, the value of this integral will depend only upon the endpoints of the integration path and not upon the particular path followed. In the following calculations we will choose an

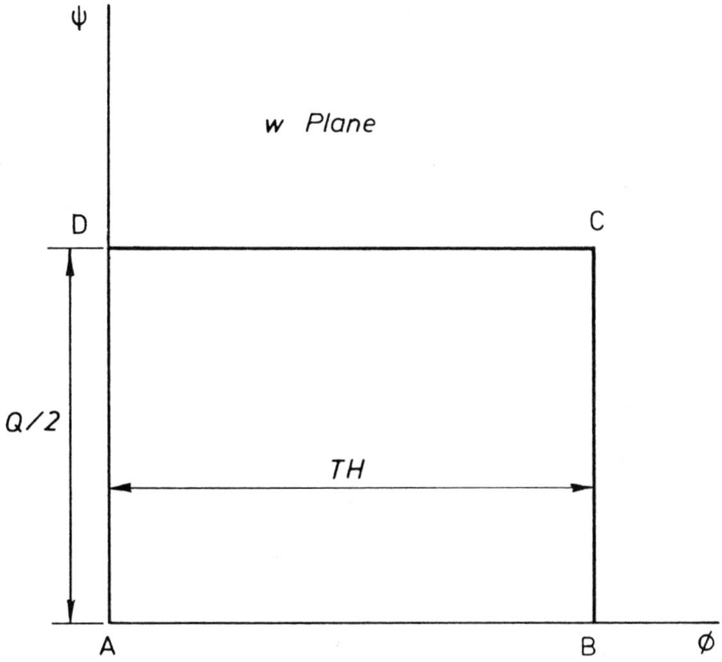

Fig. 7.17 - The w plane for the problem shown in Fig. 7.16.

integration path that coincides with the branch cut along the real t axis.

Setting w = TH + iQ/2 and iQ/2 at t = γ and 1, respectively, and using the chosen branches in the t plane to rewrite the integrand as a real function at each point along the integration path gives

$$TH + i\tfrac{1}{2}Q = -C_1 \int_0^{\gamma} \frac{d\xi}{\sqrt{\xi(\gamma - \xi)(1 - \xi)}} + TH \tag{7.135}$$

$$i\tfrac{1}{2}Q = C_1 \left[-\int_0^{\gamma} \frac{d\xi}{\sqrt{\xi(\gamma - \xi)(1 - \xi)}} - i \int_{\gamma}^{1} \frac{d\xi}{\sqrt{\xi(\xi - \gamma)(1 - \xi)}} \right] + TH \tag{7.136}$$

The simultaneous solution of Eqs. 7.135 - 7.136 for C_1 and Q gives

$$C_1 = -i \frac{Q}{2I_1(\gamma)} \tag{7.137}$$

$$\frac{Q}{TH} = 2 \frac{I_1(\gamma)}{I_2(\gamma)} \tag{7.138}$$

in which

$$I_1(\gamma) = \int_0^{\gamma} \frac{dx}{\sqrt{x(\gamma - x)(1 - x)}} \tag{7.139}$$

$$I_2(\gamma) = \int_{\gamma}^{1} \frac{dx}{\sqrt{x(x - \gamma)(1 - \gamma)}} \tag{7.140}$$

The dummy integration variable in Eqs. 7.139 - 7.140 has been changed from ξ to x, and, although the integrands contain singularities, the integrals exist. In fact, integrals listed by Gradshteyn and Ryzhik (1965) allow I_1 and I_2 to be calculated as multiples of elliptic integrals of the first kind.

$$I_1(\gamma) = 2K(\sqrt{\gamma}) \tag{7.141}$$

$$I_2(\gamma) = 2K(\sqrt{1 - \gamma}) \tag{7.142}$$

Numerical values of the elliptic integral, K(k), are tabulated extensively. See, for example, Abramowitz and Stegun (1964).

Flow to a collection gallery that parallels the river edge is shown in Fig. 7.18. Here again, flow symmetry in the multiply-connected aquifer allows us to solve the problem by using a simply-connected flow region, which is shown in Fig. 7.18(b). Boundary conditions analogous to Eqs. 7.132 lead to the construction of the w plane that is shown in Fig. 7.18(d), and the problem is solved by mapping the z and w planes upon the upper half t plane shown in Fig. 7.18(c).

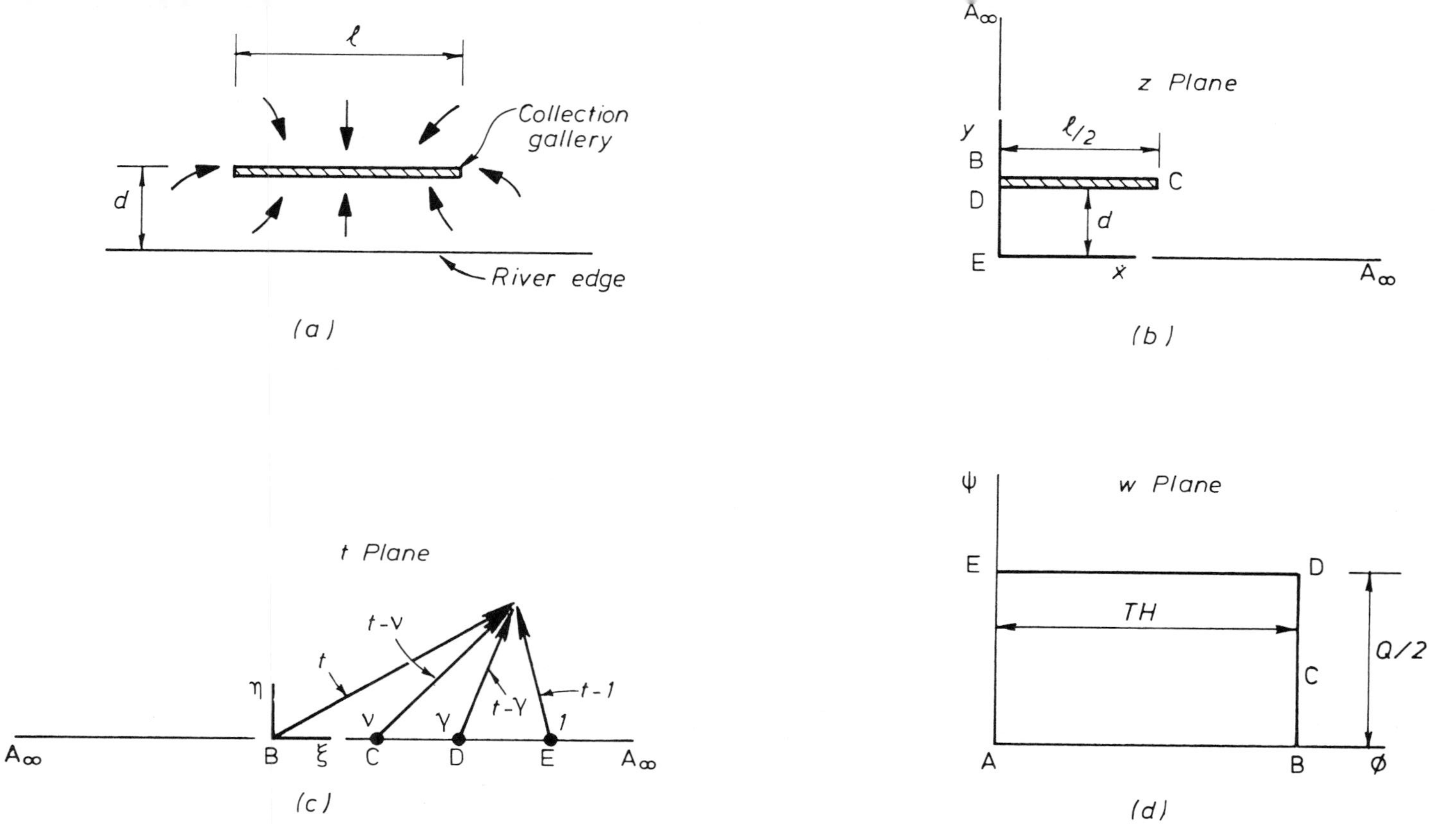

Fig. 7.18 - Flow to a collection gallery that parallels the river edge.

The mapping between the w and t planes is identical with the analogous result obtained for the problem in Figs. 7.16 - 7.17.

$$w = -\frac{iQ}{4K(\sqrt{\gamma})}\int_0^t \frac{dt}{\sqrt{t(t-\gamma)(t-1)}} + TH \tag{7.143}$$

$$\frac{Q}{TH} = 2\,\frac{K(\sqrt{\gamma})}{K(\sqrt{1-\gamma})} \tag{7.144}$$

The value of γ must be obtained from the mapping between the z and t planes, which means that γ is a function of ℓ/d.

The mapping between the z and t planes is found by integrating

$$\frac{dz}{dt} = C_1 t^{-\frac{1}{2}}\,(t-\nu)^1\,(t-\gamma)^{-\frac{1}{2}}\,(t-1)^{-\frac{1}{2}} \tag{7.145}$$

Taking the definite integral of Eq. 7.145 and setting $z = id$ at $t = 0$ gives

$$z = C_1\int_0^t \frac{t-\nu}{\sqrt{t(t-\gamma)(t-1)}}\,dt + id \tag{7.146}$$

The multiple-valued integrand will be made single-valued by choosing the arguments of t, $t-\gamma$ and $t-1$ to lie between 0 and π for values of t in the upper half plane. Thus, requiring points C, D and E to map into their correct image points gives

$$I_1(\nu, \gamma) = I_2(\nu, \gamma) \tag{7.147}$$

$$\frac{\ell}{d} = 2\,\frac{I_1(\nu, \gamma)}{I_3(\nu, \gamma)} \tag{7.148}$$

$$C_1 = \frac{\ell}{2I_1(\nu, \gamma)} \tag{7.149}$$

The integrals I_1, I_2 and I_3 are positive, real numbers that are given by

$$I_1(\nu, \gamma) = \int_0^\nu \frac{(\nu - x)\,dx}{\sqrt{x(\gamma-x)(1-x)}} \tag{7.150}$$

$$I_2(\nu, \gamma) = \int_\nu^\gamma \frac{(x-\nu)\,dx}{\sqrt{x(\gamma-x)(1-x)}} \tag{7.151}$$

$$I_3(\nu, \gamma) = \int_\gamma^1 \frac{(x-\nu)\,dx}{\sqrt{x(x-\gamma)(1-x)}} \tag{7.152}$$

Equations 7.144 and 7.147 - 7.148 show that $Q/(TH)$ is a function of ℓ/d only. This relationship can be calculated in the following way:

1. Plot $I_1(\nu, \gamma)$ and $I_2(\nu, \gamma)$ as functions of ν for constant values of γ.

2. Substitute the values of ν and γ at the intersections of the curves for I_1 and I_2, which are roots of Eq. 7.147, into Eqs. 7.144 and 7.148 to obtain corresponding values for $Q/(TH)$ and ℓ/d.

The results of these calculations are shown in Fig. 7.19. Equations 7.131 and 7.138 are also plotted in Fig. 7.19, and the results show that a maximum value of Q for fixed values of H, d and ℓ is obtained by constructing the collection gallery parallel rather than orthogonal to the river edge. The problem of computing numerical values for singular integrals like Eqs. 7.150 - 7.152 is discussed in Appendix II.

A large number of groundwater problems can be solved with Pavlovsky's method. Other applications of this technique are shown in the books by Aravin and Numerov (1965), Harr (1962) and Polubarinova-Kochina (1962). The same basic idea is used in all of these applications: the interior of simply-connected polygons in the z and w planes are mapped onto either an upper or lower half plane by using the Schwarz-Christoffel transformation. Complications in the integrations arise (a) when interior angles of the polygon vertices are no longer integral multiples of $\pi/2$ and (b) as the numbers of polygon vertices increase. These considerations mean that integrals often must be calculated numerically and that the location of some of the polygon vertex image points along the real axis must be found by solving sets of non-linear equations.

39. The Inverse Velocity Hodograph

Relatively simple, two-dimensional problems in a vertical plane that have a free surface or an interface between fluids of different densities can sometimes be solved with the use of the inverse velocity hodograph. This hodograph plane is defined as a plot of the function

$$\Omega = \frac{dz}{dw} = \frac{1}{dw/dz} \tag{7.153}$$

Setting $\Omega = \Omega_1 + i\Omega_2$ and $dw/dz = u - iv$ and separating real and imaginary parts in Eq. 7.153 shows that the real and imaginary parts of Ω are given, respectively, by

$$\Omega_1 = \frac{u}{u^2 + v^2} \qquad \text{(7.154 a, b)}$$

$$\Omega_2 = \frac{v}{u^2 + v^2}$$

Problems in which all boundaries consist of free streamlines, interfaces between fluids of different densities and reservoir and impermeable boundaries that are straight lines have an Ω plane and a w plane with boundaries that consist entirely of straight lines. Thus, the Ω and w planes can each be

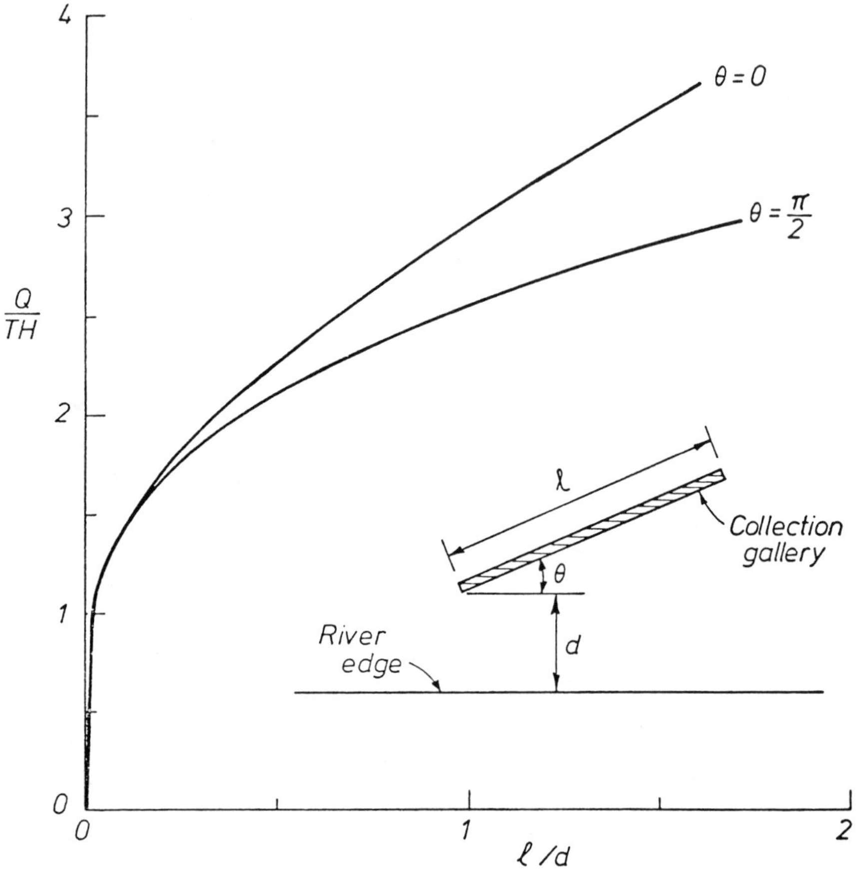

Fig. 7.19 - Dimensionless flow rates to a collection gallery along a river.

mapped onto an upper or lower half t plane with the Schwarz-Christoffel transformation to obtain $\Omega = \Omega(t)$ and $w = w(t)$. The final solution must often be written parametrically with the equations $w = w(t)$ and $z = z(t)$ in which $z(t)$ is obtained by integrating the equation

$$\frac{dz}{dt} = \frac{dz}{dw}\frac{dw}{dt} = \Omega(t)\,\frac{dw(t)}{dt} \tag{7.155}$$

The resulting integration constant is fixed by choosing a location for the coordinate origin in the z plane.

If the y axis points upward, then the following two boundary conditions must be satisfied along a free streamline:

$$\phi = -Ky \qquad \psi = \text{constant} \tag{7.156a, b}$$

Differentiation of Eqs. 7.156 with respect to arc length, s, along the streamline gives

$$\frac{\partial\phi}{\partial x}\frac{dx}{ds} + \frac{\partial\phi}{\partial y}\frac{dy}{ds} = -K\frac{dy}{ds} \qquad \frac{\partial\psi}{\partial x}\frac{dx}{ds} + \frac{\partial\psi}{\partial y}\frac{dy}{ds} = 0 \tag{7.157a, b}$$

Finally, using Eqs. 7.3 and 7.6 in Eq. 7.157 and eliminating dx/ds and dy/ds gives the result

$$\frac{v}{u^2 + v^2} = \frac{-1}{K} \tag{7.158}$$

Thus, Equations 7.154b and 7.158 give the equation of a free streamline in the inverse hodograph plane as a straight line parallel to the real axis.

$$\Omega_2 = -\frac{1}{K} \tag{7.159}$$

Equation 2.65 in Chapter II shows that an interface between fresh water and sea water in steady flow can be treated by replacing K with $-K\varepsilon$ in Eqs. 7.156-159.

An impermeable boundary has the same boundary condition as Eq. 7.156b. Thus, Eqs. 7.157b and 7.6 give the result

$$\frac{v}{u} = \frac{dy/ds}{dx/ds} = \tan(\alpha) \tag{7.160}$$

in which α = constant angle that the straight, impermeable boundary makes with the x axis. Equations 7.154 and 7.160 give the equation of the impermeable boundary in the hodograph plane as

$$\frac{\Omega_2}{\Omega_1} = \tan(\alpha) \tag{7.161}$$

Thus, the impermeable boundary in the inverse hodograph plane is a straight line through the origin with the same slope as the slope of the boundary in the physical z plane.

A reservoir boundary has the boundary condition

$$\phi = \text{constant} \tag{7.162}$$

Differentiation of Eq. 7.162 with respect to arc length gives

$$\frac{\partial\phi}{\partial x}\frac{dx}{ds} + \frac{\partial\phi}{\partial y}\frac{dy}{ds} = 0 \tag{7.163}$$

and Eqs. 7.3 and 7.163 give the result

$$\frac{v}{u} = -\frac{dx/ds}{dy/ds} = -\frac{1}{\tan(\alpha)} \tag{7.164}$$

Thus, Eqs. 7.154 and 7.164 give the equation of a reservoir boundary in the Ω plane as

$$\frac{\Omega_2}{\Omega_1} = -\frac{1}{\tan(\alpha)} \tag{7.165}$$

When the reservoir boundary is straight, Eq. 7.165 is the equation of a straight line through the origin in a direction that is orthogonal to the reservoir boundary in the physical z plane.

For a first application of the inverse hodograph, consider the interface between fresh and sea water in the infinitely deep coastal aquifer that is shown in Fig. 7.20. The boundary AB is either an impermeable boundary or a very gently sloping free surface with the boundary condition $\psi = 0$. The gently sloping sea bed is taken as the horizontal line BC, which has the boundary condition $\phi = 0$. The interface, AC, has the two boundary conditions $\phi = K\varepsilon y$ and $\psi = q$. (The first boundary condition comes from Eqs. 2.65 and 7.2, and the sign on the right side of the second boundary condition is fixed by Eq. 7.6 and the fact that $u < 0$.) The w and Ω planes are shown in Fig. 7.20(b) and (c), respectively, and traversing the boundary points in Figs. 7.20(a), (b) and (c) in the same order shows that the flow regions are contained in the interior of the semi-infinite strips in the w and Ω planes. The w and Ω planes are mapped to the upper half t plane in Fig. 7.20(d) by integrating

$$\frac{dw}{dt} = C_1\, t^{-\frac{1}{2}}\,(t-1)^{-\frac{1}{2}} \tag{7.166}$$

$$\frac{d\Omega}{dt} = C_2\, t^{-\frac{1}{2}}\,(t-1)^{-\frac{1}{2}} \tag{7.167}$$

Since the right sides of Eqs. 7.166 and 7.167 differ only by multiplicative constants,

$$\frac{d\Omega}{dt} = \frac{C_2}{C_1}\frac{dw}{dt} \tag{7.168}$$

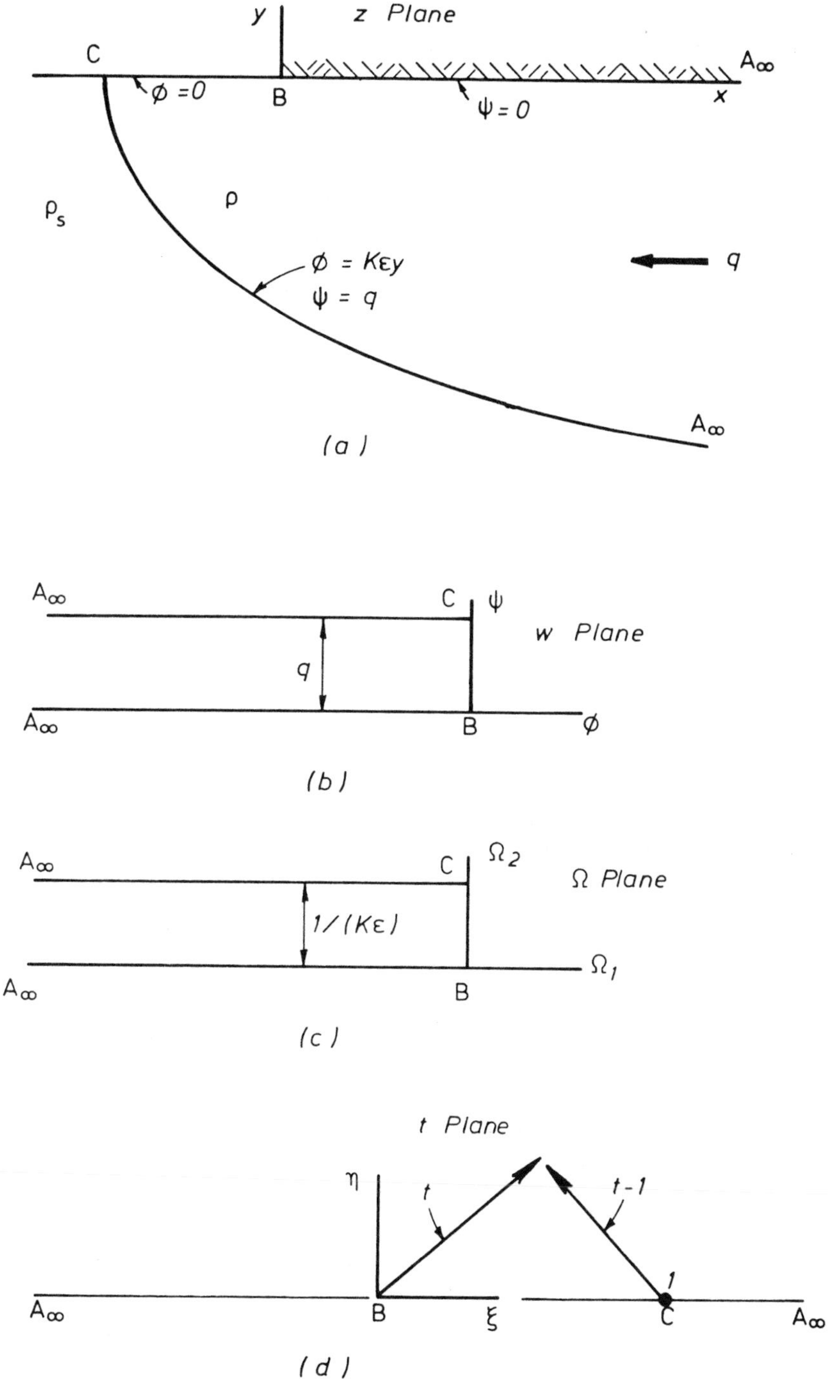

Fig. 7.20 - An interface between fresh and sea water in an infinitely deep coastal aquifer.

which can be integrated to give

$$\Omega = \frac{C_2}{C_1} w + C_3 \tag{7.169}$$

Requiring points B and C to map into their correct image points gives $C_3 = 0$ and $C_2/C_1 = 1/(K\varepsilon q)$. Thus Eq. 7.169 can be rewritten in the form

$$\Omega = \frac{dz}{dw} = \frac{w}{K\varepsilon q} \tag{7.170}$$

and the integration of Eq. 7.170 subject to the condition $z = 0$ when $w = 0$, which locates the coordinate origin in the z plane, gives

$$w^2 = 2K\varepsilon qz \tag{7.171}$$

Equation 7.171 is a solution that was initially given by Glover (1959) who obtained his solution from a solution given earlier by Kozeny (1953) for flow through a dam. (See problem 18.)

Setting $w = \phi + i\psi$, $z = x + iy$ and separating real and imaginary parts in Eq. 7.171 gives

$$\phi^2 - \psi^2 = 2K\varepsilon qx$$
$$\phi\psi = K\varepsilon qy \tag{7.172a, b}$$

The equation of the interface can be found by putting the boundary conditions for ϕ and ψ along AC into Eq. 7.172 to obtain the equation of the parabola

$$\left(\frac{K\varepsilon y}{q}\right)^2 = 1 + 2\left(\frac{K\varepsilon x}{q}\right) \tag{7.173}$$

The piezometric head along AB, or the first approximation to the elevation of a free surface along AB, can be calculated by setting $\psi = y = 0$ and $\phi = -Kh$ in Eq. 7.172 to obtain

$$\frac{Kh}{q} = \sqrt{2\frac{K\varepsilon x}{q}} \tag{7.174}$$

Setting $y = 0$ in Eq. 7.173 gives the x coordinate of point C,

$$x_C = -\frac{q}{2K\varepsilon} \tag{7.175}$$

which shows that an increase in q causes the interface to move seaward.

A more difficult application of the inverse hodograph method occurs for seepage from the triangular canal shown in Fig. 7.21. Polubarinova-Kochina (1962) states that this problem was first solved by V.V. Vedernikov. Symmetry allows us to consider flow in half the flow region, which is marked ABCA in Fig. 7.21(a). The flow regions in the w and Ω planes are the interiors of a semi-infinite strip and a triangle, respectively. These regions are shown in Fig. 7.21(b) and (c) and are mapped upon the upper half t plane in Fig. 7.21(d) with the transformations

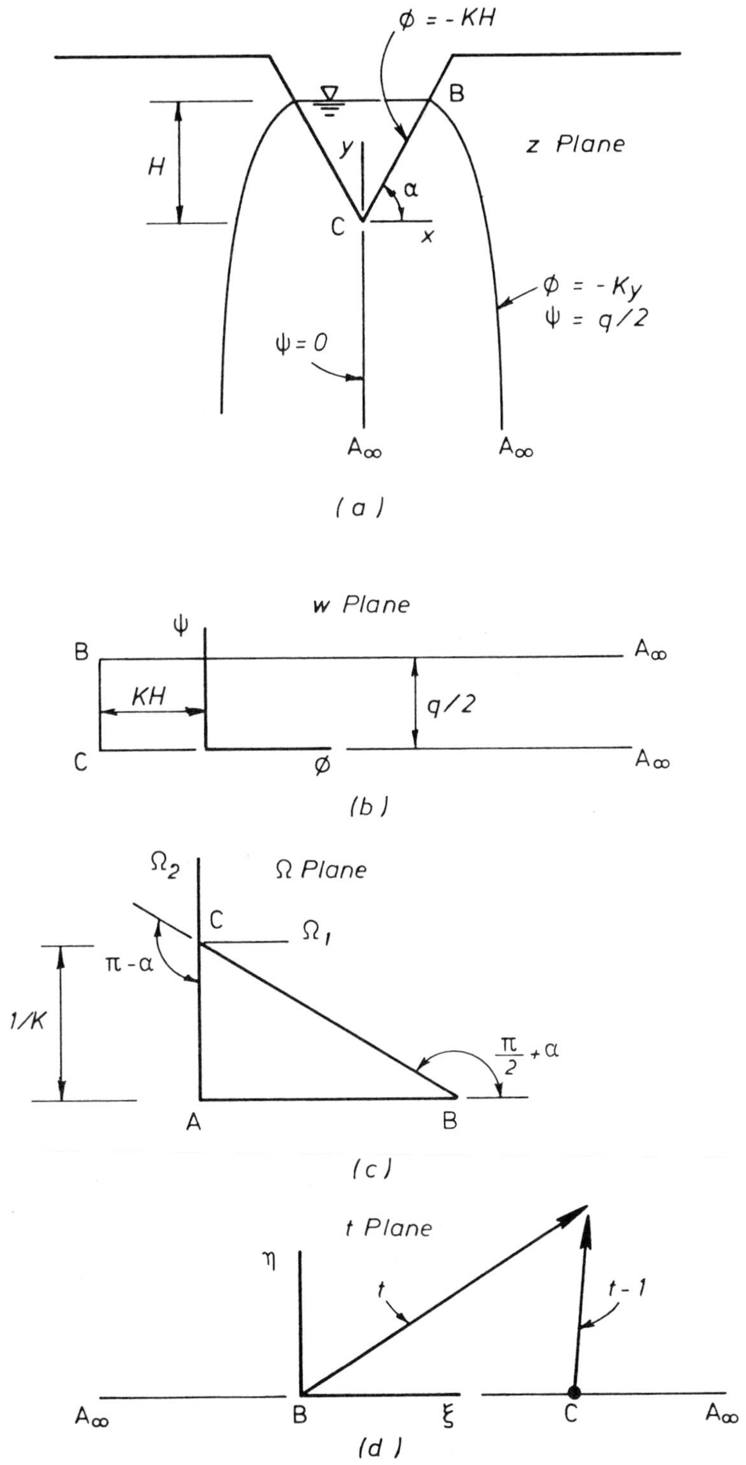

Fig. 7.21 - The free-surface seepage from a triangular canal.

$$w = \frac{q}{2\pi} \cosh^{-1} (2t - 1) - KH \tag{7.176}$$

$$\Omega = \frac{1}{K} [\tan(\alpha) - i] - \frac{i}{K \cos(\alpha) I_1(\alpha)} \int_0^t \frac{dt}{t^{\frac{1}{2}+\alpha/\pi} (t-1)^{1-\alpha/\pi}} \tag{7.177}$$

in which the arguments of t and (t - 1) vary between 0 and π for points in the upper half t plane and the integral $I_1(\alpha)$ is a Beta function that can be written as the product of two Gamma functions:

$$I_1(\alpha) = \int_0^1 \frac{dx}{x^{\frac{1}{2}+\alpha/\pi} (1 - x)^{1-\alpha/\pi}} = \frac{1}{\sqrt{\pi}} \Gamma\left(\frac{1}{2} - \frac{\alpha}{\pi}\right) \Gamma\left(\frac{\alpha}{\pi}\right) \tag{7.178}$$

Substituting Eq. 7.176 into Eq. 7.155 gives

$$\frac{dz}{dt} = \frac{q}{2\pi\sqrt{t(t - 1)}} \Omega(t) \tag{7.179}$$

Since z = 0 at t = 1, Eq. 7.179 can be integrated to obtain

$$z = - \frac{q}{2\pi} \int_t^1 \frac{\Omega(t)}{\sqrt{t(t - 1)}} dt \tag{7.180}$$

An expression for q can be obtained from Eq. 7.180 by setting $z = H[\cot(\alpha) + i]$ at t = 0:

$$H[\cot(\alpha) + i] = - \frac{q}{2\pi} \int_0^1 \frac{\Omega(t)}{\sqrt{t(t - 1)}} dt \tag{7.181}$$

The integrals in Eq. 7.181 can be put in real form by setting $(t - 1) = (1 - \xi)e^{i\pi}$ in which $0 \leq \xi \leq 1$. This eventually leads to the result

$$\frac{2\pi KH}{q \tan(\alpha)} = \int_0^1 \left[1 - \frac{1}{I_1(\alpha)} \int_0^x \frac{d\xi}{\xi^{\frac{1}{2}+\alpha/\pi} (1 - \xi)^{1-\alpha/\pi}}\right] \frac{dx}{\sqrt{x(1 - x)}} \tag{7.182}$$

However, the definition of $I_1(\alpha)$ in Eq. 7.178 allows the inner integral in Eq. 7.182 to be rewritten as

$$\int_0^x \frac{d\xi}{\xi^{\frac{1}{2}+\alpha/\pi} (1 - \xi)^{1-\alpha/\pi}} = I_1(\alpha) - \int_x^1 \frac{d\xi}{\xi^{\frac{1}{2}+\alpha/\pi} (1 - \xi)^{1-\alpha/\pi}} \tag{7.183}$$

which allows Eq. 7.182 to be simplified to

$$\frac{2\pi KH}{q \tan(\alpha)} = \frac{1}{I_1(\alpha)} \int_0^1 \left[\int_x^1 \frac{d\xi}{\xi^{\frac{1}{2}+\alpha/\pi} (1 - \xi)^{1-\alpha/\pi}}\right] \frac{dx}{\sqrt{x(1 - x)}} \tag{7.184}$$

Finally, the order of integration in Eq. 7.184 can be inverted and the inside integral can be calculated by noting that the double integral is calculated over the triangular area in the (x, ξ) plane that is shown in Fig. 7.22. This leads to the final result

$$\frac{q}{KH} = 2\pi \frac{I_1(\alpha)}{I_2(\alpha)} \cot(\alpha) \tag{7.185}$$

in which

$$I_2(\alpha) = \int_0^1 \frac{\pi/2 + \sin^{-1}(2\xi - 1)}{\xi^{\frac{1}{2}+\alpha/\pi}(1 - \xi)^{1-\alpha/\pi}} d\xi \tag{7.186}$$

Numerical values for $I_2(\alpha)$ can only be calculated after looking carefully at the behaviour of the integrand on the integration interval. If we define

$$g(\xi) = \pi/2 + \sin^{-1}(2\xi - 1) \tag{7.187}$$

then

$$\frac{dg(\xi)}{d\xi} = \frac{1}{\sqrt{\xi(1 - \xi)}} \tag{7.188}$$

Hence, $g(\xi)$ has the following behaviour near the end points of the integration interval:

$$\begin{aligned} g(\xi) &\sim 2\sqrt{\xi}\,(1 + \tfrac{1}{6}\xi + \cdots) \quad \text{for } \xi \to 0 \\ &\sim \pi - 2\sqrt{1 - \xi}\left[1 + \tfrac{1}{6}(1 - \xi) + \cdots\right] \quad \text{for } \xi \to 1 \end{aligned} \tag{7.189}$$

Thus, the integrand of $I_2(\alpha)$ has the behaviour

$$\begin{aligned} \frac{\pi/2 + \sin^{-1}(2\xi - 1)}{\xi^{\frac{1}{2}+\alpha/\pi}(1 - \xi)^{1-\alpha/\pi}} &\sim \frac{2}{\xi^{\alpha/\pi}}\left[1 + \left(\frac{7}{6} - \frac{\alpha}{\pi}\right)\xi + \cdots\right] \quad \text{for } \xi \to 0 \\ &\sim \frac{\pi}{(1 - \xi)^{1-\alpha/\pi}}\left[1 + \left(\frac{1}{2} + \frac{\alpha}{\pi}\right)(1 - \xi) + \cdots\right] \\ &\quad - \frac{2}{(1 - \xi)^{\frac{1}{2}-\alpha/\pi}}\left[1 + \left(\frac{2}{3} + \frac{\alpha}{\pi}\right)(1 - \xi) + \cdots\right] \quad \text{for } \xi \to 1 \end{aligned} \tag{7.190}$$

The fairly complicated way in which singularities appear in the integrand suggests that the method of "subtracting the singularity," which is explained in Appendix II, is probably the easiest way to computer numerical values for $I_2(\alpha)$. Application of this method gives

$$I_2(\alpha) = \int_0^1 F(\xi)\,d\xi + G(\alpha) \tag{7.191}$$

in which

$$\begin{aligned} F(\xi) = {} & \frac{\pi/2 + \sin^{-1}(2\xi - 1)}{\xi^{\frac{1}{2}+\alpha/\pi}(1 - \xi)^{1-\alpha/\pi}} - \frac{2}{\xi^{\alpha/\pi}}\left[1 + \left(\frac{7}{6} - \frac{\alpha}{\pi}\right)\xi\right] \\ & - \frac{\pi}{(1 - \xi)^{1-\alpha/\pi}}\left[1 + \left(\frac{1}{2} + \frac{\alpha}{\pi}\right)(1 - \xi)\right] + \frac{2}{(1 - \xi)^{\frac{1}{2}-\alpha/\pi}}[1 \\ & + \left(\frac{2}{3} + \frac{\alpha}{\pi}\right)(1 - \xi)] \end{aligned} \tag{7.192}$$

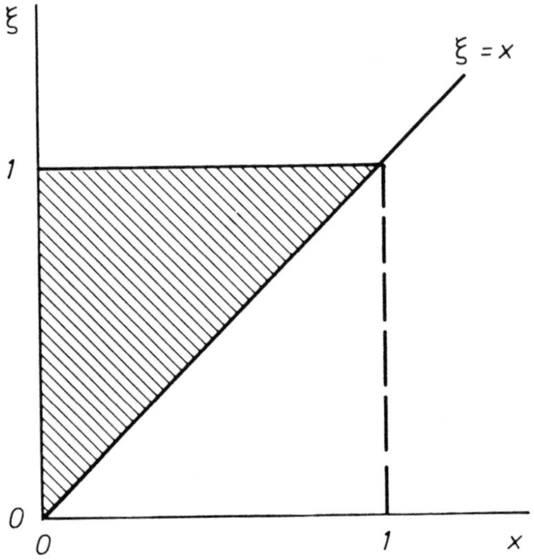

Fig. 7.22 - The triangular region that the double integral in Eq. 7.184 is calculated over in the (x, ξ) plane.

Since $F(\xi)$ is the difference between the integrand in Eq. 7.186 and the terms shown on the right side of Eq. 7.190, the function $G(\alpha)$ is obtained by integrating the terms shown on the right side of Eq. 7.190 to obtain

$$G(\alpha) = 2\left[\frac{1}{\left(1 - \frac{\alpha}{\pi}\right)} + \left(\frac{\frac{7}{6} - \frac{\alpha}{\pi}}{2 - \frac{\alpha}{\pi}}\right)\right] + \pi\left[\frac{\pi}{\alpha} + \left(\frac{\frac{1}{2} + \frac{\alpha}{\pi}}{1 + \frac{\alpha}{\pi}}\right)\right]$$

$$- 2\left[\frac{1}{\left(\frac{1}{2} + \frac{\alpha}{\pi}\right)} + \left(\frac{\frac{2}{3} + \frac{\alpha}{\pi}}{\frac{3}{2} + \frac{\alpha}{\pi}}\right)\right] \qquad (7.193)$$

Equations 7.191 - 7.193 are exact, and the numerical approximation is made when the integral on the right side of Eq. 7.191 is calculated numerically with a quadrature formula. This is relatively straight-forward since $F(\xi)$ and $dF(\xi)/d\xi$ are finite for $0 \leqq \xi \leqq 1$. The result of calculating $I_1(\alpha)$ and $I_2(\alpha)$ is shown in Fig. 7.23 with a plot of Eq. 7.185.

A number of other solution techniques have been applied to solve free-streamline problems in groundwater flow. These methods include the use of Zhukovsky's function, Cauchy integrals and the analytic theory of linear differential equations. The interested reader will find descriptions and applications of these techniques, as well as additional applications of the inverse velocity hodograph, in the books by Aravin and Numerov (1965), Harr (1962) and Polubarinov-Kochina (1962).

REFERENCES

Abramowitz, M. and Stegun, I.A. (Editors), 1964. Handbook of Mathematical Functions, U.S. National Bureau of Standards, Applied Mathematics Series, No. 55, U.S. Government Printing Office, Washington, D.C., Ch. 17.

Aravin, V.I. and Numerov, S.N. 1965. Theory of Fluid Flow in Undeformable Porous Media, Israel Program for Scientific Translations, Jerusalem.

Churchill, R.V. 1960. Complex Variables and Applications, second edition, McGraw-Hill Book Company, New York, pp. 284-291.

Glover, R.E. 1959. "The Pattern of Fresh Water in a Coastal Aquifer, "Journal of Geophysical Research, Vol. 64, pp. 457-459.

Gradshteyn, I.S. and Ryzhik, I.M. 1965. Table of Integrals, Series, and Products, Academic Press, New York, pp. 219-220.

Harr, M.E. 1962. Groundwater and Seepage, McGraw-Hill Book Co., New York.

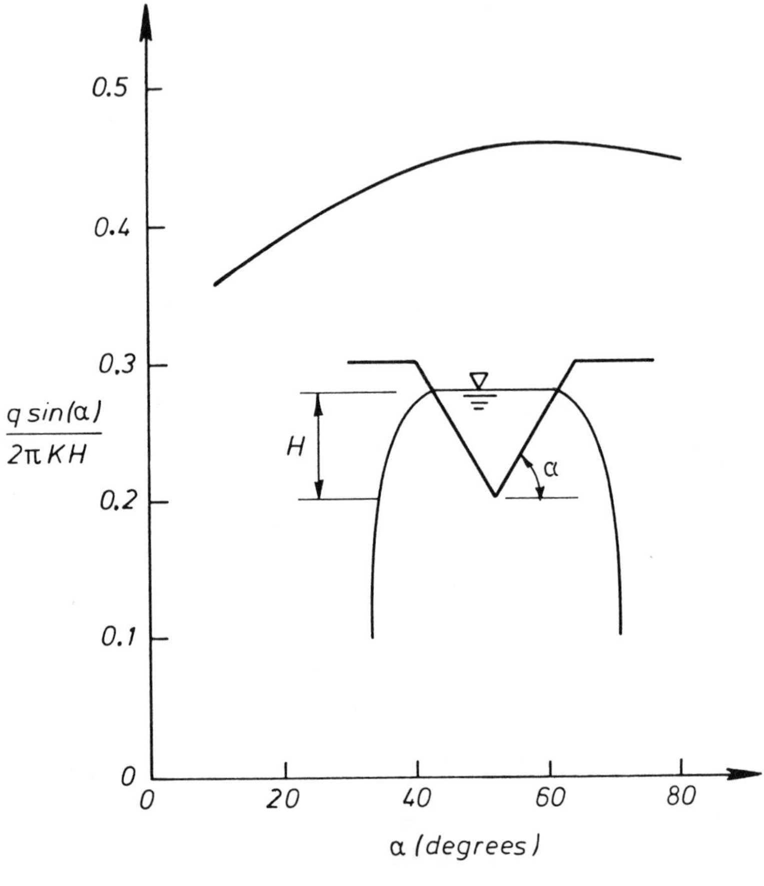

Fig. 7.23 - Dimensionless flow rates from a triangular ditch.

Kober, H. 1957. Dictionary of Conformal Representations, Dover Publications, Inc.

Kozeny, J. 1953. Hydraulik, Springer Verlag, Vienna.

Nehari, Z. 1952. Conformal Mapping, McGraw-Hill Book Company Inc., New York, pp. 102-109, 190.

Polubarinova-Kochina, P.Ya. 1962. Theory of Groundwater Movement, Translated by J.M.R. de Wiest, Princeton University Press, Princeton.

Strack, O.D.L. 1976. "A Single-Potential Solution for Regional Interface Problems in Coastal Aquifers," Water Resources Research, Vol. 12, No. 6, pp. 1165-1174.

PROBLEMS

1. Write the equations, analogous to Eqs. 7.27 - 7.30, that can be used to define ϕ and ψ for sea water intrusion in the confined aquifer that is considered in problem 7 of Chapter III.

2. Set $z = x + iy$ in the function $w(z) = z^2$ and separate real and imaginary parts to obtain expressions for ϕ and ψ. Then carry out the following calculations:

   (a) Show by differenting the expressions for ϕ and ψ that ϕ and ψ are harmonic.

   (b) Calculate dw/dz, $\partial\phi/\partial x$, $\partial\phi/\partial y$, $\partial\phi/\partial x$ and $\partial\psi/\partial y$. Then compare these expressions to show that

$$\frac{\partial\phi}{\partial x} = \frac{\partial\psi}{\partial y} \quad , \quad \frac{\partial\phi}{\partial y} = -\frac{\partial\psi}{\partial x}$$

$$\frac{dw}{dz} = \frac{\partial\phi}{\partial x} + i\,\frac{\partial\psi}{\partial x} \quad , \quad \frac{dw}{dz} = \frac{1}{i}\left(\frac{\partial\phi}{\partial y} + i\,\frac{\partial\psi}{\partial y}\right)$$

   (c) Note that the x and y axes are both lines of constant ψ that intersect at the origin. Hence, lines of constant ϕ are not orthogonal to lines of constant ψ at the origin. Why?

3. A recharge well and an abstraction well, each of strength Q, are separated a distance ℓ in a horizontal, confined aquifer. Show that the minimum time required for a pollutant to travel from the recharge well to the abstraction well is

$$t = \frac{\pi B \sigma \ell^2}{3Q}$$

in which σ = aquifer porosity and B = constant aquifer thickness. Assume that the velocity vanishes as $|z| \to \infty$ and neglect dispersion.

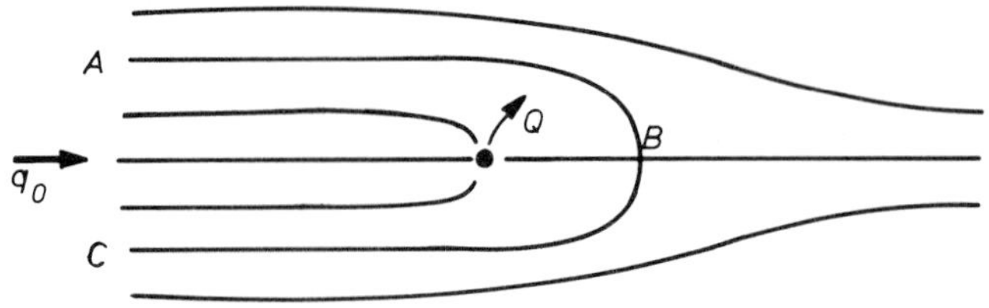

PROB. 4

4. An abstraction well is placed in a flow that is uniform at infinity. If dispersion is neglected, then the streamline ABC becomes the boundary of a region in which pollutant sources must be absent if the well water is to remain uncontaminated. Show that the distance between the well and point B is $Q/(2\pi q_0)$ and that the upstream width of this zone is Q/q_0. (Notice that the width of the zone can be calculated by equating the flow in the zone at infinity to the well flow rate.) Then show that the streamline ABC achieves 99% of this asymptotic width at a distance upstream from the well of 15.75 Q/q_0.

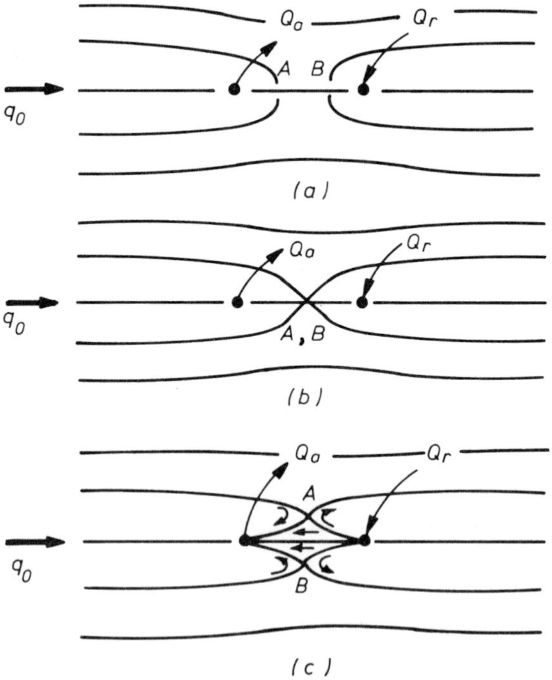

PROB. 5

5. A recharge well, of strength Q_r, is spaced a horizontal distance ℓ downstream from an abstraction well, of strength Q_a, in a flow that is uniform at infinity. Sketches (a), (b) and (c) suggest, when dispersion is neglected, that water from the recharge well will not enter the abstraction well as long as the two stagnation points, A and B, remain separate and distinct points on the straight line joining the two wells. Use this fact to show that Q_a and Q_r must satisfy the following inequality if the abstraction well is to remain unpolluted with water from the recharge well:

$$Q_r < (\sqrt{(\pi q_0 \ell} - \sqrt{Q_a})^2$$

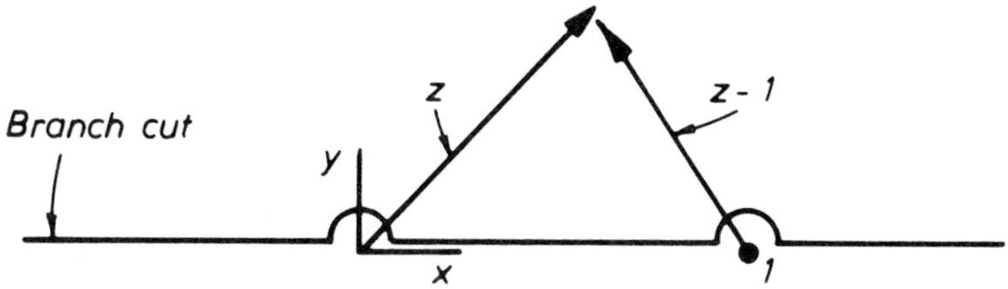

PROB. 6

6. Let the function $w(z) = \sqrt{z(z-1)}$ have the real axis as a branch cut so that $w(z)$ is analytic for values of z in the upper half z plane. Choose the branch for this function for which the arguments of z and $z - 1$ vanish when $z \to x + i0$ for $1 < x < \infty$. Show that $w(z)$ has the following values when z approaches points along the real axis:

$$\begin{aligned} w(x) &= \sqrt{x(x-1)} \quad \text{for } 1 < x < \infty \\ &= i\sqrt{x(1-x)} \quad \text{for } 0 < x < 1 \\ &= -\sqrt{x(x-1)} \quad \text{for } -\infty < x < 0 \end{aligned}$$

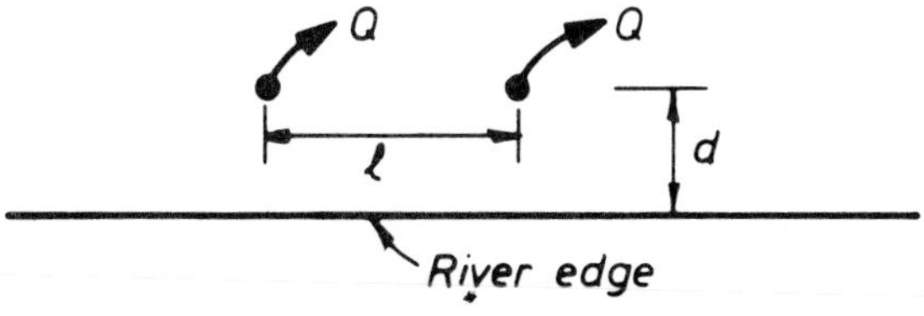

PROB. 7

7. Two abstraction wells, spaced a distance ℓ apart and a distance d from a river edge, each withdraw the same flow. Use Eq. 7.84 to calculate values of ℓ/δ, in which δ = well radius, for values of d/δ in the range $10 \leqq d/\delta \leqq 1000$ at which well interference effects become negligibly small.

8. A circular aquifer of radius R has a well at its geometrical center. The unconfined aquifer is recharged by a constant rainfall, P, and is surrounded by the sea. Use the theory of Strack to show that the critical well abstraction rate, Q, that will be just sufficient to cause sea-water intrusion is a root of the following transcendental equation:

$$\frac{Q}{\pi}\left[2\ \ln\left(\frac{R}{\sqrt{\frac{Q}{\pi P}}}\right) + 1\right] = PR^2 - 2K\varepsilon(1 + \varepsilon)D^2$$

Hint: Since the potential satisfies $\nabla^2\phi = P$, set $\phi = \frac{P}{4}(r^2 - R^2) + \phi_1(r)$. Since the first term satisfies the inhomogeneous equation and vanishes at the boundary r = R, then $\phi_1(r)$ must be a harmonic function that has a logarithmic singularity at r = 0 and also vanishes at r = R. Thus, $\phi_1(r)$ is given by

$$\phi_1(r) = -\frac{Q}{2\pi}\ \ln\left(\frac{r}{R}\right)$$

Notice that you will not need to use complex variables to solve this problem. However, calculation of the harmonic function, ϕ_1, is often simplified by using complex variables when aquifer geometries are more complicated.

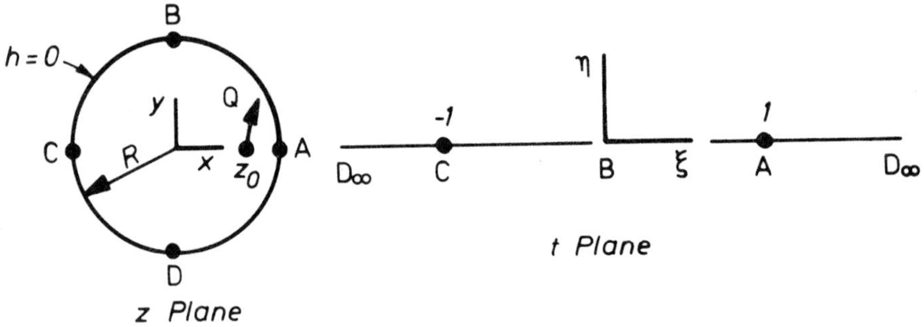

PROB. 9

9. Set $z = Re^{i\theta}$ and verify that the transformation

$$t = \frac{R + iz}{z + iR}$$

maps the boundary of the circle $|t| = R$ into the real t axis, as shown in the sketch. An abstraction well of radius $|\Delta| = \delta$ at $z_0 = r_0 + i0$ removes a flow rate Q from the circular aquifer when the piezometric head inside the well bore is h = -H. Show that Q is given by the following expression if the aquifer is confined and δ is relatively small:

$$\frac{Q}{TH} = \frac{2\pi}{\ln\left(\frac{R^2 - r_0^2}{R\delta}\right)}$$

The transformation between the z and t planes was obtained by applying the bilinear transformation that is discussed in most introductory texts on conformal mapping.

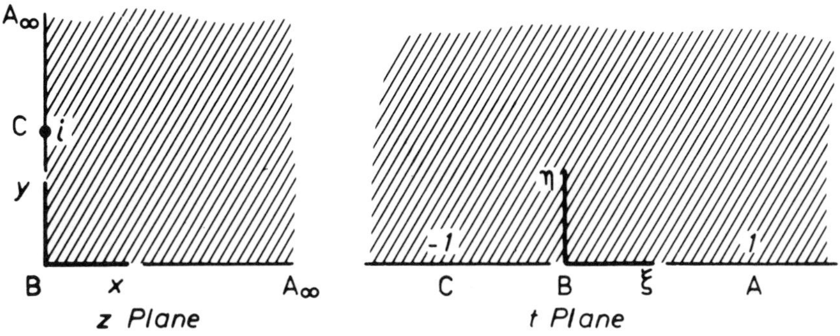

PROB. 10

10. Use the Schwarz-Christoffel transformation to obtain the following equation for mapping the cross hatched regions upon each other:

$$z = i\sqrt{\frac{2t}{t - 1}}$$

Hint: The value of α at vertex A is the angle that the unit tangent to the boundary rotates in traversing the boundary at A in the positive direction. As a check, the sum of α at A and B must equal 2π. Also, the arguments of t and t - 1 have been taken as zero when t approaches a point on the real t axis for which $1 < \xi < \infty$.

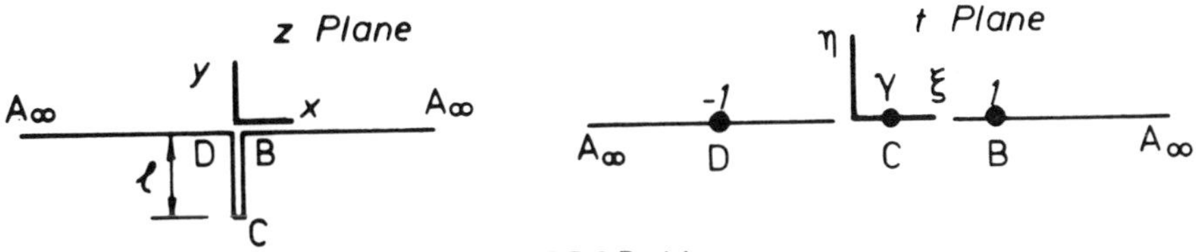

PROB. 11

11. Use the Schwarz-Christoffel transformation to map the polygon in the z plane to the lower half t plane. Symmetry suggests that $\gamma = 0$, but you should be able to prove this from the transformation equation. Show that the answer can be put in the form

$$z = \ell\sqrt{t^2 - 1}$$

12. Hold two mirrors in vertical planes at right angles to each other. Place a vertical pencil in the enclosed right angle and note the location of the three image points that are shown in Fig. 7.7(c) and (d). Repeat the experiment when the two mirrors are held in vertical,

parallel planes and note the "infinite" number of images that result.

13. Locate the image points and indicate whether a source or sink should be placed at each image point in order to model flow to a well in a wedge-shaped region with an interior angle of $\pi/4$. Do this for the case when both boundaries are reservoir boundaries, then repeat the exercise for the case when one boundary is a reservoir boundary and the other boundary is a streamline. These types of problems are more easily solved with conformal mapping than with the method of images when the flow is steady and governed by the Laplace equation. However, the method of images can also be used to solve unsteady flow problems when the coefficients of the linear "heat conduction" equation are constant. Conformal mapping can not be used in these cases.

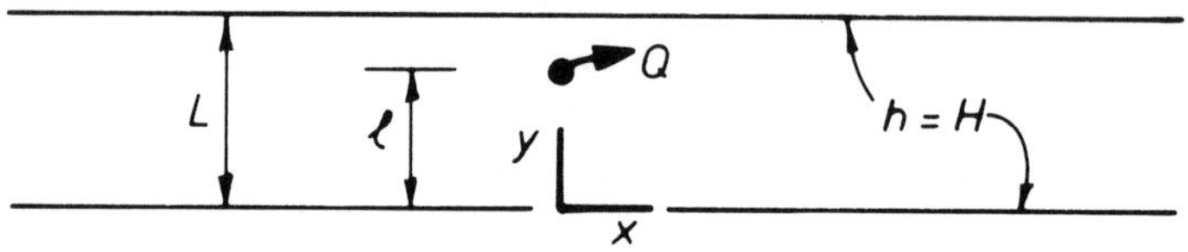

PROB. 14

14. Show that the flow from the well in the infinitely long, unconfined aquifer is given by

$$\frac{K(H^2 - h_0^2)}{Q} = \frac{1}{\pi} \ln\left[\frac{2\sin(\pi\ell/L)}{\pi(\delta/L)}\right]$$

in which h_0 = piezometric head at the well and δ = well radius.

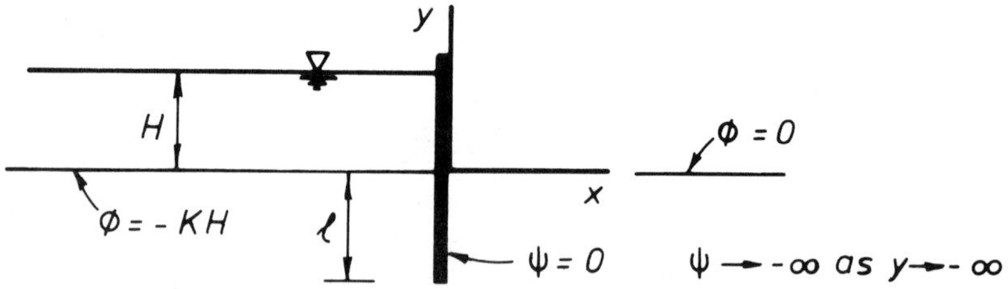

PROB. 15

15. Use Pavlovsky's method and the transformation in problem 11 to calculate the complex velocity

$$\frac{dw}{dz} = -\frac{KH}{\pi\ell}\frac{i}{\sqrt{1 + (x/\ell)^2}} \text{ on } y = 0 \text{ for } 0 < x < \infty$$

for flow beneath the sheetpile.

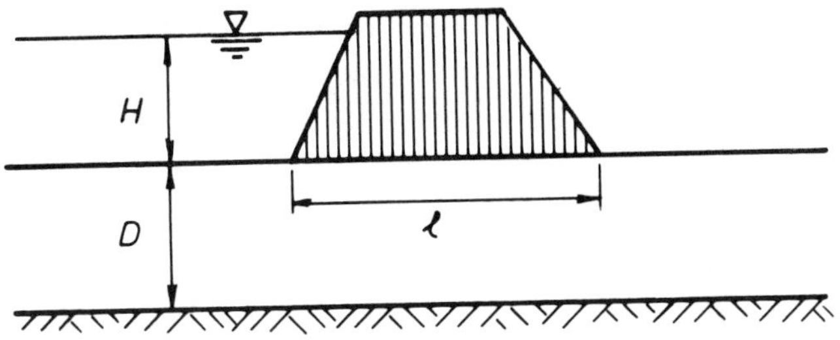

PROB. 16

16. Use Pavlovsky's method to show that the flow rate beneath the dam is given by

$$\frac{q}{KH} = \frac{K(\sqrt{\gamma})}{K(\sqrt{1-\gamma})}, \quad \gamma = e^{-\pi \ell/D}$$

in which K = coefficient of permeability and K(k) = complete elliptic integral of the first kind with modulus k.

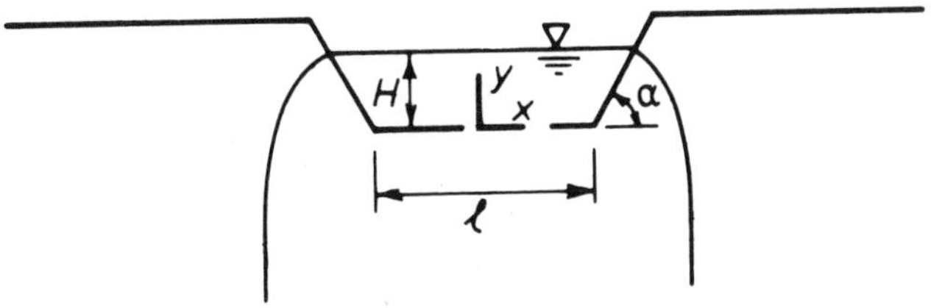

PROB. 17

17. Show the complex planes that you would map upon each other to calculate an expression for the flow rate lost by the trapezoidal canal through seepage.

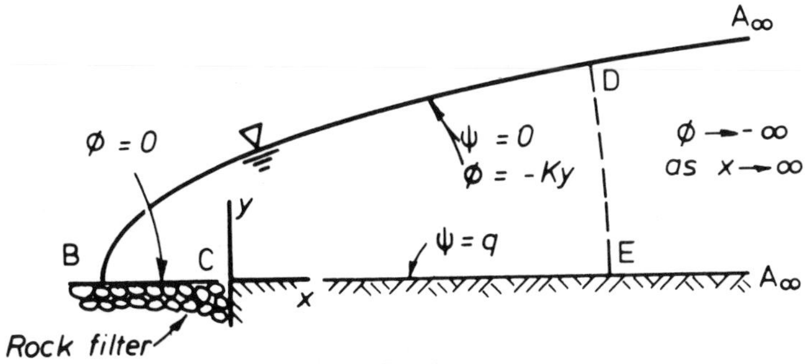

PROB. 18

18. Use the inverse velocity hodograph to show that the Kozeny (1953) solution for flow through a dam is

$$z = \frac{w^2}{2Kq} - \frac{iw}{K} - \frac{q}{2K}$$

The curved equipotential line DE in the infinite flow domain is used to approximate the upstream embankment that forms the reservoir boundary. If the free surface of the reservoir has an elevation of H, show that the equations of BD and DE, respectively, are

$$2\left(\frac{Kx}{q}\right) = \left(\frac{Ky}{q}\right)^2 - 1, \quad 2\left(\frac{Kx}{q}\right) = \left(\frac{KH}{q}\right)^2 - \left(\frac{y}{H}\right)^2$$

Thus, if $x = \ell$ at point E, the flow rate through the dam is given by

$$\frac{q}{KH} = \frac{1}{2}\left(\frac{H}{\ell}\right)$$

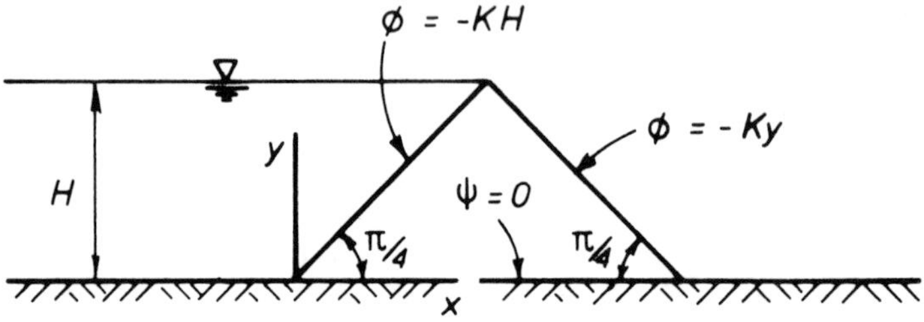

PROB. 19

19. The velocity hodograph, which is a plot of $dw/dz = u - iv$, is sometimes useful for problems without a free surface. Construct the velocity hodograph for seepage through the triangular dam core and map it upon the z plane to obtain the solution

$$w(z) = \frac{Kz^2}{4H} - KH$$

Thus, show that the discharge through the dam is given by

$$\frac{q}{KH} = \frac{1}{2}$$

Harr (1962) states that this solution was given first by B.B. Davison in 1937.

8 The Exact Solution of Unsteady Problems

40. The Laplace Transform

In this chapter we will be concerned with exact solutions of the linear equation

$$T\nabla^2 h = S\frac{\partial h}{\partial t} + \frac{K'}{B'}h - R \qquad (8.1)$$

in which the coefficients T, S and K'/B' are constants and R represents a recharge flux velocity from rainfall, irrigation, recharge ponds, etc. Equation 8.1 is a linear equation derived from the Dupuit approximation, and we will interpret h as the change in piezometric head that results from a perturbation in either R, boundary conditions or initial conditions.

An exact solution of Eq. 8.1 for flow to a well, for R = 0, has already been obtained in Chapter V. Since the coefficients of Eq. 8.1 are constant, superposition and the method of images discussed in the previous chapter can also be used to obtain solutions for unsteady flow to wells in more complicated solution domains. The locations of image points are found by replacing straight-line boundaries with mirrors and placing either an abstraction or recharge well at each image point. Solution domain geometries that require an infinite number of image wells are approximated, in practice, by using a finite number of wells. The process can become laborious for more complicated geometries than an infinite half plane, but it is a straightforward extension of methods introduced in the previous chapter and will not be discussed further in this chapter.

Conformal mapping techniques are not useful in solving Eq. 8.1 because h no longer satisfies the Laplace equation. Thus, transform techniques and infinite series of orthogonal functions furnish the main tools of analysis for solving Eq. 8.1. Of these techniques, probably the most general for our purposes is the Laplace transform:

$$\phi(x, y, p) = \int_0^\infty e^{-pt}\, h(x, y, t)\, dt \qquad (8.2)$$

Equation 8.2 is applied by multiplying Eq. 8.1 and all boundary conditions by e^{-pt} and integrating with respect to t from t = 0 to t = ∞. The real part of the complex variable, p, is chosen to be positive and large enough to insure the existence of all integrals. The integral containing ∂h/∂t is integrated by parts once, a process which allows that integral to be replaced with a linear combination of ϕ and an initial condition. Thus, the initial-valued, boundary-value problem for h is replaced with a boundary-value problem for ϕ in which p is a complex parameter. The solution of this simpler problem for ϕ then allows h to be calculated from the inversion formula

$$h(x, y, t) = \frac{1}{2\pi i}\int_{\lambda-i\infty}^{\lambda+i\infty} e^{pt}\,\phi(x, y, p)\,dp \tag{8.3}$$

in which λ is a real, positive constant. The integration path in Eq. 8.3 is a straight line that is parallel to the imaginary p axis and which lies to the right of all singularities and branch points of $\phi(x, y, p)$ in the p plane.

The application of Eq. 8.3 usually proceeds by calculating the integral

$$\frac{1}{2\pi i}\int_{\Gamma} e^{pt}\,\phi(x, y, p)\,dp = \Sigma a_{-1} \tag{8.4}$$

in which Σa_{-1} is the sum of all residues within the closed curve Γ. The curve Γ is chosen so that (a) the function $e^{pt}\,\phi(x, y, p)$ is an analytic function of p within Γ, (b) a portion of Γ coincides with the integration path of Eq. 8.3 and (c) the contributions to Eq. 8.4 from all portions of Γ that consist of circular arcs at infinity vanish. The usual choices for Γ are shown in Fig. 8.1, in which the straight line AB lies to the right of all singularities and branch points of $\phi(x, y, p)$. The branch cuts to the left of AB are only needed when ϕ has branch points. The mapping transformation

$$p = z + \lambda \tag{8.5}$$

in which λ is a positive, real constant transforms Eq. 8.4 to

$$\frac{e^{\lambda t}}{2\pi i}\int_{\Gamma_z} e^{zt}\,\phi(x, y, z+\lambda)\,dz = \Sigma a_{-1} \tag{8.6}$$

in which Γ_z has the same geometry shown in Fig. 8.1 except that the line AB coincides with the imaginary z axis. (i.e. - Eq. 8.5 merely shifts the coordinate origin to the right by the distance λ.) Hildebrand (1962) shows that contributions to Eq. 8.6 (and, thus, Eq. 8.4) from the arc BCA vanish when $t > 0$ provided that $\phi \to 0$ uniformly on BCA as $|p| \to \infty$.* Likewise, it is shown that contributions from the arc ADB vanish when $t < 0$ if $\phi \to 0$ uniformly on ADB as $|p| \to \infty$. Thus, we will choose ABCA for Γ when $t > 0$ and BADB for Γ when $t < 0$. Most advanced calculus texts give introductions to the Laplace transform. The author also finds the book by Spiegel (1965) to be helpful for its inclusion of many elementary examples and problems, and more advanced applications are covered thoroughly in the two books by Carslaw and Jaeger (1948, 1959).

41. Some One-Dimensional Examples

For our first example, we will consider the following problem for the

---

*$\phi \to 0$ uniformly on a given arc if $|\phi| \to 0$ as $|p| \to \infty$ for all positions of point p on the arc.

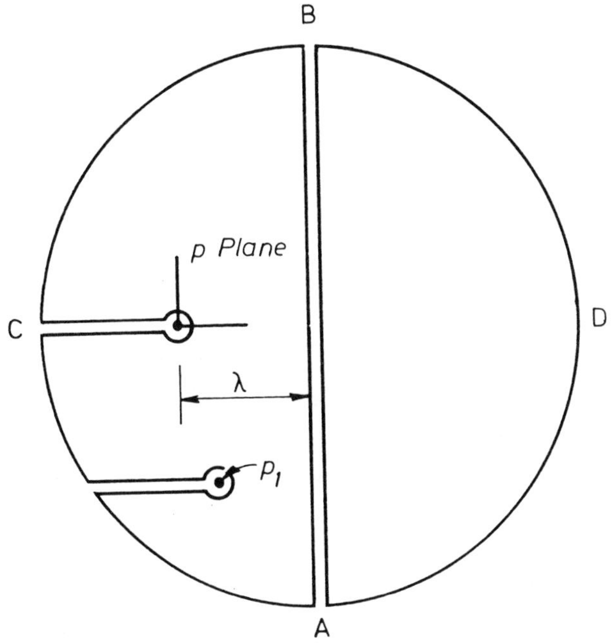

Fig. 8.1 - Contours in the p plane for Eq. 8.4. Contour ABC is used for t > 0, contour BAD for t < 0 and the line AB lies to the right of all singularities and branch points of $\phi(x, y, p)$.

aquifer geometry shown in Fig. 4.9:

$$T \frac{\partial^2 h}{\partial x^2} = S \frac{\partial h}{\partial t}, \quad (0 < x < \ell,\ 0 < t < \infty) \tag{8.7}$$

$$h(0,\ t) = h_0 f(t), \quad (0 \leqq t < \infty) \tag{8.8}$$

$$h(\ell,\ t) = 0, \quad (0 \leqq t < \infty) \tag{8.9}$$

$$h(x,\ 0) = 0, \quad (0 \leqq x \leqq \ell) \tag{8.10}$$

Multiplying Eqs. 8.7 - 8.9 by e^{pt} and integrating from t = 0 to t = ∞ gives:

$$T \frac{\partial^2 \phi}{\partial x^2} = S \int_0^\infty e^{-pt} \frac{\partial h}{\partial t}\, dt \tag{8.11}$$

$$\phi(0,\ p) = h_0\ F(p), \quad \left[F(p) = \int_0^\infty e^{-pt}\ f(t)\ dt\right] \tag{8.12}$$

$$\phi(\ell,\ p) = 0 \tag{8.13}$$

Integrating Eq. 8.11 once, by parts, and using the initial condition given by Eq. 8.10 gives

$$T \frac{\partial^2 \phi}{\partial x^2} = Sp\phi \tag{8.14}$$

The solution of Eq. 8.14, subject to the boundary conditions given by Eqs. 8.12 - 8.13, is

$$\phi(x,\ p) = h_0\ F(p)\ \frac{\sinh(\ell - x)\sqrt{Sp/T}}{\sinh(\ell\sqrt{Sp/T})} \tag{8.15}$$

Thus, Eq. 8.3 gives the solution

$$h(x,\ t) = \frac{h_0}{2\pi i} \int_{\lambda - i\infty}^{\lambda + i\infty} e^{pt}\ F(p)\ \frac{\sinh(\ell - x)\sqrt{Sp/T}}{\sinh(\ell\sqrt{Sp/t})}\, dp \tag{8.16}$$

The final calculation of the integral in Eq. 8.16 requires that the coefficient of e^{pt} in Eq. 8.16 approach zero uniformly as $|p| \to \infty$, and it is not obvious from the definition of F(p) in Eq. 8.12 that this condition will always be satisfied. However, one integration, by parts, in the definition of F(p) and substitution into Eq. 8.16 gives

$$\frac{h(x,\ t)}{h_0} = \frac{1}{2\pi i} \int_{\lambda - i\infty}^{\lambda + i\infty} \frac{e^{pt}}{p} \left[f(0) + \int_0^\infty e^{-p\tau}\ f'(\tau)\, d\tau\right] \frac{\sinh(\ell - x)\nu}{\sinh(\ell\nu)}\, dp \tag{8.17}$$

in which $\nu = \sqrt{Sp/T}$. Thus, calculating F(p) in this alternative way insures that the integrals exist when the order of integration is inverted to obtain

$$\frac{h(x,\ t)}{h_0} = f(0)\ I(t) + \int_0^{\infty} f'(\tau)\ I(t - \tau)d\tau \tag{8.18}$$

in which

$$I(t) = \frac{1}{2\pi i}\int_{\lambda-i\infty}^{\lambda+i\infty} \frac{e^{pt}}{p}\,\frac{\sinh(\ell - x)\nu}{\sinh(\ell\nu)}\,dp, \quad \left[\nu = \sqrt{\frac{Sp}{T}}\right] \tag{8.19}$$

Since $\sinh(\ell - x)\nu / [p \sinh(\ell\nu)] \to 0$ uniformly on the circular contour ADBCA in Fig. 8.1 as $|p| \to \infty$, the application of Eq. 8.4 gives

$$I(t) = 0 \text{ for } t < 0 \tag{8.20}$$

Since $\sinh(\ell - x)\nu/\sinh(\ell\nu)$ has an expansion near $p = 0$ of the form

$$\frac{\sinh(\ell - x)\nu}{\sinh(\ell\nu)} \sim \frac{1 + \frac{1}{6}\nu^2(\ell - x)^2 + \cdots\cdot}{1 + \frac{1}{6}\nu^2\ell^2 + \cdots\cdot}\left(1 - \frac{x}{\ell}\right) \tag{8.21}$$

we see that the integrand of Eq. 8.4 has no branch point but, rather, a residue of $(1 - x/\ell)$ at $p = 0$. Since the other zeros of $\sinh(\ell\nu)$ occur at $\nu = \pm\, in\pi/\ell$ for $n = 1, 2, 3, \ldots.$ (where $p = -n^2\pi^2T/S\ell^2$), we see that the remaining residues of the integrand in Eq. 8.4 are

$$\begin{aligned} a_{-1} &= \lim_{\substack{\nu \to \pm\, in\pi/\ell \\ p \to -n^2\pi^2T/(S\ell^2)}} \frac{e^{pt}}{p}\,\frac{\sinh(\ell - x)\nu}{\frac{d\nu}{dp}\frac{d}{d\nu}\sinh(\ell\nu)} \\ &= (-1)^n\,\frac{2}{\pi}\,\frac{\sin n\pi(1 - x/\ell)}{n}\,e^{-n^2\pi^2Tt/(S\ell^2)} \end{aligned} \tag{8.22}$$

Hence, the calculation of Eq. 8.4 around the contour ABCA in Fig. 8.1 gives the following result for $I(t)$ when $t > 0$:

$$I(t) = 1 - x/\ell + \frac{2}{\pi}\sum_{n=1}^{\infty}(-1)^n\,\frac{\sin n\pi(1 - x/\ell)}{n}\,e^{-n^2\pi^2Tt/(S\ell^2)} \tag{8.23}$$

Thus, Eqs. 8.18 and 8.20 show that the final solution is

$$\frac{h(x,\ t)}{h_0} = f(0)\ I(t) + \int_0^{t} f'(\tau)\ I(t - \tau)\ d\tau \tag{8.24}$$

in which $I(t)$ is given by Eq. 8.23.

For a second example, which introduces a branch point in the inversion integral, we will consider the result of letting $\ell \to \infty$ in Eqs. 8.7 - 8.10

$$T\frac{\partial^2 h}{\partial x^2} = S\frac{\partial h}{\partial t}, \quad (0 < x < \infty,\ 0 < t < \infty) \tag{8.25}$$

$$h(0,\ t) = h_0 f(t), \quad (0 \leqq t < \infty) \tag{8.26}$$

$$h(\infty,\ t) = 0, \quad (0 \leqq t < \infty) \tag{8.27}$$

$$h(x,\ 0) = 0, \quad (0 \leqq x < \infty) \tag{8.28}$$

Repeating the same steps that were followed in the first example leads to the following problem for ϕ:

$$T \frac{\partial^2 \phi}{\partial x^2} = Sp\phi \tag{8.29}$$

$$\phi(0, p) = h_0 F(p) \tag{8.30}$$

$$\phi(\infty, p) = 0 \tag{8.31}$$

The solution of Eqs. 8.29 - 8.31 for ϕ gives

$$\phi(x, p) = h_0 F(p)\, e^{-x\nu}, \quad \left[\nu = \sqrt{\frac{Sp}{T}}\right] \tag{8.32}$$

Thus, Eq. 8.3 gives the solution

$$\frac{h(x, t)}{h_0} = \frac{1}{2\pi i} \int_{\lambda - i\infty}^{\lambda + i\infty} e^{pt - x\nu}\, F(p)\, dp \tag{8.33}$$

Integrating F(p) once, by parts, and substituting the result into Eq. 8.33 gives

$$\frac{h(x, t)}{h_0} = \frac{1}{2\pi i} \int_{\lambda - i\infty}^{\lambda + i\infty} \frac{e^{pt - x\nu}}{p} \left[f(0) + \int_0^\infty e^{-p\tau} f'(\tau)\, d\tau \right] dp \tag{8.34}$$

Thus, interchanging the integration order gives

$$\frac{h(x, t)}{h_0} = f(0)\, I(t) + \int_0^\infty f'(\tau)\, I(t - \tau)\, d\tau \tag{8.35}$$

in which

$$I(t) = \frac{1}{2\pi i} \int_{\lambda - i\infty}^{\lambda + i\infty} e^{pt - x\nu} \frac{dp}{p}, \quad \left[\nu = \sqrt{\frac{Sp}{T}}\right] \tag{8.36}$$

Since $pt - x\nu \sim pt$ as $|p| \to \infty$, we see that contributions to the inversion integral from any portion of the circular arc ADBCA in Fig. 8.1 will vanish in the limit as the radius of this circle becomes infinite. This, of course, is the reason for integrating the expression for F(p) by parts before substituting it into Eq. 8.33 to obtain Eq. 8.34.

Applying Eq. 8.4 to Eq. 8.36 around the contour ADBA in Fig. 8.1 gives

$$I(t) = 0 \text{ for } t < 0 \tag{8.37}$$

Thus, Eq. 8.35 reduces to

$$\frac{h(x, t)}{h_0} = f(0)\, I(t) + \int_0^t f'(\tau)\, I(t - \tau)\, d\tau \tag{8.38}$$

The integrand of I(t) has a branch point at p = 0. Thus, we will choose a branch cut along the negative real axis and consider the integral

$$\frac{1}{2\pi i}\int_{\Gamma} e^{pt-x\nu}\frac{dp}{p} = 0 \text{ for } t > 0 \tag{8.39}$$

in which the closed curve Γ is shown in Fig. 8.2. As noted before, the contributions to Eq. 8.39 from the arcs BC and FA vanish as the radius of these arcs becomes infinite. Thus, Eqs. 8.36 and 8.39 give

$$I(t) + \frac{1}{2\pi i}\int_{CDEF} e^{pt-x\nu}\frac{dp}{p} = 0 \text{ for } t > 0 \tag{8.40}$$

Setting $p = re^{i\pi}$ on CD, $p = \varepsilon e^{i\theta}$ on DE and $p = re^{-i\pi}$ on EF in Eq. 8.40 gives

$$I(t) + \frac{1}{2\pi i}\int_{\infty}^{\varepsilon} e^{-rt-ix\sqrt{Sr/T}}\frac{dr}{r} + \frac{1}{2\pi}\int_{\pi}^{-\pi} e^{\varepsilon te^{i\theta} - x\sqrt{S\varepsilon/T}\,e^{i\theta/2}}\,d\theta$$

$$+ \frac{1}{2\pi i}\int_{\varepsilon}^{\infty} e^{-rt+ix\sqrt{Sr/T}}\frac{dr}{r} = 0 \tag{8.41}$$

Hence, taking the limit $\varepsilon \to 0$ gives

$$I(t) = 1 - \frac{1}{\pi}\int_0^{\infty} e^{-rt}\sin\left(x\sqrt{\frac{Sr}{T}}\right)\frac{dr}{r} \tag{8.42}$$

The substitution $r = u^2$ reduces the integral in Eq. 8.42 to an integral calculated by Hildebrand (1962):

$$I(t) = \operatorname{erfc}\left(\frac{x}{2}\sqrt{\frac{S}{Tt}}\right) \text{ for } t > 0 \tag{8.43}$$

The complementary error function, erfc, has already been met in Chapter VI, and it is easily verified that Eqs. 8.38 and 8.43 give the solution to Eqs. 8.25 - 8.28.

42. A Groundwater Recharge Problem

If water percolates downward to an unconfined aquifer, then the formation of a groundwater mound beneath the recharge area, A, is described by the solution to the following problem:

$$T\nabla^2 h = S\frac{\partial h}{\partial t} + \frac{K'}{B'}h - Rf(t), \quad (-\infty < x < \infty,\ -\infty < y < \infty,\ 0 < t < \infty) \tag{8.44}$$

$$h \to 0 \text{ as } (x^2 + y^2) \to \infty \text{ for } 0 < t < \infty \tag{8.45}$$

$$h(x, y, 0) = 0, \quad (-\infty < x < \infty,\ -\infty < y < \infty) \tag{8.46}$$

The function f(t) is a dimensionless function of time, and R, with dimensions of a flux velocity, is constant for (x, y) within A and zero for (x, y) outside A.

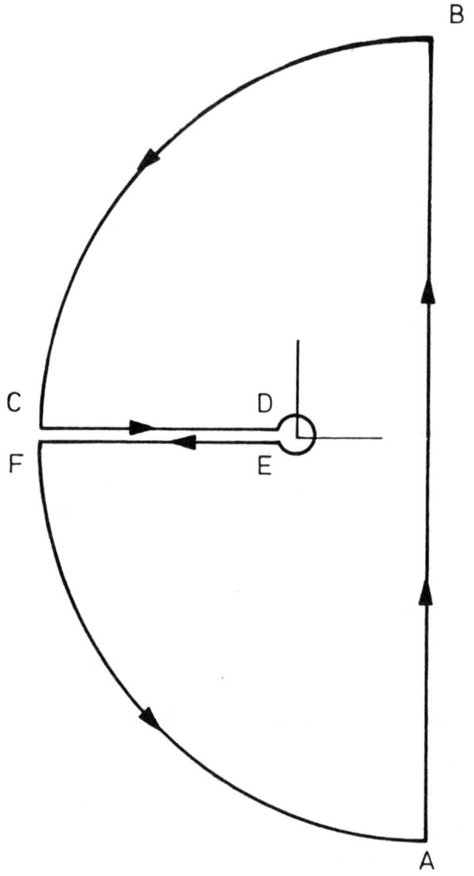

Fig. 8.2 - The integration path in the p plane for Eq. 8.39.

It is easily verified that a solution of Eqs. 8.44 - 8.46 is given by

$$h(x, y, t) = \int_A H(x - \xi, y - \eta, t)\, d\xi d\eta \qquad (8.47)$$

in which H(x, y, t) is a solution of the following problem:

$$T\nabla^2 H = S \frac{\partial H}{\partial t} + \frac{K'}{B'} H - Rf(t)\, \delta(x)\delta(y) \qquad (8.48)$$

$$H \to 0 \text{ as } (x^2 + y^2) \to \infty \qquad (8.49)$$

$$H(x, y, 0) = 0 \qquad (8.50)$$

Equations 8.48 - 8.50 are identical with the equations that describe flow to a well with a flow rate of Rf(t). Thus, the solution for H is axisymmetric and satisfies the system of equations

$$\frac{T}{r}\frac{\partial}{\partial r}\left(r \frac{\partial H}{\partial r}\right) = S \frac{\partial H}{\partial t} + \frac{K'}{B'} H, \ (0 < r < \infty,\ 0 < t < \infty) \qquad (8.51)$$

$$H(\infty, t) = 0, \ (0 < t < \infty) \qquad (8.52)$$

$$H(r, 0) = 0, \ (0 < r < \infty) \qquad (8.53)$$

$$\underset{r \to 0}{\text{Limit}} \left(r \frac{\partial H}{\partial r}\right) = - \frac{R}{2\pi T} f(t), \ (0 < t < \infty) \qquad (8.54)$$

Equation 8.54 is obtained by integrating Eq. 8.48 throughout a small circular area around the origin, using the divergence theorem and taking the limit as the radius of this circle goes to zero. The function $H(x - \xi, y - \eta, t)$ in Eq. 8.47 is obtained from H(r, t) by setting

$$r = \sqrt{(x - \xi)^2 + (y - \eta)^2} \qquad (8.55)$$

Taking the Laplace transform of the system of equations for H gives

$$\frac{T}{r}\frac{\partial}{\partial r}\left(r \frac{\partial \phi}{\partial r}\right) = \left(Sp + \frac{K'}{B'}\right)\phi \qquad (8.56)$$

$$\phi(\infty, p) = 0 \qquad (8.57)$$

$$\underset{r \to 0}{\text{Limit}} \left(r \frac{\partial \phi}{\partial r}\right) = - \frac{R}{2\pi T} F(p) \qquad (8.58)$$

in which F(p) is the transform of f(t).

$$F(p) = \int_0^\infty e^{-pt} f(t)\, dt \qquad (8.59)$$

In order to insure the existence of the inverse transform, it is necessary to integrate 8.59 once, by parts, to obtain

$$F(p) = \frac{1}{p}\left[f(0) + \int_0^\infty e^{-p\tau} f'(\tau)\, d\tau\right] \qquad (8.60)$$

The solution of Eqs. 8.56 - 8.58 and 8.60 is

$$\phi(r,\ p) = \frac{R}{2\pi Tp}\left[f(0) + \int_0^\infty e^{-p\tau}\ f'(\tau)\,d\tau\right] K_0\left(r\sqrt{\frac{Sp + K'/B'}{T}}\right) \qquad (8.61)$$

in which K_0 is the zero-order, modified Bessel function of the second kind.

Equation 8.61 can be inverted by using the known, definite integral

$$K_0(z) = \frac{1}{2}\int_0^\infty e^{-\frac{1}{u} - \frac{u}{4}z^2}\ \frac{du}{u} \text{ for } \mathrm{Re}\{z^2\} > 0 \qquad (8.62)$$

Thus, Eqs. 8.3, 8.61 and 8.62 lead to the result

$$H(r,\ t) = \frac{R}{4\pi T}\left[f(0)\int_0^\infty e^{-\frac{1}{u} - \frac{ur^2K'/B'}{4T}}\ I(u,\ t)\ \frac{du}{u}\right.$$

$$\left. + \int_0^\infty f'(\tau)\int_0^\infty e^{-\frac{1}{u} - \frac{ur^2K'/B'}{4T}}\ I(u,\ t - \tau)\ \frac{du}{u}\ d\tau\right] \qquad (8.63)$$

in which

$$I(u,\ t) = \frac{1}{2\pi i}\int_{\lambda - i\infty}^{\lambda + i\infty} e^{p\left(t - \frac{ur^2S}{4T}\right)}\ \frac{dp}{p} = 0 \text{ for } t < \frac{ur^2S}{4T}$$

$$= 1 \text{ for } t > \frac{ur^2S}{4T} \qquad (8.64)$$

The integral in Eq. 8.64 was calculated by using the two contours ADBA and ABCA in Fig. 8.1, and the use of this result in Eq. 8.63 gives

$$H(r,\ t) = \frac{R}{4\pi T}\left[f(0)\int_0^{\frac{4Tt}{Sr^2}} e^{-\frac{1}{u} - \frac{ur^2K'/B'}{4T}}\ \frac{du}{u}\right.$$

$$\left. + \int_0^t f'(\tau)\int_0^{\frac{4T(t - \tau)}{Sr^2}} e^{-\frac{1}{u} - \frac{ur^2K'/B'}{4T}}\ \frac{du}{u}\ d\tau\right] \qquad (8.65)$$

The substitution $u = 1/\xi$ in both integrals in Eq. 8.65 allows the result to be written in terms of the leaky aquifer function. A simpler result, though, is obtained by carrying out an integration by parts in the second term on the right side of Eq. 8.65 to obtain

$$H(r,\ t) = \frac{R}{4\pi T}\int_0^t f(\tau)\ e^{-\frac{Sr^2}{4T(t - \tau)} - \frac{(t - \tau)K'/B'}{S}}\ \frac{d\tau}{(t - \tau)} \qquad (8.66)$$

Equation 8.47, 8.55 and 8.66 give the solution to the problem.

For a particular application, let A be the rectangle $-a < x < a$, $-b < y < b$. Then Eqs. 8.47, 8.55 and 8.66 give

$$\frac{4\pi Th}{R} = \int_0^t f(\tau)\, e^{-\frac{(t-\tau)K'/B'}{S}} \left[\int_{-b}^{b}\int_{-a}^{a} e^{-\frac{S(x-\xi)^2}{4T(t-\tau)} - \frac{S(y-\eta)^2}{4T(t-\tau)}}\, d\xi d\eta\right] \frac{d\tau}{(t-\tau)} \qquad (8.67)$$

in which the order of the time and spatial integrations has been interchanged. Calculation of the two spatial integrals gives

$$\frac{4Sh}{R} = \int_0^t f(\tau)\, e^{-\frac{(t-\tau)K'/B'}{S}} \left\{\operatorname{erf}\left[(a-x)\sqrt{\frac{S}{4T(t-\tau)}}\right] + \operatorname{erf}\left[(a+x)\sqrt{\frac{S}{4T(t-\tau)}}\right]\right\} \left\{\operatorname{erf}\left[(b-y)\sqrt{\frac{S}{4T(t-\tau)}}\right] + \operatorname{erf}\left[(b+y)\sqrt{\frac{S}{4T(t-\tau)}}\right]\right\} d\tau \qquad (8.68)$$

The integral in Eq. 8.68 is in a form that is relatively easy to compute numerically with standard quadrature formulas.

43. The Well Storage Effect

The classical solutions for flow toward a well, which are derived and used in Chapter V, assume that flow into the well from the aquifer equals the constant flow that is pumped from the well bore. In actual fact, this is not generally true since the difference between these two flow rates equals the change in storage within the well as the free surface in the well changes its level. A more complete solution to this problem, which includes the well storage effect, can be calculated by solving the following set of equations:

$$\frac{T}{r}\frac{\partial}{\partial r}\left(r\frac{\partial h}{\partial r}\right) = S\frac{\partial h}{\partial t}, \quad (r_0 < r < \infty,\ 0 < t < \infty) \qquad (8.69)$$

$$h(r, 0) = 0, \quad (r_0 \leq r < \infty) \qquad (8.70)$$

$$h(\infty, t) = 0, \quad (0 \leq t < \infty) \qquad (8.71)$$

$$\frac{\partial h(r_0, t)}{\partial r} = \frac{Q(t)}{2\pi T r_0}, \quad (0 \leq t < \infty) \qquad (8.72)$$

$$Q(t) - Q_0 = \pi r_0^2 \frac{\partial h(r_0, t)}{\partial t}, \quad (0 \leq t < \infty) \qquad (8.73)$$

These equations contain two unknowns: the piezometric head, $h(r, t)$, in the aquifer and the flow rate, $Q(t)$, from the aquifer into the well. The radius of the well is r_0, Q_0 is the constant flow rate pumped from the well bore and Eq. 8.73 is a mass conservation statement for the water within the well bore.

Taking the Laplace transform of Eqs. 8.69 - 8.73 gives

$$\frac{T}{r}\frac{\partial}{\partial r}\left(r\frac{\partial \phi}{\partial r}\right) = Sp\phi \qquad (8.74)$$

$$\phi(\infty, p) = 0 \tag{8.75}$$

$$\frac{\partial\phi(r_0, p)}{\partial r} = \frac{F(p)}{2\pi T r_0}$$

$$F(p) - \frac{Q_0}{p} = \pi r_0^{\,2} p\phi(r_0, p) \tag{8.77}$$

in which $\phi(r, p)$ and $F(p)$ are the Laplace transforms of $h(r, t)$ and $Q(t)$, respectively. A straightforward solution of Eqs. 8.74 - 8.77 yields

$$F(p) = \frac{Q_0\, K_1(r_0\sqrt{pS/T})}{P[K_1(r_0\sqrt{pS/T}) + \frac{r_0}{2}\sqrt{\frac{P}{ST}}\, K_0\,(r_0\sqrt{pS/T})]} \tag{8.78}$$

$$\phi(r, p) = \frac{-Q_0\, K_0(r\sqrt{pS/T})}{2\pi r_0\sqrt{STp^3}\,[K_1(r_0\sqrt{pS/T}) + \frac{r_0}{2}\sqrt{\frac{p}{ST}}\, K_0(r_0\sqrt{pS/T})]} \tag{8.79}$$

The expressions for $F(p)$ and $\phi(r, p)$ have a single branch point at $p = 0$ and vanish uniformly as $|p| \to \infty$ for all values of the argument of p between $+\pi$ and $-\pi$. Thus, we will use the inversion contour shown in Fig. 8.2 and note that the arcs BC and FA will not make a contribution to the contour integral in Eq. 8.4. Furthermore, the following analysis shows that $F(p)$ and $\phi(r, p)$ have no residues within the contour ABCDEFA of Fig. 8.2.

The residues of Eqs. 8.78 and 8.79 can only occur at the zeros of the equation

$$K_1(\alpha) + \frac{\alpha}{2S} K_0(\alpha) = 0 \tag{8.80}$$

in which the argument of α varies between $\pi/2$ and $-\pi/2$ as the argument of p in Eq. 8.79 varies between π and $-\pi$. Since the coefficients of Eq. 8.80 are real, the complex conjugate, $\bar{\alpha}$, of α is also a root of Eq. 8.80.

$$K_1(\bar{\alpha}) + \frac{\bar{\alpha}}{2S} K_0(\bar{\alpha}) = 0 \tag{8.81}$$

But an integral given by Gradshteyn and Ryzhik (1965) leads to the result

$$(\alpha^2 - \bar{\alpha}^2) \int_1^\infty x\, K_0(\alpha x) K_0(\bar{\alpha} x)\, dx = \alpha K_0(\bar{\alpha}) K_1(\alpha) - \bar{\alpha} K_0(\alpha) K_1(\bar{\alpha}) \tag{8.82}$$

Setting $\alpha = \alpha_1 + i\alpha_2$ and using Eqs. 8.80 and 8.81 in Eq. 8.82 gives

$$4i\alpha_1\alpha_2 \left[\int_1^\infty x\, K_0(\alpha x) K_0(\bar{\alpha} x)\, dx + \frac{1}{2S} K_0(\alpha) K_0(\bar{\alpha}) \right] = 0 \tag{8.83}$$

Thus, Eq. 8.83 show that the roots of Eq. 8.80 can only lie along the real axis ($\alpha_1 = 0$) or the imaginary axis ($\alpha_2 = 0$). But each term in Eq. 8.80 is greater than zero for real, positive values of α. Thus, zeros of Eq. 8.83

can only lie along the negative real axis, which does not concern us, or along the imaginary axis. Setting $\alpha = re^{\pm i\pi/2}$ in Eq. 8.80 gives

$$[J_0'(r) + \frac{r}{2S} J_0(r)] \pm i[Y_0'(r) + \frac{r}{2S} Y_0(r)] = 0 \qquad (8.84)$$

in which J_0 and Y_0 are Bessel functions of the first and second kinds, respectively. Since the real and imaginary parts of Eq. 8.84 must vanish simultaneously, any roots of Eq. 8.80 along the imaginary axis must satisfy

$$J_0(r)Y_0'(r) - J_0'(r)Y_0(r) = 0 \qquad (8.85)$$

But a well known identity gives

$$J_0(r)Y_0'(r) - J_0'(r)Y_0(r) = \frac{2}{\pi r} > 0 \qquad (8.86)$$

Thus, Eq. 8.80 has no roots along the imaginary α axis, and Eqs. 8.78 and 8.79 have no poles within the contour ABCDEFA in Fig. 8.2.

It is possible to use Eqs. 8.78, 8.79, 8.3 and 8.4, with $\Sigma a_{-1} = 0$, together with the contour shown in Fig. 8.2 to obtain expressions for h(r, t) and Q(t) in the form of definite integrals. This process leads to the result

$$h(r, t) = \frac{1}{2\pi i} \int_{-\infty}^{(0+)} e^{pt} \phi(r, p)\, dp \qquad (8.87)$$

$$Q(t) = \frac{1}{2\pi i} \int_{-\infty}^{(0+)} e^{pt} F(p)\, dp \qquad (8.88)$$

in which the limits on the integrals indicate that the integrals are to be calculated over the path FEDC in Fig. 8.2. The integrals in Eqs. 8.87 - 8.88 can be reduced to real variable form in the same way that was used to obtain Eq. 8.43, but the result is too complicated to be useful. Instead, we will use Eqs. 8.78, 8.79, 8.87 and 8.88 to obtain the first few terms in the asymptotic expansions of h(r, t) and Q(t) for large values of t.

Carslaw and Jaeger (1948) show that asymptotic expansions for h(r, t) and Q(t) can be calculated by expanding $\phi(r, p)$ and F(p) about p = 0 and calculating the integrals in Eqs. 8.87 and 8.88 for each term in the series. The resulting series may not converge because the expansions for ϕ and F may not converge for all values of p upon the integration path, but these asymptotic expansions are usually capable of giving extremely accurate results with only the first few terms when t is large enough. The first few terms in the expansions of ϕ and F about the origin are

$$\frac{4\pi T\phi(r, p)}{Q_0} = \frac{1}{p} \ln\left(\frac{pSr^2 e^{2\gamma}}{4T}\right) + \frac{Sr^2}{4T} \ln\left[\frac{pSr^2 e^{2(\gamma-1)}}{4T}\right]$$
$$- \frac{Sr_0^2}{4T} \ln\left(\frac{pSr^2 e^{2\gamma}}{4T}\right) \left[\ln\left(\frac{pSr_0^2 e^{2\gamma-1}}{4T}\right) - \frac{1}{S} \ln\left(\frac{pSr_0^2 e^{2\gamma}}{4T}\right)\right]$$
$$+ \ldots. \qquad (8.89)$$

$$\frac{F(p)}{Q_0} = 1 + \frac{pr_0^2}{4T} \ln\left(\frac{pSr_0^2 e^{2\gamma}}{4T}\right) + \ldots. \tag{8.90}$$

in which γ = Euler's constant = 0.5772 .... Carslaw and Jaeger (1959) give values for the different integrals that result when Eqs. 8.89 and 8.90 are inserted in Eqs. 8.87 and 8.88. (These integrals are also given in problem 5 at the end of this chapter.) The final result of integrating Eqs. 8.89 and 8.90 is

$$\frac{4\pi Th(r,\ t)}{Q_0} \sim \ln\left(\frac{Sr^2e^{\gamma}}{4Tt}\right) - \frac{Sr_0^2}{4Tt}\left\{1 + \left(\frac{r}{r_0}\right)^2 + \left(\frac{1}{S} - 1\right) \ln\left[\left(\frac{Sr^2e^{\gamma}}{4Tt}\right)\left(\frac{Sr_0^2e^{\gamma}}{4Tt}\right)\right]\right\} + \ldots. \tag{8.91}$$

$$\frac{Q(t)}{Q_0} \sim 1 - \frac{r_0^2}{4Tt} + \ldots.. \tag{8.92}$$

The significance of the storage effect becomes apparent when Eqs. 8.91 and 8.92 are compared with the corresponding result for zero well storage. This result, which can be obtained either from Eq. 5.31 or by letting $r_0 \to 0$ in Eqs. 8.91 and 8.92, is

$$\frac{4\pi Th(r,\ t)}{Q_0} \sim \ln\left(\frac{Sr^2e^{\gamma}}{4Tt}\right) - \frac{Sr^2}{4Tt} + \ldots. \tag{8.93}$$

$$\frac{Q(t)}{Q_0} = 1 \tag{8.94}$$

Equations 8.92 and 8.94 show that the steady flow rate, Q_0, is approached very rapidly. Furthermore, since the first term in Eq. 8.91 is identical with the first term in Eq. 8.93, and since only this first term is used in Jacob's straight-line method for analyzing pump test data (as described in section 25), we see that well storage effects have no influence upon the analysis of pump test data when Jacob's method is used. This is true because Jacob's method can only be used after t has become large enough to allow the neglect of every term except for the first in either Eq. 8.91 or 8.93. Since a similar statement can not be made about the match-point method, we conclude that Jacob's method is probably the better method for analyzing pump test data.

44. Two Well Recovery Problems

Equation 5.49 in Chapter V can be derived by solving the following problem, which describes flow from a well after an instantaneous water level rise of H_0 within the well:

$$\frac{T}{r}\frac{\partial}{\partial r}\left(r\frac{\partial h}{\partial r}\right) = S\frac{\partial h}{\partial t},\quad (r_0 < r < \infty,\ 0 < t < \infty) \tag{8.95}$$

$$h(r_0, t) = H(t), \quad (0 < t < \infty) \tag{8.96}$$

$$h(r, 0) = 0, \quad (r_0 < r < \infty) \tag{8.97}$$

$$h(\infty, t) = 0, \quad (0 < t < \infty) \tag{8.98}$$

$$\frac{dH(t)}{dt} - 2\frac{T}{r_0}\frac{h(r_0, t)}{\partial r} = 0, \quad (0 < t < \infty) \tag{8.99}$$

$$H(0) = H_0 \tag{8.100}$$

The radius of the well is r_0, and Eq. 8.99 requires that a decrease in storage within the well be balanced by a corresponding flow into the aquifer. The two unknowns in these equations are the piezometric heads, $h(r, t)$ and $H(t)$, in the aquifer and well, respectively.

Taking the Laplace transform of Eqs. 8.95 - 8.100 gives

$$\frac{T}{r}\frac{\partial}{\partial r}\left(r\frac{\partial\phi}{\partial r}\right) = Sp\phi \tag{8.101}$$

$$\phi(r_0, p) = \psi(p) \tag{8.102}$$

$$\phi(\infty, p) = 0 \tag{8.103}$$

$$p\psi(p) - H_0 - 2\frac{T}{r_0}\frac{\partial\phi(r_0, p)}{\partial r} = 0 \tag{8.104}$$

in which $\phi(r, p)$ and $\psi(p)$ are the transforms of $h(r, t)$ and $H(t)$, respectively. The solution of Eqs. 8.101 - 8.104 is given by

$$\phi(r, p) = \frac{H_0 K_0(r\sqrt{pS/T})}{pK_0(r_0\sqrt{pS/T}) + \frac{2}{r_0}\sqrt{pST}\,K_1(r_0\sqrt{pS/T})} \tag{8.105}$$

$$\psi(p) = \frac{H_0 K_0(r_0\sqrt{pS/T})}{pK_0(r_0\sqrt{pS/T}) + \frac{2}{r_0}\sqrt{pST}\,K_1(r_0\sqrt{pS/T})} \tag{8.106}$$

The poles in Eqs. 8.105 - 8.106 can only occur at the zeros of

$$\frac{r_0}{2}\sqrt{\frac{p}{ST}}\,K_0(r_0\sqrt{pS/T}) + K_1(r_0\sqrt{pS/T})$$

However, it was shown in section 43 that this function has no zeros within the contour ABCDEFA in Fig. 8.2. Thus, since $\psi(p) \sim H_0/P$ as $|p| \to \infty$, $H(t)$ is given by

$$H(t) = \frac{1}{2\pi i}\int_{-\infty}^{(0+)} e^{pt}\,\psi(p)\,dp \tag{8.107}$$

in which the integration limits indicate that the integral is to be calculated over the contour FEDC in Fig. 8.2. Since $\psi(p) \sim -H_0 r_0^2 \ln(p)/(4T)$ as $|p| \to 0$, the contribution around the small circle ED vanishes as the circle radius goes to zero. Thus, H(t) is given by

$$H(t) = -\frac{1}{\pi}\int_0^\infty e^{-\xi t}\ \text{Im}\{\psi(\xi e^{i\pi})\}d\xi \tag{8.108}$$

in which $\xi = |p|$ and Im{} denotes the imaginary part of a function. The identities

$$K_\nu(ze^{i\pi/2}) = i\frac{\pi}{2}e^{-i\pi\nu/2}\ [-J_\nu(z) + iY_\nu(z)] \tag{8.109}$$

$$J_{\nu+1}(z)\ Y_\nu(z) - J_\nu(z)\ Y_{\nu+1}(z) = \frac{2}{\pi z} \tag{8.110}$$

and Eq. 8.106 allow Eq. 8.108 to be reduced to the form

$$\frac{H(t)}{H_0} = \frac{4S}{\pi^2}\int_0^\infty \frac{e^{-\xi t}}{\xi\Delta\left(r_0\sqrt{\frac{\xi S}{T}}\right)}d\xi \tag{8.111}$$

in which

$$\Delta(x) = [xJ_0(x) - 2SJ_1(x)]^2 + [xY_0(x) - 2SY_1(x)]^2 \tag{8.112}$$

Finally, the substitution $x = r_0\sqrt{\xi S/T}$ in Eq. 8.111 puts the result in the form of Eq. 5.49.

The analogous solution for flow toward the open end of a pipe is found by solving the following problem, which has been written in terms of the spherical coordinate system shown in Fig. 6.2(b):

$$\frac{K}{r^2}\frac{\partial}{\partial r}\left(r^2\frac{\partial h}{\partial r}\right) = S_s\frac{\partial h}{\partial t},\ (r_0 < r < \infty,\ 0 < t < \infty) \tag{8.113}$$

$$h(r_0,\ t) = H(t),\ (0 < t < \infty) \tag{8.114}$$

$$h(r,\ 0) = 0,\ (r_0 < r < \infty) \tag{8.115}$$

$$h(\infty,\ t) = 0,\ (0 < t < \infty) \tag{8.116}$$

$$\frac{dH(t)}{dt} - 4K\frac{\partial h(r_0,\ t)}{\partial r} = 0,\ (0 < t < \infty) \tag{8.117}$$

$$H(0) = H_0 \tag{8.118}$$

in which K = coefficient of permeability and S_s = specific storage. A procedure very similar to the one used to solve Eqs. 8.95 - 8.100 leads to the following solution for the Laplace transform, $\psi(p)$, of H(t):

$$\psi(p) = \frac{H_0}{p + \frac{4K}{r_0} + 4\sqrt{KS_s p}} \tag{8.119}$$

Since $\psi(p)$ has no poles within the contour ABCDEFA in Fig. 8.2*, and since $\psi(p) \sim H_0/p$ as $|p| \to \infty$, taking the inverse of Eq. 8.119 gives

$$H(t) = \frac{1}{2\pi i} \int_{-\infty}^{(0+)} e^{pt} \psi(p)\, dp \qquad (8.120)$$

Since $\psi(p) \sim r_0 H_0/(4K)$ as $|p| \to 0$, the contribution from the small circle ED vanishes as the circle radius goes to zero and Eq. 8.120 gives the result

$$H(t) = -\frac{1}{\pi} \int_0^{\infty} e^{-\xi t}\, \mathrm{Im}\{\psi(\xi e^{i\pi})\} d\xi \qquad (8.121)$$

Finally, using Eq. 8.119 in Eq. 8.121 and setting $x = \xi r_0/K$ puts the result in the form of Eq. 5.51.

Since the product $S_s r_0$ is normally a very small, dimensionless number, Eq. 8.119 can be expanded in ascending powers of S_s to obtain

$$\frac{\psi(p)}{H_0} = \frac{1}{\left(p + \frac{4K}{r_0}\right)} - \frac{4\sqrt{KS_s p}}{\left(p + \frac{4K}{r_0}\right)^2} + \frac{16KS_s p}{\left(p + \frac{4K}{r_0}\right)^3} + \ldots \qquad (8.122)$$

The method of residues can be used to show that the inverses of the first and third terms on the right side of Eq. 8.122 are given by

$$\frac{1}{2\pi i} \int_{\lambda - i\infty}^{\lambda + i\infty} \frac{e^{pt}}{p + \frac{4K}{r_0}} dp = e^{-4\tau}, \quad \left(\tau = \frac{Kt}{r_0}\right) \qquad (8.123)$$

$$\frac{1}{2\pi i} \int_{\lambda + i\infty}^{\lambda + i\infty} \frac{16KS_s p e^{pt}}{\left(p + \frac{4K}{r_0}\right)^3} dp = \underset{p \to -\frac{4K}{r_0}}{\mathrm{Limit}} \frac{1}{2!} \frac{\partial^2}{\partial p^2} 16KS_s p e^{pt}$$

$$= 16\varepsilon\tau(1-2\tau)e^{-4\tau}, \quad (\varepsilon = S_s r_0) \qquad (8.124)$$

The second term on the right side of Eq. 8.122 can be inverted by first making the substitution $z = p + 4K/r_0$ to obtain

$$\frac{1}{2\pi i} \int_{\lambda - i\infty}^{\lambda + i\infty} \frac{4\sqrt{KS_s p}}{\left(p + \frac{4K}{r_0}\right)^2} e^{pt}\, dp = \frac{1}{2\pi i} \int_{\lambda' - i\infty}^{\lambda' + i\infty} \frac{4\sqrt{KS_s (z - 4K/r_0)}}{z^2} e^{(z - 4K/r_0)t}\, dz \qquad (8.125)$$

in which $\lambda' = \lambda + 4K/r_0$. An expansion of the square root in descending powers of z gives

---

*A pole of $\psi(p)$ can only occur within ABCDEFA in Fig. 8.2 if $|\mathrm{Arg}\{\sqrt{p}\}| < \pi/2$ at the pole. However, the poles of $\psi(p)$ occur at the two points $\sqrt{p} = -2\sqrt{KS_s} \pm 2i\sqrt{K(1/r_0 - S_s)}$, and $|\mathrm{Arg}\{\sqrt{p}\}| > \pi/2$ at both of these points. In other words the poles of $\psi(p)$ occur upon a sheet of the Riemann surface that is different from the one shown in Fig. 8.2.

$$\frac{1}{2\pi i}\int_{\lambda-i\infty}^{\lambda+i\infty} \frac{4\sqrt{KS_s}p}{\left(p+\frac{4K}{r_0}\right)^2} e^{pt}\, dp = \frac{4\sqrt{KS_s}\, e^{-4\tau}}{2\pi i}\int_{\lambda'-i\infty}^{\lambda'+i\infty} e^{zt}\left[\frac{1}{z^{3/2}} - \frac{1}{2\sqrt{\pi}}\sum_{n=1}^{\infty}\frac{\Gamma(n-1/2)}{n!}\frac{(4K/r_0)^n}{z^{n+3/2}}\right] dz \tag{8.126}$$

The infinite series expansion only converges for $4K/|r_0 z| < 1$, but λ' can always be chosen large enough to insure that the expansion converges for all points on the integration path. Finally, since the definition and properties of the Gamma function give

$$\frac{1}{2\pi i}\int_{\lambda'-i\infty}^{\lambda'+i\infty} e^{zt}\frac{dz}{z^{\nu}} = \frac{t^{\nu-1}}{\Gamma(\nu)}, \quad (\nu > 0) \tag{8.127}$$

$$\Gamma(n+3/2) = (n^2-1/4)\,\Gamma(n-1/2) \tag{8.128}$$

the result of integrating each term in Eq. 8.126 gives the final result

$$\frac{1}{2\pi i}\int_{\lambda-i\infty}^{\lambda+i\infty} \frac{4\sqrt{KS_s}p}{\left(p+\frac{4K}{r_0}\right)^2} e^{pt}\, dp = 8e^{-4\tau}\sqrt{\frac{\epsilon\tau}{\pi}}\left[1 - \sum_{n=1}^{\infty}\frac{(4\tau)^n}{n!\,(4n^2-1)}\right] \tag{8.129}$$

The infinite series in Eq. 8.129 is absolutely convergent for all finite values of τ, and the use of Eqs. 8.123, 8.124 and 8.129 in Eq. 8.122 gives Eq. 5.52 in Chapter V.

REFERENCES

Abramowitz, M. and Stegun, I.A. (Editors), 1964. Handbook of Mathematical Functions, U.S. National Bureau of Standards, Applied Mathematics Series, No. 55, U.S. Government Printing Office, Washington, D.C., p. 488.

Carslaw, H.S. and Jaeger, J.C. 1948. Operational Methods in Applied Mathematics, second edition, Oxford University Press.

Carslaw, H.S. and Jaeger, J.C. 1959. Conduction of Heat in Solids, second edition, Oxford at the Clarendon Press, Chs. XII and XIII.

Gradshteyn, I.S. and Ryzhik, I.M. 1965. Tables of Integrals, Series and Products, Fourth Edition, Academic Press, New York, p. 634.

Hildebrand, F.B. 1962. Advanced Calculus for Applications, Prentice-Hall, Inc., Englewood Cliffs, pp. 464, 556-558.

Spiegel, M.R. 1965. Theory and Problems of Laplace Transforms, Schaum's Outline Series, McGraw-Hill Book Company, New York.

PROBLEMS

1. Calculate the solution of Eqs. 8.7 - 8.10 if Eq. 8.9 is replaced with

$$\frac{\partial h(\ell, t)}{\partial x} = 0, \quad (0 \leqq t < \infty)$$

Use the Laplace transform to show that the answer is

$$\frac{h(x, t)}{h_0} = f(0)\ I(t) + \int_0^t f'(\tau)\ I(t-\tau)\ d\tau$$

$$I(t) = 1 + 2 \sum_{n=1}^{\infty} \frac{(-1)^n}{\alpha_n} e^{-\frac{Tt}{S\ell^2}\alpha_n^2} \cos\left[\alpha_n\left(1 - \frac{x}{\ell}\right)\right]$$

in which $\alpha_n = (2n - 1)\pi/2$.

2. Use an integration by parts to show that the solution given by Eqs. 8.38 and 8.43 can be written in the alternative forms

$$\frac{h(x, t)}{h_0} = \frac{x}{2}\sqrt{\frac{S}{\pi T}} \int_0^t f(\tau)\ e^{-\frac{Sx^2}{4T(t - \tau)}} \frac{d\tau}{(t - \tau)^{3/2}}$$

$$= \frac{2}{\sqrt{\pi}} \int_{\frac{x}{2}\sqrt{\frac{S}{Tt}}}^{\infty} f\left(t - \frac{Sx^2}{4T\xi^2}\right) e^{-\xi^2}\ d\xi$$

Although this procedure puts the answer in a simpler form, the reader should be advised that it can occasionally lead to difficulties when I(t) is given by an infinite series. This is because I(t) is differentiated in the process, and differentiation always slows the convergence rate of an infinite series. In fact, differentiating an infinite series can sometimes cause the series to diverge!

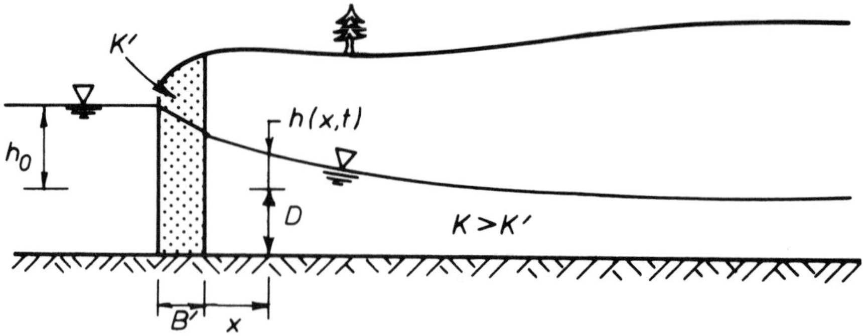

PROB. 3

3. Use the Laplace transform to solve the following problem:

$$T\frac{\partial^2 h}{\partial x^2} = S\frac{\partial h}{\partial t}, \quad (0 < x < \infty,\ 0 < t < \infty,\ T = KD)$$

$$h(x, 0) = 0, \quad (0 \leqq x < \infty)$$

$$h(\infty, t) = 0, \quad (0 \leqq t < \infty)$$

$$\frac{\partial h(0, t)}{\partial x} - \varepsilon h(0, t) = -\varepsilon h_0, \quad (0 \leqq t \leqq \infty,\ \varepsilon = \frac{K'/K}{B'},$$

$$h_0 = \text{constant})$$

Explain the physical significance of the boundary condition at $x = 0$ and show that the answer can be written in the form

$$\frac{h(x, t)}{h_0} = 1 - (\varepsilon x)\frac{2}{\pi}\int_0^\infty \frac{e^{-\left(\frac{Tt}{Sx^2}\right)\xi^2}}{\xi(\xi^2 + \varepsilon^2 x^2)}[(\varepsilon x)\sin(\xi) + \xi\cos(\xi)]\,d\xi$$

4. The rainfall recharge of an aquifer with the shape of an infinite strip is described by the solution of the following problem:

$$T\frac{\partial^2 h}{\partial x^2} = S\frac{\partial h}{\partial t} - R, \quad (-\ell < x < \ell,\ 0 < t < \infty,\ R = \text{constant})$$

$$h(\pm\ell, t) = 0, \quad (0 \leqq t < \infty)$$

$$h(x, 0) = 0, \quad (-\ell \leqq x \leqq \ell)$$

Use the Laplace transform to obtain the solution in the form

$$h = \frac{R}{2T}(\ell^2 - x^2) + \frac{2R\ell^2}{T}\sum_{n=1}^{\infty}\frac{(-1)^n}{\alpha_n^3}e^{-\left(\frac{Tt}{S\ell^2}\right)\alpha_n^2}\cos\left(\frac{x}{\ell}\alpha_n\right)$$

in which $\alpha_n = (2n - 1)\pi/2$.

5. As indicated in the text, Eq. 8.61 gives the Laplace transform of the leaky aquifer well function if R and f(t) are replaced with the constants Q and unity, respectively. If one also sets $K'/B' = 0$, it becomes the Laplace transform of the exponential integral (the "well function" in hydrological literature). The expansion of $K_0(z)$ about the origin is given by

$$K_0(z) = -\tfrac{1}{2}\left(1 + \frac{z^2}{4} + \cdots\right) \ln\left(\frac{z^2 e^{2\gamma}}{4}\right) + \frac{z^2}{4} + \cdots\cdots$$

in which γ = Euler's constant = 0.5772 $\cdots$. In addition, $K_0(z)$ is known to be an anlytic function that vanishes as $|z| \to 0$ and has no poles for $|\mathrm{Arg}\{z\}| < \pi/2$. Hence, use the integration path shown in Fig. 8.2 to obtain an asymptotic expansion, for large t, for the well function. You will need to use some of the following integrals, which are given by Carslaw and Jaeger (1959):

$$\frac{1}{2\pi i}\int_{-\infty}^{(0+)} e^{pt}\,\frac{dp}{p^{n+1}} = \frac{t^n}{n!} \quad \text{for } n = 0, 1, 2, 3 \cdots$$

$$= 0 \quad \text{for } n = -1, -2, -3, \cdots$$

$$\frac{1}{2\pi i}\int_{-\infty}^{(0+)} p^n e^{pt} \ln(Ap)\,dp = \frac{(-1)^{n+1} n!}{t^{n+1}} \quad \text{for } n = 0, 1, 2, \cdots$$

$$= \ln\left(\frac{A}{te^{\gamma}}\right) \quad \text{for } n = -1$$

$$\frac{1}{2\pi i}\int_{-\infty}^{(0+)} e^{pt}\,[\ln(Ap)]^n\,dp = \frac{2}{t}\ln\left(\frac{te^{\gamma}}{A}\right) \quad \text{for } n = 2$$

$$= \frac{\pi^2}{2t} - \frac{3}{t}\left[\ln\left(\frac{te^{\gamma}}{A}\right)\right]^2 \quad \text{for } n = 3$$

The integrals are calculated around the path FEDC in Fig. 8.2, and your answer should agree with Eq. 8.93.

6. Set $T = .1\ \mathrm{m^2/s}$, $S = 10^{-4}$ and $r = r_0 = .2$ m in Eq. 8.91. Calculate the value of t, in seconds, for which the second term is 1% of the magnitude of the first term. Then repeat the calculation using $S = .1$ and comment upon the results.

Appendix 1
Trigonometric and Hyperbolic Functions with Complex Arguments

Equation 7.48 can be used to obtain the result

$$\cos(z) = \frac{1}{2}(e^{iz} + e^{-iz}) \tag{1}$$

$$\sin(z) = \frac{1}{2i}(e^{iz} - e^{-iz}) \tag{2}$$

This leads in a natural way to the following definitions for the hyperbolic functions:

$$\cosh(z) = \frac{1}{2}(e^{z} + e^{-z}) \tag{3}$$

$$\sinh(z) = \frac{1}{2}(e^{z} - e^{-z}) \tag{4}$$

The other hyperbolic functions, tanh(z), sech(z), csch(z) and coth(z) are defined as sinh(z)/cosh(z), 1/cosh(z), 1/sinh(z) and 1/tanh(z), respectively.

Replacing z with iz in Eq. 1 - 2 and using Eqs. 3 - 4 gives

$$\cos(iz) = \cosh(z) \tag{5}$$

$$\sin(iz) = i \sinh(z) \tag{6}$$

Replacing z with iz in Eqs. 5 - 6 also gives

$$\cosh(iz) = \cos(z) \tag{7}$$

$$\sinh(iz) = i \sin(z) \tag{8}$$

Equations 5 - 6 can be used to derive a large number of identities for the hyperbolic functions from the more familiar trigonometric identities. For example, Eqs. 5 - 6 and the identities

$$\cos(x + iy) = \cos(x) \cos(iy) - \sin(x) \sin(iy)$$

$$\sin(x + iy) = \sin(x) \cos(iy) + \sin(iy) \cos(x)$$

lead to the new identities

$$\cos(x + iy) = \cos(x) \cosh(y) - i \sin(x) \sinh(y) \tag{9}$$

$$\sin(x + iy) = \sin(x) \cosh(y) + i \sin(y) \cos(x) \tag{10}$$

Likewise, Eqs. 5 - 6 give

$$\cosh(x + iy) = \cos(-y + ix) = \cos(y) \cosh(x) + i \sin(y) \sinh(x) \tag{11}$$

$$\sinh(x + iy) = \frac{1}{i} \sin(-y + ix) = \cos(y) \sinh(x) + i \sin(y) \cosh(x) \tag{12}$$

Equations 9 - 12 make it possible to compute the trigonometric and hyperbolic functions with complex arguments in terms of trigonometric and hyperbolic functions with real arguments.

Other identities for the hyperbolic functions can be obtained by replacing z with iz in trigonometric identities. For example, this procedure causes the identities

$$\cos^2(z) + \sin^2(z) = 1$$

$$\cos(z_1 + z_2) = \cos(z_1)\cos(z_2) - \sin(z_1)\sin(z_2)$$

$$\sin(z_1 + z_2) = \sin(z_1)\cos(z_2) + \sin(z_2)\cos(z_1)$$

to take the form

$$\cosh^2(z) - \sinh^2(z) = 1 \qquad (13)$$

$$\cosh(z_1 + z_2) = \cosh(z_1)\cosh(z_2) + \sinh(z_1)\sinh(z_2) \qquad (14)$$

$$\sinh(z_1 + z_2) = \sinh(z_1)\cosh(z_2) + \sinh(z_2)\cosh(z_1) \qquad (15)$$

Obviously, the possibilities are endless, and the reader is advised to remember Eqs. 1 - 4 and the method of derivation rather than specific formulae like Eqs. 5 - 15.

The inverse trigonometric and hyperbolic functions are multiple-valued and can be calculated in terms of logarithms. For example,

$$t = \cosh^{-1}(z)$$

can be rewritten as

$$z = \cosh(t) = \frac{1}{2}(e^t + e^{-t})$$

Multiplying both sides by $2e^t$ gives

$$(e^t)^2 - 2z(e^t) + 1 = 0$$

This is a quadratic equation that can be solved, by completing the square, to obtain

$$e^t = z + \sqrt{z^2 - 1}$$

Thus, taking the logarithm of both sides gives

$$t = \cosh^{-1}(z) = \ln(z + \sqrt{z^2 - 1}) \qquad (16)$$

The correct sign in front of $\sqrt{z^2 - 1}$ for real values of z must always be fixed by choosing a branch for any particular problem.

The same procedure can be applied to the inverse trigonometric functions. For example, it can be shown that

$$\sin^{-1}(z) = \frac{1}{i}\ln(iz + \sqrt{1 - z^2}) \qquad (17)$$

Here again, the reader is advised to derive these formulae as he needs them rather than trying to memorize the results.

Appendix 2
The Numerical Calculation of Singular Integrals

We will consider the calculation of integrals that have singularities in either the integrand or first derivative of the integrand at end points of finite integration intervals. Examples include

$$\int_0^{\pi/2} \frac{\cos(x)}{x^{1/3}}\,dx, \quad \int_0^2 \frac{\ln(x)}{\sqrt{4+x}}\,dx, \quad \int_1^2 x^2\sqrt{2-x}\,dx$$

The first two integrals have singularities in the integrand at the lower integration limit, and the third integral has a singularity in the first derivative of the integrand at the upper integration limit. These integrals have integrands that vary extremely rapidly in the neighborhood of the singularities, and an attempt to calculate their value with standard quadrature formulas, such as the trapezoidal rule or Simpson's rule, can lead to diasterous results.

Series expansions about the singularity furnish one way to compute these integrals. As an example, if f(x) has the expansion

$$f(x) = \sum_{n=0}^{\infty} \frac{f^n(a)}{n!}(x-a)^n \tag{1}$$

then multiplying both sides of Eq. 1 by $(x - a)^{-\alpha}$ and integrating gives

$$\int_a^b \frac{f(x)}{(x-a)^{\alpha}}\,dx = \sum_{n=0}^{\infty} \frac{f^n(a)}{n!}\,\frac{(b-a)^{n+1-\alpha}}{(n+1-\alpha)} \quad \text{for } \alpha < 1 \tag{2}$$

This method is most useful when the expansion for f(x) is readily calculated and when this expansion converges rapidly. The rate of convergence can be judged qualitatively by replacing x with the complex variable z and noting the location in the z plane of the singularities and branch points of f(z). A circle which has point z_0 as its center and which passes through the singularity or branch point of f(z) that lies closest to point z_0 defines the region in which the expansion of f(z) in powers of $(z - z_0)$ converges*, and the integration interval $a \leq x \leq b$ must lie well within this circle of convergence in order to obtain a rapidly convergent expansion for the integral. The method has the advantage of allowing the accuracy of approximation to be estimated from the magnitude of succeeding terms in the series.

As a numerical example, replacing x with z in

*See Nehari (1952) for a proof of this fact.

$$f(x) = \frac{1}{\sqrt{2 + x^2}} \tag{3}$$

shows that f(z) has two singularities at the branch points $z = \pm 2i$. Thus, the expansion

$$f(x) = \frac{1}{\sqrt{2}}\left[1 + \sum_{n=1}^{\infty} (-1)^n \frac{1.3.5\cdots(2n-1)}{n!}\left(\frac{x}{2}\right)^{2n}\right] \tag{4}$$

has a circle of convergence with its center at the origin and which passes through the points $z = \pm 2i$. Since the interval $0 \leq x \leq 1$ lies well within this circle of convergence, a termwise integration of the expansion for $f(x)/\sqrt{x}$ should give a rapidly converging series expansion for the following integral:

$$\int_0^1 \frac{dx}{\sqrt{x(2 + x^2)}} = \sqrt{2}\left[1 + \sum_{n=1}^{\infty} (-1)^n \frac{1.3.5\cdots(2n-1)}{n!2^{2n}\ (4n+1)}\right] \tag{5}$$

The first few terms of this expansion give

$$\int_0^1 \frac{dx}{\sqrt{x(2 + x^2)}} = \sqrt{2}\left(1 - \frac{1}{20} + \frac{1}{96} - \cdots\right) \simeq 1.36 \tag{6}$$

The fourth term in the expansion has a magnitude of $.00425\cdots$, which shows that three terms give a result that is in error by about 0.3 per cent. Difficulties with this method occur when f(x) is complicated or when an endpoint of the integration interval lies relatively close to the circle of convergence for the expansion of f(z).

A second computational method is known as "subtracting the singularity." For example, if f(x) has the behaviour near x = a of

$$f(x) \sim a_1\phi_1(x) + a_2\phi_2(x) + a_3\phi_3(x) + \cdots \tag{7}$$

in which a singularity occurs in $\phi_1(x)$ or its first derivative, then

$$\int_a^b f(x)\,dx = \int_a^b F(x)\,dx + \sum_{n=1}^{N} a_n \int_a^b \phi_n(x)\,dx \tag{8}$$

in which

$$F(x) = f(x) - \sum_{n=1}^{N} a_n\phi_b(x) \tag{9}$$

The value of N is chosen so that ϕ_{N+1} is the first term with a zero derivative at the singularity. A standard quadrature formula can be used to compute the first integral since F(x) and its first derivative are finite for $a \leq x \leq b$. (The computer must be told, however, to set F(a) = 0 or error messages will result!) The integrals that appear in the sum are usually of simple form and are computed exactly.

For a numerical example, consider the computation of

$$I = \int_0^1 \frac{dx}{\sqrt{(1 - x^2)(1 - .5x^2)}} = K(\sqrt{.5}) = 1.8541\cdots\cdots \qquad (10)$$

in which K = complete elliptic integral of the first kind. The integrand has a singularity at x = 1, where it has an expansion of the form

$$\frac{1}{\sqrt{(1 - x^2)(1 - .5x^2)}} \sim (1 - x)^{-\frac{1}{2}} - \frac{3}{4}(1 - x)^{\frac{1}{2}} + \cdots\cdots \qquad (11)$$

Since the third term in the expansion has a zero derivative at x = 1, we will write the identity

$$I = \int_0^1 F(x)\,dx + \int_0^1 [(1 - x)^{-\frac{1}{2}} - \frac{3}{4}(1 - x)^{\frac{1}{2}}]\,dx \qquad (12)$$

in which

$$F(x) = \frac{1}{\sqrt{(1 - x^2)(1 - .5x^2)}} - [(1 - x)^{-\frac{1}{2}} - \frac{3}{4}(1 - x)^{\frac{1}{2}}] \qquad (13)$$

Values of F(x) at x = 0, .5 and 1 are .7500, .3505 and 0, respectively. Thus, computing the integral of F(x) with Simpson's rule and evaluating the singular integrals exactly gives

$$I \simeq \frac{.5}{3}[.7500 + 4(.3505) + 0] + 2 - \frac{1}{2} = 1.8587 \qquad (14)$$

The answer overestimates I by about .25 per cent. An improvement in accuracy can be obtained by using more nodes with a smaller spacing in calculating the integral of F(x).

A third computational method consists of constructing specialized quadrature formulas. In general terms, a quadrature formula for the integral

$$\int_a^b f(x)\,\phi(x)\,dx$$

in which $\phi(x)$ contains any singularities and f(x) is well-behaved for $a \leq x \leq b$ is obtained by approximating f(x) with a polynomial. The following integral comes from the definition of the Beta function and is sometimes helpful in constructing these quadrature formulas:

$$\int_a^b \frac{dx}{(x - a)^{\alpha}(b - x)^{\beta}} = \frac{\Gamma(1 - \alpha)\Gamma(1 - \beta)}{\Gamma(2 - \alpha - \beta)}(b - a)^{1-\alpha-\beta} \qquad (15)$$

It has been assumed in this equation that $\alpha < 1$ and $\beta < 1$, and Γ is the Gamma function that is widely tabulated in handbooks.

For an example, we will obtain a quadrature formula for

$$\int_0^1 \frac{f(x)}{\sqrt{x}}\,dx$$

in which f(x) is a well-behaved function that we will approximate with a second-degree polynomial for $0 \leq x \leq 1$. We will obtain the quadrature formula in the form

$$\int_0^1 \frac{f(x)}{\sqrt{x}}\,dx = W_1 f(0) + W_2 f(.5) + W_3 f(1) \qquad (16)$$

in which the W_i's are weighting constants that must be calculated. Since f(x) appears linearly on both sides of Eq. 16, setting f(x) = 1 gives an equation that holds when f(x) is any constant multiple of 1. Likewise, setting f(x) = x on both sides of Eq. 16 gives an equation that holds when f(x) is any constant multiple of x, and setting f(x) = x(1 - x) gives a result that holds when f(x) is any constant multiple of x(1 - x). Thus, setting f(x) = 1, x and x(1 - x) in both sides of Eq. 16 gives three equations to determine values for W_1, W_2 and W_3:

$$2 = W_1 + W_2 + W_3. \qquad (17a, b, c)$$

$$\frac{2}{3} = \frac{1}{2} W_2 + W_3$$

$$\frac{4}{15} = \frac{1}{4} W_2$$

The particular functions chosen for f(x) were selected so that they contained a zero, first and second-degree polynomial and so that the right side of the equations for W_i have a form that allows the equations to be solved easily. The solution for W_i gives

$$(W_1, W_2, W_3) = \left(\frac{12}{15}, \frac{16}{15}, \frac{2}{15}\right) \qquad (18)$$

Thus, the quadrature formula is

$$\int_0^1 \frac{f(x)}{\sqrt{x}}\,dx = \frac{1}{15}\,[12f(0) + 16f(.5) + 2f(1)] \qquad (19)$$

Since Eq. 19 holds for all constant multiples of 1, x and x(1 - x), and since all second-degree polynomials can be constructed from a linear combination of these polynomials, we see that the quadrature formula becomes exact when f(x) is any second-degree polynomial. Application of this quadrature formula to the first example, with $f(x) = (2 + x^2)^{-1/2}$, gives

$$\int_0^1 \frac{dx}{\sqrt{x(2 + x^2)}} \simeq \frac{1}{15}\,[12(.707) + 16(.667) + 2(.577)] = 1.354 \qquad (20)$$

The series expansion method considered initially gives a more accurate value of 1.355...., which means that the error of approximation is about .07 per cent.

REFERENCE

Nehari, Z. 1952. Conformal Mapping, McGraw-Hill Book Company Inc., New York, pp. 98-101.

Appendix 3
Worked Example Problems

CHAPTER I

1. Calculate $\overline{\nabla}\phi$ for $\phi = x^2 + y^2 + z^2$ and show that $\overline{\nabla}\phi$ is perpendicular to surfaces of constant ϕ. Then use Eq. 1.8 to calculate $\partial\phi/\partial r$ and check this result by noting that $\phi = r^2$.

Solution:

$$\overline{\nabla}\phi = \left(\hat{i}\frac{\partial}{\partial x} + \hat{j}\frac{\partial}{\partial y} + \hat{k}\frac{\partial}{\partial z}\right)(x^2 + y^2 + z^2) = 2x\hat{i} + 2y\hat{j} + 2z\hat{k} = 2\overline{r}$$

in which $\overline{r} = x\hat{i} + y\hat{j} + z\hat{k}$. Thus, since surfaces of constant ϕ are concentric spheres with centers at the origin, and since $\overline{r}$ is a vector drawn from the origin to any point in space, then $\overline{\nabla}\phi = 2\overline{r}$ is a vector that is perpendicular to the surfaces of constant ϕ. Furthermore, since $\hat{e}_r = \overline{r}/|\overline{r}|$, Eq. 1.8 gives

$$\frac{d\phi}{dr} = \hat{e}_r.\overline{\nabla}\phi = \frac{\overline{r}}{|\overline{r}|}\ .\ 2\overline{r} = 2\,\frac{\overline{r}.\overline{r}}{|\overline{r}|} = 2|\overline{r}| = 2\sqrt{x^2 + y^2 + z^2} = 2r$$

Since $\phi = r^2$, this same result can be obtained by calculating $d\phi/dr$ by direct differentiation.

2. Let $\forall$ be a spherical volume with its center at the origin and with an external surface, S, given by the equation $r = R$ in which R is the constant sphere radius. Show that Eq. 1.21 holds when $\overline{F} = \overline{r}$.

Solution:

Since $\hat{e}_n = \overline{r}/|\overline{r}|$ on $r = R$,

$$\int_S \overline{F}.\hat{e}_n\, dS = \int_S \overline{r}\,.\,\frac{\overline{r}}{|\overline{r}|}\, dS = \int_S |\overline{r}|\, dS = R\int_S dS = 4\pi R^3$$

Also, since

$$\overline{\nabla}.\overline{F} = \left(\hat{i}\frac{\partial}{\partial x} + \hat{j}\frac{\partial}{\partial y} + \hat{k}\frac{\partial}{\partial z}\right).\ (x\hat{i} + y\hat{j} + z\hat{k}) = 1 + 1 + 1 = 3$$

the volume integral is easily calculated to be

$$\int_\forall \overline{\nabla}.\overline{F}\, d\forall = \int_\forall 3\, d\forall = 3\left(\frac{4}{3}\pi R^3\right) = 4\pi R^3$$

Hence, Eq. 1.21 is satisfied with this particular choice for $\overline{F}$, $\forall$ and S.

3. Use Eq. 1.20 and Fig. 1.4 to derive the equation of hydrostatic equilibrium in a motionless fluid.

3. Ctd. <u>Solution</u>:

Let P = fluid pressure (positive when P is a compressive stress), ρ = fluid mass density and $\bar{g}$ = vector with a magnitude equal to the gravitational constant, g, and which points towards the center of the earth. Then the condition that the summation of external forces on $\forall$ in Fig. 1.4 equals zero is expressed by the mathematical equation

$$-\int_S P\hat{e}_n \, dS + \int_{\forall} \rho\bar{g} \, d\forall = 0$$

The minus sign in front of the first integral comes from the fact that a compressive stress on S creates a force on S in the direction of the inward normal. Thus, using Eq. 1.20 upon the first integral allows this equation to be rewritten in the form

$$\int_{\forall} (-\bar{\nabla}P + \rho\bar{g}) \, d\forall = 0$$

Since this equation must hold for all arbitrary choices for $\forall$, the integrand must vanish at all points within $\forall$ and

$$-\bar{\nabla}P + \rho\bar{g} = 0$$

CHAPTER II

1. A particular kind of pollutant is known to decay to a harmless level in 30 days. What is the minimum distance from the seacoast to a point where the pollutant can be buried safely in the sand if $\sigma = 0.005$, K = 0.1 mm/s and the constant water-table gradient is 0.002 m/m toward the seacoast. Neglect dispersion in this calculation.

<u>Solution</u>:

Darcy's law gives a flux velocity of

$$u = -K\frac{\partial h}{\partial x} = (0.0001)\ (0.002) = 2 \times 10^{-7} \text{ m/s} = 0.0173 \text{ m/day}$$

This gives a constant pore velocity of

$$\frac{dx}{dt} = u/\sigma = 0.0173/0.005 = 3.46 \text{ m/day}$$

Hence, an easy integration gives

$$\Delta x = \frac{dx}{dt}\Delta t = (3.46)\ (30) = 104 \text{ m}$$

2. A field experiment gives $T = 0.002\ m^2/s$ and $S = 1 \times 10^{-4}$ for a confined aquifer that is 4 m thick. Calculate the average permeability for the aquifer and the ratio of α/β, which is a measure of the compressibility of the aquifer relative to the compressibility of the water. Assume that $\sigma = 0.1$, $\rho = 1000\ Kg/m^3$, $g = 9.81\ m/s^2$, $\beta = 4.74 \times 10^{-10}\ m^2/N$ and that the total change in aquifer thickness, B', is negligible.

Solution:

Eq. 2.38 shows that the average permeability of the aquifer is given by

$$K_{ave} = \frac{T}{B} = \frac{0.002}{4} = 0.0005\ m/s = 0.5\ mm/s$$

Since B' = 0, Eqs. 2.33, 2.34 and 2.49 show that

$$S = B_0 S_s = B_0 \rho_0 g\ [\alpha(1 - \sigma_0) + \sigma_0 \beta]$$

Hence, solution for α gives

$$\alpha = \left[\frac{S}{B_0 \rho_0 g} - \sigma_0 \beta\right] / (1 - \sigma_0)$$

$$= \left[\frac{1 \times 10^{-4}}{4(1000)\ 9.81} - .1(4.74 \times 10^{-10})\right] / (1 - .1) = 2.78 \times 10^{-9}\ m^2/N$$

This gives the ratio $\alpha/\beta = 5.87$, which suggests that the aquifer is almost six times more compressible than water.

3. Derive Eq. 2.62 for the case R = 0 by making use of the fact that a fluid particle on a free surface remains on the free surface for a finite time period when R = 0.

Solution:

Since the particle upon the free surface remains at atmospheric pressure,

$$h(x, y, z, t) = y$$

in which the fluid particle has the coordinates [x(t), y(t), z(t)]. Thus, using the chain rule to differentiate both sides with respect to time gives

$$\frac{\partial h}{\partial x}\frac{dx}{dt} + \frac{\partial h}{\partial y}\frac{dy}{dt} + \frac{\partial h}{\partial z}\frac{dz}{dt} + \frac{\partial h}{\partial t} = \frac{dy}{dt}$$

or

$$(\overline{\nabla} h - \hat{j}) \cdot \frac{d\overline{r}}{dt} + \frac{\partial h}{\partial t} = 0$$

But $d\overline{r}/dt$ is the pore velocity of the fluid particle, and Eqs. 2.2 and 2.11 can be used to rewrite this equation in the form

$$-\frac{K}{\sigma}\overline{\nabla} h \cdot (\overline{\nabla} h - \hat{j}) + \frac{\partial h}{\partial t} = 0$$

Thus, multiplying through by σ and computing the dot products gives

$$K\ \overline{\nabla} h \cdot \overline{\nabla} h = K \frac{\partial h}{\partial y} + \sigma \frac{\partial h}{\partial t}$$

4. Use the Ghyben-Herzberg approximation to estimate the maximum elevation of the free surface above mean sea level for a lens of fresh water beneath an oceanic island if the lens has a maximum total thickness of 10 m. The drawing for problem 6 at the end of chapter III would be applicable for this problem if points A and B coincided and if the bottom impermeable boundary were not present.

Solution:

Since $\varepsilon \simeq 1/40$, Eq. 2.65 gives

$$h = D/40$$

in which D = distance between mean sea level and the bottom boundary of the fresh-water lens. Since the total thickness of the lens is

$$B = h + D$$

the variable D can be eliminated between these two equations to obtain

$$h = B/41 = 10/41 = 0.243 \text{ m}$$

CHAPTER III

1. Work problem 6 at the end of chapter III for the axisymmetric case in which the aquifer lies beneath a circular island of radius r_0.

Solution:

(a) Since T and R are constants and Q = 0, and since

$$\bar{\nabla}.\bar{\nabla}h = \nabla^2 h = \frac{1}{r}\frac{\partial}{\partial r}\left(r\frac{\partial h}{\partial r}\right) + \frac{1}{r^2}\frac{\partial^2 h}{\partial \theta^2}$$

in polar coordinates, Eqs. 3.34 and 3.35 reduce to

$$\frac{1}{r}\frac{\partial}{\partial r}\left(r\frac{\partial h}{\partial r}\right) = -\frac{R}{T} \text{ for } 0 \leqq r < r_0$$

$$h = 0 \text{ on } r = r_0$$

Integration of the first equation gives

$$h = -\left(\frac{R}{T}\right)\frac{r^2}{4} + C_1 \ln(r) + C_2$$

Since h must be finite at r = 0, we must set $C_1 = 0$. Then the boundary condition at $r = r_0$ gives

$$0 = -\left(\frac{R}{T}\right)\frac{r_0^2}{4} + C_2$$

which can be solved for C_2 and substituted into the solution for h to obtain

1. Ctd.

$$h = \left(\frac{R}{T}\right) \frac{\left(r_0^2 - r^2\right)}{4} \quad \text{for } 0 \leqq r \leqq r_0$$

Finally, the interface toes will meet at r = 0 when Eq. 3.37 is satisfied:

$$\varepsilon D = \left(\frac{R}{T}\right) \frac{r_0^2}{4}, \quad \left(\varepsilon = \frac{\rho_s}{\rho} - 1 \simeq \frac{1}{40}\right)$$

Thus, the solution of the linearized problem is

$$R = \frac{4\varepsilon DT}{r_0^2}$$

(b) The non-linear formulation given by Eqs. 3.41 and 3.42 reduces for this problem to

$$\frac{1}{r}\frac{\partial}{\partial r}\left(r\frac{\partial\phi}{\partial r}\right) = -\frac{R}{K} \quad \text{for } 0 \leqq r < r_0$$

$$\phi = 0 \text{ on } r = r_0$$

This is identical with the problem solved for part (a), so that ϕ is given by

$$\phi = \left(\frac{R}{K}\right) \frac{\left(r_0^2 - r^2\right)}{4} \quad \text{for } 0 \leqq r \leqq r_0$$

The interface toes will meet at r = 0 when Eq. 3.44 is satisfied:

$$\frac{\varepsilon}{2}(1 + \varepsilon)\, D^2 = \left(\frac{R}{K}\right) \frac{r_0^2}{4}$$

Thus, the solution of the non-linear problem is

$$R = \frac{1}{2} \frac{4\varepsilon D\ (KD)}{r_0^2}(1 + \varepsilon)$$

Since $1 + \varepsilon \simeq 1$, and since we would normally set $T \simeq KD$ in the linearized solution, we see that the linearized solution over estimates R by a factor of about two.

2. Carry out a uniqueness proof for the problem described by Eqs. 3.53 - 3.57 and Fig. 3.3.

Solution:

Assume that two different solutions, h_1 and h_2, satisfy equations 3.53 - 3.57. Since the equations are linear, and since K, σ and R are identical for the solutions h_1 and h_2, the difference $h = h_1 - h_2$ satisfies the homogeneous set of equations

$$\overline{\nabla}.(K\overline{\nabla}h) = 0$$

$$h(x, 0, 0) = 0$$

2.
Ctd.

$$\frac{\partial h}{\partial y} = 0 \text{ on } y = -D$$

$$K \frac{\partial h}{\partial y} + \sigma \frac{\partial h}{\partial t} = 0 \text{ on } y = 0 \text{ for } \frac{L}{2} < |x| < \infty$$

$$(K + R) \frac{\partial h}{\partial y} + \sigma \frac{\partial h}{\partial t} = 0 \text{ on } y = 0 \text{ for } 0 \leqq |x| < \frac{L}{2}$$

The uniqueness proof must show that h = 0 is the only possible solution of this set of equations.

Multiplication of the first equation by h and manipulation of the deriva- gives

$$\overline{\nabla}.(Kh\ \overline{\nabla}h) - K\ \overline{\nabla}h.\overline{\nabla}h = 0$$

Thus, integrating throughout the solution domain and using the divergence theorem gives

$$\int_{\Gamma} Kh \frac{dh}{dn}\, ds = \int_{A} K\ \overline{\nabla}h.\overline{\nabla}h \quad dA$$

in which the boundary Γ consists of the four straight lines y = 0, y = -D and $x = \pm d$ as $d \to \infty$. If we assume that $h = h_1 - h_2$ vanishes as $|x| \to \infty$, then the contribution to the line integral from the boundaries $x = \pm d$ vanishes as $d \to \infty$ and we are left with the result

$$\left[\int_{-\infty}^{\infty} Kh \frac{\partial h}{\partial y}\, dx\right]_{y=0} - \left[\int_{-\infty}^{\infty} Kh \frac{\partial h}{\partial y}\, dx\right]_{y=-D} = \int_{A} K\ \overline{\nabla}h.\overline{\nabla}h\ dA$$

Thus, use of the boundary conditions on y = 0 and y = -D gives

$$-\int_{-\infty}^{-L/2} \sigma h \frac{\partial h}{\partial t}\, dx - \int_{-L/2}^{L/2} \frac{K\sigma}{K + R}\, h \frac{\partial h}{\partial t}\, dx - \int_{L/2}^{\infty} \sigma h \frac{\partial h}{\partial t}\, dx = \int_{A} K\ \overline{\nabla}h.\overline{\nabla}h\ dA$$

in which the integrals on the left are calculated along the path y = 0. Finally, since σ and K do not depend upon t, integrating both sides from 0 to t and using the initial condition h(x, 0, 0) = 0 gives

$$\int_{-\infty}^{-L/2} \frac{\sigma}{2} h^2(x, 0, t)\, dx + \int_{-L/2}^{L/2} \frac{K\sigma}{2(K + R)} h^2(x, 0, t)\, dx + \int_{L/2}^{\infty} \frac{\sigma}{2} h^2\ (x, 0, t)\, dx$$

$$+ \int_0^t \int_A K\ \overline{\nabla}h.\overline{\nabla}h\ dA\ dt = 0$$

Since the integrand of each integral is positive definite, we see that

$$\overline{\nabla}h = 0 \text{ for all } x, y \text{ in A with } 0 < t < \infty$$

$$h(x, 0, t) = 0 \text{ for } -\infty < x < \infty \text{ and } 0 < t < \infty$$

2. Ctd. The first equation can be integrated to obtain $h = h_1 - h_2$ = constant for all x, y in A with $0 < t < \infty$, and the second equation shows that the integration constant vanishes for $0 < t < \infty$. Thus, $h_1 = h_2$ for all x, y in A over the time period $0 < t < \infty$, and uniqueness has been shown. Note that it has been necessary to specify the initial condition, Eq. 3.54, only along the straight line $y = 0$ rather than throughout the entire solution domain. It has also been assumed that h vanishes fast enough as $|x| \to \infty$ to insure the existance of the integrals along the boundary $y = 0$ and over the region A.

CHAPTER IV

1. Use finite differences with five equally-spaced nodes to solve the following boundary-value problem:

$$\frac{\partial^2 h}{\partial x^2} = h - 1 \text{ for } 0 < x < 1$$

$$h(0) = 0$$

$$\frac{\partial h(1)}{\partial x} = 0$$

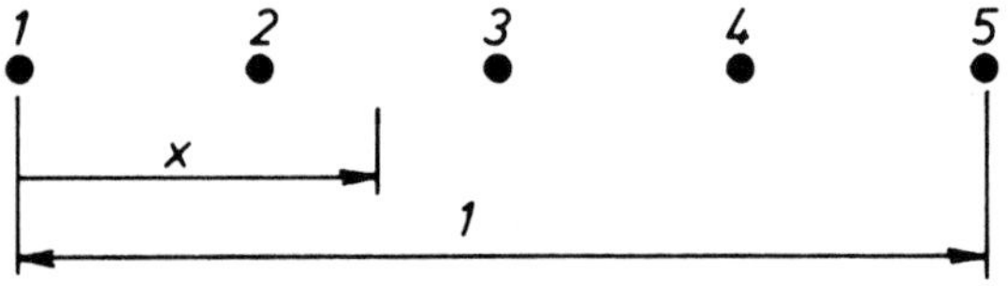

Solution:

We have a node spacing of $\Delta = 0.25$ and five unknown values of h at nodes 1 through 5. At nodes 2 through 4 we will write second-order, central-difference approximations to the ordinary differential equation, and at nodes 1 and 5 we will write exact and second-order approximations, respectively, for the two boundary conditions to close the system of five equations with five unknowns. For example, the central difference approximation given by Eq. 4.6 allows us to approximate the differential equation at node 2 with

$$\frac{h_3 + h_1 - 2h_2}{\Delta^2} = h_2 - 1, \quad (\Delta = 0.25)$$

and the boundary condition at $x = 0$ gives

$$h_1 = 0$$

At node 5 we will use the second-order, backward-difference approximation

1. Ctd. introduced in problem 1 of chapter IV to obtain

$$\frac{h_3 + 3h_5 - 4h_4}{2\Delta} = 0$$

In this way we obtain the following system of five equations in five unknowns:

$$\begin{bmatrix} 1 & 0 & 0 & 0 & 0 \\ -1 & 2.0625 & -1 & 0 & 0 \\ 0 & -1 & 2.0625 & -1 & 0 \\ 0 & 0 & -1 & 2.0625 & -1 \\ 0 & 0 & 1 & -4 & 3 \end{bmatrix} \begin{bmatrix} h_1 \\ h_2 \\ h_3 \\ h_4 \\ h_5 \end{bmatrix} = \begin{bmatrix} 0 \\ .0625 \\ .0625 \\ .0625 \\ 0 \end{bmatrix}$$

A solution by iteration can be obtained by putting the off-diagonal terms on the right side and starting with the initial approximation vector (0, 0, 0, 0, 0) for the unknown values of h_i. Thus, the first equation, $h_1=0$, does not change any of the approximations for h_i. The second equation gives

$$h_2 = (.0625 + h_1 + h_3)/2.0625 = (.0625 + 0 + 0)/2.0625 = .0303$$

and the corresponding approximation for h_i becomes

(0, .0303, 0, 0, 0)

The third equation gives

$$h_3 = (.0625 + h_2 + h_4)/2.0625 = (.0625 + .0303 + 0)/2.0625 = .0450$$

and the corresponding approximation for h_i becomes

(0, .0303, .0450, 0, 0)

Continuing in this way, we find the fifth approximation for h_i, which completes one cycle of the iteration, to be

(0, .0303, .0450, .0521, .0545) (cycle 1)

Additional computations give approximations for h_i at the end of further cycles of

(0, .0521, .0808, .0959, .1009) (cycle 2)

(0, .0695, .1105, .1328, .1403) (cycle 3)

(0, .0839, .1354, .1639, .1735) (cycle 4)

(0, .1331, .2209, .2713, .2881) (cycle 10)

(0, .1557, .2603, .3207, .3409) (cycle 20)

(0, .1599, .2675, .3297, .3505) (cycle 30)

(0, .1606, .2688, .3314, .3522) (cycle 40)

(0, .1607, .2691, .3317, .3526) (cycle 50)

1. Ctd. The exact solution of the problem is given by

$$h = 1 - \frac{\cosh(1 - x)}{\cosh(1)}$$

which gives the exact solution vector

(0, .1610, .2692, .3316, .3519)

The relatively slow convergence rate is typical of problems in which the normal derivative is specified at one or more boundaries.

2. The alluvial plain in Fig. A.1 is bounded on the north and west by impermeable clay embankments. The other boundaries are along rivers that carry a constant flow, and the aquifer beneath the plain has an average transmissivity of 0.01 m^2/s. Water from the rivers is to be used to irrigate the plain, and the excess irrigation water will recharge the aquifer and cause a rise in the water table. What is the maximum, long-term recharge rate, R, that can be allowed if the water table is to rise not more than 1 m above its present position?

Solution:

The problem can be formulated as follows:

$$\nabla^2 h = -\frac{R}{T} \text{ for x, y in the aquifer}$$

$$\frac{dh}{dn} = 0 \text{ for x, y along the clay embankments}$$

$$h = 0 \text{ for x, y along the river boundary}$$

Note that the aquifer has been assumed homogeneous, that h will be the change in water level created by the uniformly distributed recharge rate R and that the problem is an inverse problem in which the unknown is the value for R that will create a specified maximum for h. This inverse problem, however, is easily solved because the problem is linear in both h and R. Thus, multiplying all of the equations by the same constant, λ, shows that if a recharge rate of R creates a water level change of h, then a recharge rate of λR will create a water level change of λh. The computer program shown in Figs. A.2 and A.3 gives the water table rise contours shown in Fig. A.4 for a recharge rate of $R = .1 \times 10^{-6}$ m/s = 263 mm/month. Since the maximum computed rise is 0.315 m at node 12, the value for R that will give a maximum water level rise of 1 m is

$$R = \frac{1 \text{ m}}{0.315 \text{ m}} \times 263 \text{ mm/month} = 835 \text{ mm/month}$$

This, of course, is the maximum rate at which recharge water could actually be allowed to enter the saturated portion of the aquifer. Evapotranspiration would probably permit the application of irrigation water on the ground surface to be two or three times this rate.

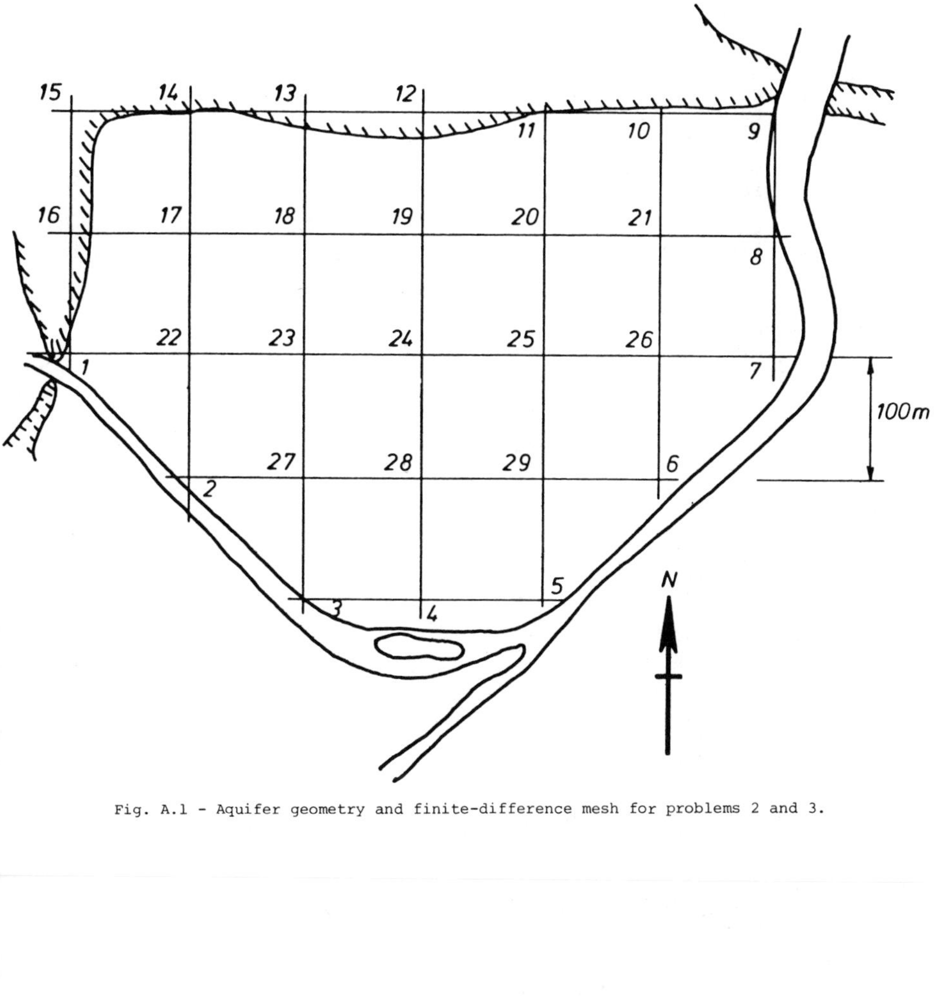

Fig. A.1 - Aquifer geometry and finite-difference mesh for problems 2 and 3.

```
C THIS PROGRAM SOLVES STEADY FLOW PROBLEMS
      DIMENSION ID(100,4),D(100),F(100),T(100),A0(100),A1(100),A2(100),
     1 A3(100),A4(100),B0(100),H(100),BK(100),HP(100),R(100),Q(100)
C DATA IS ENTERED AND COEFFICIENTS ARE CALCULATED
      READ(5,1000) NB,N,ERR

      WRITE(6,1500) NB,N,ERR

      DO 100 I=1,NB
      READ(5,2000) ID(I,1),ID(I,2),ALPHA,BETA,XN,YN,DP,D(I),F(I),T(I)
      WRITE(6,2500) I,ID(I,1),ID(I,2),ALPHA,BETA,XN,YN,DP,D(I),F(I),T(I)
      A=(ALPHA+BETA*DP)/D(I)
      A1(I)=A*YN
      A2(I)=A*XN
      A0(I)=A1(I)+A2(I)+BETA
  100 H(I)=0.
      N1=NB+1
      WRITE(6,2700)
      DO 200 I=N1,N
      READ(5,3000)  (ID(I,J),J=1,4),T(I),D(I),BK(I),HP(I),R(I),Q(I)
  200 WRITE(6,3500) I,(ID(I,J),J=1,4),T(I),D(I),BK(I),HP(I),R(I),Q(I)
      DO 300 I=N1,N
      I1=ID(I,1)
      I2=ID(I,2)
      I3=ID(I,3)
      I4=ID(I,4)
      A1(I)=.5*(T(I)+T(I1))
      A2(I)=.5*(T(I)+T(I2))
      A3(I)=.5*(T(I)+T(I3))
      A4(I)=.5*(T(I)+T(I4))
      A0(I)=A1(I)+A2(I)+A3(I)+A4(I)+BK(I)*D(I)**2
      B0(I)=-Q(I)+(BK(I)*HP(I)+R(I))*D(I)**2
  300 H(I)=0.
C THE EQUATIONS ARE SOLVED BY ITERATION
   50 ERRC=0.
      DO 400 I=1,NB
      I1=ID(I,1)
      I2=ID(I,2)
      A=(F(I)+A1(I)*H(I1)+A2(I)*H(I2))/A0(I)
      ERRC=ERRC+(A-H(I))**2
  400 H(I)=A
      DO 500 I=N1,N
      I1=ID(I,1)
      I2=ID(I,2)
      I3=ID(I,3)
      I4=ID(I,4)
      A=(B0(I)+A1(I)*H(I1)+A2(I)*H(I2)+A3(I)*H(I3)+A4(I)*H(I4))/A0(I)
      ERRC=ERRC+(A-H(I))**2
  500 H(I)=A
      ERRC=SQRT(ERRC)
      IF(ERRC.GT.ERR) GO TO 50
C THE SOLUTION IS WRITTEN OUT
      WRITE(6,4000) (I,I=1,10)
      DO 600 IJ=1,N,10
      ILAST=MIN0(N,(IJ+9))
      IFIRST=IJ-1
  600 WRITE(6,5000) IFIRST,(H(I),I=IJ,ILAST)
 1000 FORMAT(2I10,F10.8)
 1500 FORMAT(1H1,6H NB = ,I4,9X,4HN = ,I4,9X,6HERR = ,F10.8
     1///5H    I,8H ID(I,1),8H ID(I,2),6H ALPHA,6H  BETA,6H    XN,
     26H    YN,7H     DP,7H   D(I),7H   F(I),7H   T(I)/)
 2000 FORMAT(2I5,8F5.3)
 2500 FORMAT(1X,I4,2I8,2F6.2,2F6.3,3F7.1,F7.4)
 2700 FORMAT(1H1,4H   I,8H ID(I,1),8H ID(I,2),8H ID(I,3),8H ID(I,4),
     17H   T(I),7H   D(I),11H      BK(I),11H      HP(I),11H       R(I),
     211H       Q(I)/)
 3000 FORMAT(4I5,2F10.3,4E10.3)
 3500 FORMAT(1X,I4,4I8,F7.4,F7.1,4E11.3)
 4000 FORMAT(1H1,26H SOLUTION FOR H(I) FOLLOWS///1X,10I10)
 5000 FORMAT(1X,I4,10F10.4)
      END
```

Fig. A.2 - A computer program for steady flow written from the flow chart in Figs. 4.7 and 4.8.

```
NB =    16          N =    29          ERR = 0.00001000
```

I	ID(I,1)	ID(I,2)	ALPHA	BETA	XN	YN	DP,	D(I)	F(I)	T(I)
1	16	22	0.00	1.00	0.707	0.707	10.0	100.0	0.0	0.0100
2	22	27	0.00	1.00	0.682	0.731	10.0	100.0	0.0	0.0100
3	27	4	0.00	1.00	0.707	0.707	0.0	100.0	0.0	0.0100
4	28	5	0.00	1.00	0.000	1.000	23.0	100.0	0.0	0.0100
5	29	4	0.00	1.00	0.616	0.788	10.0	100.0	0.0	0.0100
6	26	29	0.00	1.00	0.707	0.707	19.0	100.0	0.0	0.0100
7	8	26	0.00	1.00	0.788	0.616	18.0	100.0	0.0	0.0100
8	7	21	0.00	1.00	0.940	0.342	7.0	100.0	0.0	0.0100
9	8	10	0.00	1.00	0.996	-0.087	0.0	100.0	0.0	0.0100
10	21	11	1.00	0.00	-0.035	0.999	3.0	100.0	0.0	0.0100
11	20	12	1.00	0.00	-0.105	0.995	0.0	100.0	0.0	0.0100
12	19	13	1.00	0.00	-0.174	0.985	-19.0	100.0	0.0	0.0100
13	18	14	1.00	0.00	0.087	0.995	-12.0	100.0	0.0	0.0100
14	17	15	1.00	0.00	0.000	1.000	0.0	100.0	0.0	0.0100
15	16	14	1.00	0.00	0.707	0.707	-29.0	100.0	0.0	0.0100
16	1	17	1.00	0.00	0.999	0.052	-15.0	100.0	0.0	0.0100

I	ID(I,1)	ID(I,2)	ID(I,3)	ID(I,4)	T(I)	D(I)	BK(I)	HP(I)	R(I)	Q(I)
17	18	14	16	22	0.0100	100.0	0.	0.	.100E-06	0.
18	19	13	17	23	0.0100	100.0	0.	0.	.100E-06	0.
19	20	12	18	24	0.0100	100.0	0.	0.	.100E-06	0.
20	21	11	19	25	0.0100	100.0	0.	0.	.100E-06	0.
21	8	10	20	26	0.0100	100.0	0.	0.	.100E-06	0.
22	23	17	1	2	0.0100	100.0	0.	0.	.100E-06	0.
23	24	18	22	27	0.0100	100.0	0.	0.	.100E-06	0.
24	25	19	23	28	0.0100	100.0	0.	0.	.100E-06	0.
25	26	20	24	29	0.0100	100.0	0.	0.	.100E-06	0.
26	7	21	25	6	0.0100	100.0	0.	0.	.100E-06	0.
27	28	23	2	3	0.0100	100.0	0.	0.	.100E-06	0.
28	29	24	27	4	0.0100	100.0	0.	0.	.100E-06	0.
29	6	25	28	5	0.0100	100.0	0.	0.	.100E-06	0.

SOLUTION FOR H(I) FOLLOWS

	1	2	3	4	5	6	7	8	9	10
0	0.0273	0.0187	0.0000	0.0306	0.0107	0.0282	0.0162	0.0106	0.0000	0.1663
10	0.2617	0.3149	0.3111	0.2846	0.2783	0.2719	0.2846	0.3134	0.3143	0.2673
20	0.1697	0.1686	0.2437	0.2614	0.2237	0.1344	0.1315	0.1638	0.1316	

Fig. A.3 - Output for the steady-flow computer program in Fig. A.2.

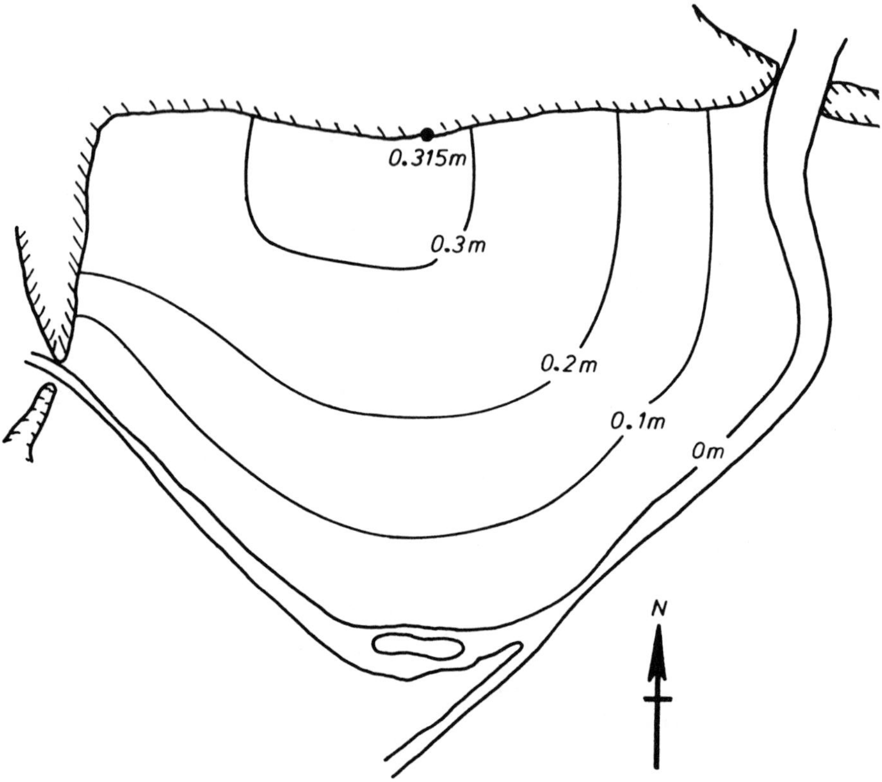

Fig. A.4 - Contours of water level rises calculated for a uniformly distributed recharge rate of $R = .1 \times 10^{-6}$ m/s = 263 mm/month.

3. Obtain the unsteady solution for the previous problem if the storage coefficient, S, is 0.05. Use this solution to estimate the time that is required for the maximum water level rise of 1 m to be approached at node 12.

Solution:

The problem can be formulated as follows:

$$\nabla^2 h = \frac{S}{T}\frac{\partial h}{\partial T} - \frac{R}{T} \text{ for x, y in the aquifer}$$

$$\frac{dh}{dn} = 0 \text{ for x, y along the clay embankments}$$

$$h = 0 \text{ for x, y along the river boundary}$$

$$h = 0 \text{ for } t = 0 \text{ for x, y in the aquifer}$$

The computer program shown in Figs. A.5 and A.6 gives the variation in h at node 12 that is shown in Fig. A.7. Thus, Fig. A.7 shows that the water level at node 12 is within 3% of its steady-state value of 1 m at the end of one week.

4. A Hele-Shaw model similar to the one shown in Fig. 4.13 is used to model flow through an embankment when the reservoir level next to the embankment is suddenly decreased to zero. The maximum free surface elevation within the embankment requires 6 minutes in the model to drop to a negligibly small value. What is the corresponding time for the prototype if the model-prototype scale ratio is 1:20 and the porosity of the prototype is 0.02. Permeabilities for the model and prototype are 5 mm/s and 0.03 mm/s, respectively.

Solution:

Equation 4.83 shows that similarity between model and prototype requires

$$\left(\frac{Kt}{\sigma L}\right)_{model} = \left(\frac{Kt}{\sigma L}\right)_{prototype}$$

Using the subscripts m and p to denote model and prototype, respectively, allows this equation to be manipulated into the form

$$t_p = \frac{\sigma_p}{\sigma_m}\frac{L_p}{L_m}\frac{K_m}{K_p} t_m$$

Thus, substitution gives the result

$$t_p = \frac{.02}{1}\frac{20}{1}\frac{5}{.03} 6 = 400 \text{ minutes} = 6.7 \text{ hrs}$$

Note that the porosity of the model has been taken as unity.

```
C THIS PROGRAM SOLVES UNSTEADY FLOW

      DIMENSION ID(100,4),D(100),F(100),T(100),A0(100),A1(100),A2(100),
     1 A3(100),A4(100),B0(100),H(100),BK(100),HP(100),R(100),Q(100),
     2 S(100),HLAST(100),TIMES(50)
C DATA IS ENTERED AND COEFFICIENTS ARE CALCULATED
      READ(5,1000) NB,N,NT,ERR

      WRITE(6,1500) NB,N,NT,ERR

      DO 100 I=1,NB
      READ(5,2000) ID(I,1),ID(I,2),ALPHA,BETA,XN,YN,DP,D(I),T(I)
      WRITE(6,2500) I,ID(I,1),ID(I,2),ALPHA,BETA,XN,YN,DP,D(I),T(I)
      A=(ALPHA+BETA*DP)/D(I)
      A1(I)=A*YN
      A2(I)=A*XN
      A0(I)=A1(I)+A2(I)+BETA
      H(I)=0.
      HLAST(I)=0.
  100 F(I)=0.
      WRITE(6,2700)
      N1=NB+1
      DO 200 I=N1,N
      READ(5,3000) (ID(I,J),J=1,4),T(I),D(I),BK(I),S(I)
      WRITE(6,3500) I,(ID(I,J),J=1,4),T(I),D(I),BK(I),S(I)
      H(I)=0.
      HLAST(I)=0.
      HP(I)=0.
      R(I)=0.
  200 Q(I)=0.
      DO 300 I=N1,N
      I1=ID(I,1)
      I2=ID(I,2)
      I3=ID(I,3)
      I4=ID(I,4)
      A1(I)=.5*(T(I)+T(I1))
      A2(I)=.5*(T(I)+T(I2))
      A3(I)=.5*(T(I)+T(I3))
      A4(I)=.5*(T(I)+T(I4))
  300 A0(I)=A1(I)+A2(I)+A3(I)+A4(I)+BK(I)*D(I)**2
C THE EQUATIONS ARE SOLVED BY ITERATION AFTER EACH TIME STEP
      TIMES(1)=0.
      READ(5,4000) (TIMES(K),K=2,NT)
      DO 400 K=2,NT
      DT=TIMES(K)-TIMES(K-1)
C INSERT NONZERO VALUES OF F(I),R(I),Q(I) AND HP(I) FOR TIME(K)
      DO 10 I=N1,N
   10 R(I)=.000000317
      DO 500 I=N1,N
  500 B0(I)=-Q(I)+(BK(I)*HP(I)+R(I))*D(I)**2
   50 ERRC=0.
      DO 600 I=1,NB
      I1=ID(I,1)
      I2=ID(I,2)
      A=(F(I)+A1(I)*H(I1)+A2(I)*H(I2))/A0(I)
      ERRC=ERRC+(A-H(I))**2
  600 H(I)=A
      DO 700 I=N1,N
      I1=ID(I,1)
      I2=ID(I,2)
```

Fig. A.5 - A computer program for unsteady flow written from the flow chart in Figs. 4.10 and 4.11.

```
      I3=ID(I,3)
      I4=ID(I,4)
      S1=S(I)*D(I)**2/DT
      A=(A1(I)*H(I1)+A2(I)*H(I2)+A3(I)*H(I3)+A4(I)*H(I4)+B0(I)
     1 +S1*HLAST(I))/(A0(I)+S1)
      ERRC=ERRC+(A-H(I))**2
  700 H(I)=A
      ERRC=SQRT(ERRC)
      IF(ERRC.GT.ERR) GO TO 50
C THE SOLUTION IS WRITTEN OUT AFTER EACH TIME STEP
      WRITE(6,5000) TIMES(K),(I,I=1,10)
      DO 800 IJ=1,N,10
      ILAST=MIN0(N,(IJ+9))
      IFIRST=IJ-1
  800 WRITE(6,6000) IFIRST,(H(I),I=IJ,ILAST)
      DO 400 I=1,N
  400 HLAST(I)=H(I)
 1000 FORMAT(3I10,F10.8)
 1500 FORMAT(1H1,6H NB = ,I4,9X,4HN = ,I4,9X,5HNT = ,I4,9X,6HERR = ,
    1 F10.8///5H    I,8H ID(I,1),8H ID(I,2),6H ALPHA,6H BETA,6H    XN,
    2 6H    YN,7H     DP,7H   D(I),7H   T(I)/)
 2000 FORMAT(2I5,7F5.3)
 2500 FORMAT(1X,I4,2I8,2F6.2,2F6.3,2F7.1,F7.4)
 2700 FORMAT(1H1,4H   I,8H ID(I,1),8H ID(I,2),8H ID(I,3),8H ID(I,4),
    1 7H   T(I),7H   D(I),11H      BK(I),11H       S(I)/)
 3000 FORMAT(4I5,2F10.6,2E10.3)
 3500 FORMAT(1X,I4,4I8,F7.4,F7.1,2E11.3)
 4000 FORMAT(8F10.3)
 5000 FORMAT(1H1,11HTIMES(K) = ,F10.1///1X,10I10)
 6000 FORMAT(1X,I4,10F10.4)
      END
```

Fig. A.5 - (Continued).

NB = 16 N = 29 NT = 13 ERR = 0.00001000

I	ID(I,1)	ID(I,2)	ALPHA	BETA,	XN	YN	DP	D(I)	T(I)
1	16	22	0.00	1.00	0.707	0.707	10.0	100.0	0.0100
2	22	27	0.00	1.00	0.682	0.731	10.0	100.0	0.0100
3	27	4	0.00	1.00	0.707	0.707	0.0	100.0	0.0100
4	28	5	0.00	1.00	0.000	1.000	23.0	100.0	0.0100
5	29	4	0.00	1.00	0.616	0.788	10.0	100.0	0.0100
6	26	29	0.00	1.00	0.707	0.707	19.0	100.0	0.0100
7	8	26	0.00	1.00	0.788	0.616	18.0	100.0	0.0100
8	7	21	0.00	1.00	0.940	0.342	7.0	100.0	0.0100
9	8	10	0.00	1.00	0.996	-0.087	0.0	100.0	0.0100
10	21	11	1.00	0.00	-0.035	0.999	3.0	100.0	0.0100
11	20	12	1.00	0.00	-0.105	0.995	0.0	100.0	0.0100
12	19	13	1.00	0.00	-0.174	0.985	-19.0	100.0	0.0100
13	18	14	1.00	0.00	0.087	0.995	-12.0	100.0	0.0100
14	17	15	1.00	0.00	0.000	1.000	0.0	100.0	0.0100
15	16	14	1.00	0.00	0.707	0.707	-29.0	100.0	0.0100
16	1	17	1.00	0.00	0.999	0.052	-15.0	100.0	0.0100

I	ID(I,1)	ID(I,2)	ID(I,3)	ID(I,4)	T(I)	D(I)	BK(I)	S(I)
17	18	14	16	22	0.0100	100.0	0.	.500E-01
18	19	13	17	23	0.0100	100.0	0.	.500E-01
19	20	12	18	24	0.0100	100.0	0.	.500E-01
20	21	11	19	25	0.0100	100.0	0.	.500E-01
21	8	10	20	26	0.0100	100.0	0.	.500E-01
22	23	17	1	2	0.0100	100.0	0.	.500E-01
23	24	18	22	27	0.0100	100.0	0.	.500E-01
24	25	19	23	28	0.0100	100.0	0.	.500E-01
25	26	20	24	29	0.0100	100.0	0.	.500E-01
26	7	21	25	6	0.0100	100.0	0.	.500E-01
27	28	23	2	3	0.0100	100.0	0.	.500E-01
28	29	24	27	4	0.0100	100.0	0.	.500E-01
29	6	25	28	5	0.0100	100.0	0.	.500E-01

TIMES(K) = 604800.0

	1	2	3	4	5	6	7	8	9	10
0	0.0840	0.0576	0.0000	0.0948	0.0333	0.0873	0.0501	0.0328	0.0000	0.5150
10	0.8077	0.9696	0.9564	0.8749	0.8553	0.8357	0.8749	0.9636	0.9672	0.8248
20	0.5253	0.5196	0.7511	0.8061	0.6913	0.4168	0.4072	0.5070	0.4081	

Fig. A.6 - Typical output for the unsteady computer program in Fig. A.5.

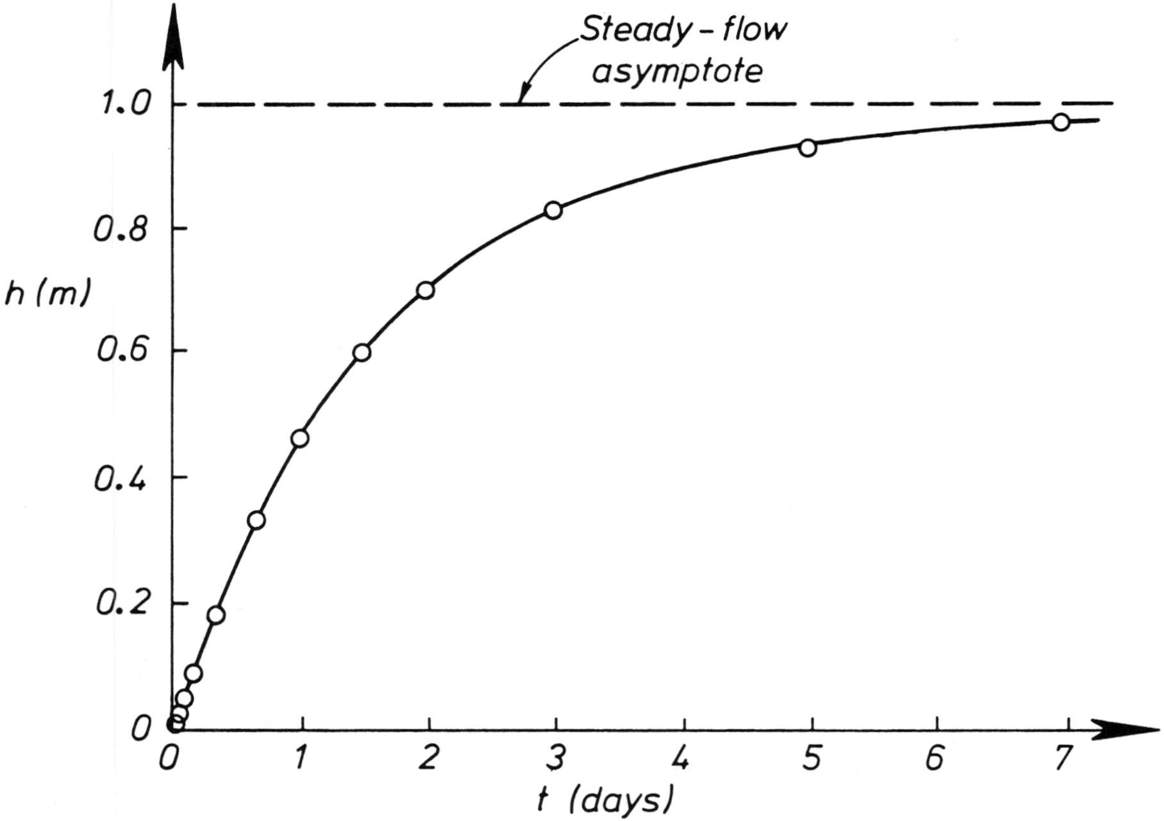

Fig. A.7 - The approach to steady-state conditions calculated at node 12 for R = 835 mm/month.

CHAPTER V

1. Use the expression for h in problem 1 at the end of chapter V

$$h = \frac{1}{2}(x^2 - y) + 1$$

and the initial condition

$$T = 1 \text{ along the line } y = 0 \text{ for } -\infty < x < \infty$$

to calculate the solution for T(x, y). Show that the initial condition for T gives one and only one, starting value for T upon each and every streamline in the entire (x, y) plane.

Solution:

Writing Eq. 5.2 in scalar form gives

$$\frac{\partial T}{\partial x}\frac{\partial h}{\partial x} + \frac{\partial T}{\partial y}\frac{\partial h}{\partial y} + T\left(\frac{\partial^2 h}{\partial x^2} + \frac{\partial^2 h}{\partial x^2}\right) = 0$$

Thus, substituting the given expression for h gives a first-order, partial differential equation for T:

$$x\frac{\partial T}{\partial x} - \frac{1}{2}\frac{\partial T}{\partial y} + T = 0$$

The characteristic form of a first-order equation is not unique, and we will use a characteristic form for this equation that is obtained by multiplying both sides of the equation by -2.

$$-2x\frac{\partial T}{\partial x} + \frac{\partial T}{\partial y} = 2T$$

Thus, this equation reduces to the ordinary differential equation

$$\frac{dT}{dy} = 2T$$

if we choose to integrate the equation along the characteristic curve

$$\frac{dx}{dy} = -2x$$

The simultaneous integration of these two ordinary differential equations gives

$$Te^{-2y} = C_1 \text{ along the characteristics } x\,e^{2y} = C_2$$

The characteristic curves coincide with streamlines, and since all of these streamlines intersect the x axis once, and only once, specifying the initial values for T along the x axis is enough to insure a unique solution for T. Using these initial values for T to calculate C_1 and C_2 at the point $(x_0, 0)$ gives

$$C_1 = 1 \text{ and } C_2 = x_0 \text{ for } -\infty < x_0 < \infty$$

1. Ctd. Thus, the solution for T is given by

$$T = e^{2y}$$

and the equations of the streamline are

$$xe^{2y} = x_0 \text{ for } -\infty < x_0 < \infty$$

2. Two wells, spaced a distance of 500 m along a line normal to the sea-coast, have a time lag of 4 minutes between maximum piezometric head values created by the rise and fall of the tide. Use a tidal period of 12 hours to calculate the ratio of T/S for this aquifer.

Solution:

Solution of Eq. 5.13 for T/S gives

$$\frac{T}{S} = \frac{1}{2\omega}\left(\frac{dx}{dt}\right)^2 = \frac{12 \times 3600}{2(2\pi)}\left(\frac{500}{4 \times 60}\right)^2 = 1.49 \times 10^4 \text{ m/s}$$

3. Use the series expansions given in Table 2 to verify the values $E_1(0.1) = 1.8229$ and $E_1(5) = 0.0011$.

Solution:

Using the expansion for $x \to 0$ gives

$$E_1(0.1) = \ln\left(\frac{0.56145948}{0.1}\right) + 0.1 - \frac{1}{4}(0.1)^2 + \frac{1}{18}(0.1)^3 - \frac{1}{96}(0.1)^4 + \ldots$$

$$= 1.72537 + 0.1 - 0.00250 + 0.00006 \ldots = 1.82293$$

Using the expansion for $x \to \infty$ gives

$$E_1(5) \sim \frac{\exp(-5)}{5}\left(1 - \frac{1}{5} + \frac{2}{5^2} - \frac{6}{5^3} + \ldots\right)$$

$$= .00135\ (1 - 0.2 + 0.08 - 0.048 + \ldots) = 0.00112$$

4. A well test gave the following match-point data:

$$\frac{4Tt}{Sr^2} = 100, \quad \frac{4\pi T|h|}{Q} = 1$$

$$|h| = 0.1 \text{ m}, \quad \frac{t}{r^2} = 0.3 \text{ s/m}^2$$

Calculate T and S if the pumped well had a flow rate of $.02 \text{ m}^3/\text{s}$.

Solution:

$$T = \frac{Q}{4\pi|h|} = \frac{.02}{4\pi(.1)} = 0.0159 \text{ m}^2/\text{s}$$

$$S = \frac{4T}{100}\frac{t}{r^2} = \frac{4(0.0159)}{100} 0.3 = 1.91 \times 10^{-4}$$

CHAPTER VI

1. Use the solution for an instantaneous point source in one dimension without radioactive decay ($\lambda = 0$)

$$c(x, t) = \frac{M_1}{2\sigma\sqrt{\pi t D_1}} \exp\left[-\frac{\left(x - \frac{u_x}{\sigma} t\right)^2}{4tD_1}\right]$$

to calculate the maximum value of c (a) in the aquifer at any fixed time and (b) at a fixed point in the aquifer over the time period $0 < t < \infty$. Note that questions (a) and (b) are two entirely different questions and that the same ideas are also applicable in two and three dimensions.

Solution

(a) When t is fixed, c is only a function of x. Since x appears only in the argument of the exponential function, and since this argument is always positive definite, the maximum value of c at any fixed value of t occurs when the argument of the exponential function vanishes. This occurs at the point

$$x = \frac{u_x}{\sigma} t$$

at which point c takes on the value

$$c = \frac{M_1}{2\sigma\sqrt{\pi t D_1}}$$

(b) When an observer remains at a fixed point in the aquifer and monitors values of c, then x is a constant and c becomes a function of t. Thus, the condition for c to become a maximum is

$$\frac{\partial c}{\partial t} = 0$$

A rather involved calculation of this derivative leads to the following value of t at which c becomes a maximum:

$$t = \frac{D_1\sigma^2}{u_x^2}\left[-1 + \sqrt{1 + \left(\frac{xu_x}{\sigma D_1}\right)^2}\right]$$

This expression for t, when substituted into the expression for c, gives a maximum value for c of

$$c = \frac{M_1}{2\sigma\sqrt{\pi t D_1}} \exp\left(-\frac{\varepsilon}{2}\right)$$

in which ε is given by

1. Ctd.

$$\varepsilon = -\frac{xu_x}{\sigma D_1} + \sqrt{\left(\frac{xu_x}{\sigma D_1}\right)^2 + 1} > 0$$

This shows that the maximum value of c, with x fixed, is always less than the maximum value of c in the aquifer at any fixed time and that the difference between these two values vanishes as

$$\frac{xu_x}{\sigma D_1} \to \infty$$

2. Twenty kilograms of conservative tracer (I = 0) are released suddenly in a well that completely penetrates a 10 m thick aquifer. The aquifer has a pore velocity of 3 m/day, lateral and longitudinal dispersivities of 1 m and 8 m, respectively, and a porosity of 0.1. What will be the maximum values of c at points 100 m and 500 m downstream from the well?

Solution:

The use of Eq. 6.50, with $\lambda = 0$, to carry out computations similar to those carried out for the previous problem show that the maximum value of c, for fixed values of x and y, is given by

$$c = \frac{M_2}{4\sigma\pi t\sqrt{D_1 D_2}} \exp\left[\frac{xu_x}{2\sigma D_1} - \sqrt{1 + \left(\frac{Ru_x}{2\sigma D_1}\right)^2}\right]$$

at the time

$$t = 2\frac{D_1\sigma^2}{u_x^2}\left[-1 + \sqrt{1 + \left(\frac{Ru_x}{2\sigma D_1}\right)^2}\right]$$

in which

$$R = \sqrt{x^2 + y^2\frac{D_1}{D_2}}$$

Since the maximum value of c occurs along y = 0, we will set R = x = 100 m and 500 m. Since concentrations are usually calculated in gms/m^3, we will set M_2 = 20,000/10 = 2,000 gms/m. Other numerical values are

$$\frac{u_x}{\sigma} = 3\ \frac{m}{day}\ \frac{day}{86,400\ s} = 3.47 \times 10^{-5}\ m/s$$

$$D_1 = \alpha_1\frac{u_x}{\sigma} = 8(3.47 \times 10^{-5}) \doteq 2.78 \times 10^{-4}\ m^2/s$$

$$D_2 = \alpha_2\frac{u_x}{\sigma} = 1(3.47 \times 10^{-5}) = 3.47 \times 10^{-5}\ m^2/s$$

$$\sigma = 0.1$$

2. Ctd. The use of these numbers leads to the following result:

x	100 m	500 m
t	28.5 days	161.8 days
c	6.08 gm/m^3	1.14 gm/m^3

3. Use Eq. 6.80 to calculate W(1, 3). Then check this result with Eq. 6.74.

Solution:

Equation 6.78 gives

$$\gamma = \left(\frac{1}{3} - \frac{1}{2}\right)\frac{3}{1} = -0.50000$$

Thus, Eqs. 6.82, 6.84 - 6.85 and the series expansion in problem 9 give

$$I_0(\gamma) = \frac{\sqrt{\pi}}{2}\,\mathrm{erfc}(\gamma) = \frac{\sqrt{\pi}}{2}\,[1 - \mathrm{erf}(\gamma)] = \frac{\sqrt{\pi}}{2}\left[1 - \frac{2}{\sqrt{\pi}}\sum_{n=0}^{\infty}\frac{(-1)^n\,\gamma^{2n+1}}{n!\,(2n+1)}\right]$$

Hence, since γ is a negative number,

$$I_0(\gamma) = \frac{\sqrt{\pi}}{2}\left[1 - \frac{2}{\sqrt{\pi}}(-0.50000 + 0.04167 - 0.00312 + 0.00019 - 0.00001 + \ldots)\right] = 1.34750$$

Thus, application of the recurrence formula given by Eq. 6.83 gives

$$I_2(\gamma) = \frac{1}{2}\,I_0(\gamma) + \frac{\gamma}{2}\exp(-\gamma^2) = 0.47905$$

$$I_4(\gamma) = \frac{3}{2}\,I_2(\gamma) + \frac{\gamma^3}{2}\exp(-\gamma^2) = 0.66990$$

$$I_6(\gamma) = \frac{5}{2}\,I_4(\gamma) + \frac{\gamma^5}{2}\exp(-\gamma^2) = 1.66258$$

Finally, Eq. 6.80 gives

$$W(1,\ 3) \sim \sqrt{\frac{2}{3}}\exp(-3)\ [1.34750 - 0.03992 + 0.00698 - 0.00241 + \ldots] = 0.05334$$

The next term in the asymptotic series contributes a value of 0.00005, which suggests that the error of approximation is about 0.00005/0.05334 = 0.09%.

The use of Eq. 6.74 requires values for

$$K_0(3) = 0.03474$$

$$E_1\left(\frac{9}{4}\right) = 0.03476$$

These numerical values have been obtained from tabulation in mathematical tables, such as the reference by Abramowitz and Stegun (1970) at the end of Chapter VI. Additional values of $E_n(9/4)$ are computed from the

3. Ctd. recurrence formula given by Eq. 5.30:

$$E_2\left(\frac{9}{4}\right) = 0.02719$$

$$E_3\left(\frac{9}{4}\right) = 0.02211$$

$$E_4\left(\frac{9}{4}\right) = 0.01855$$

$$E_5\left(\frac{9}{4}\right) = 0.01592$$

$$E_6\left(\frac{9}{4}\right) = 0.01392$$

$$E_7\left(\frac{9}{4}\right) = 0.01235$$

Thus, Eq. 6.74 gives

$$W(1, 3) = 2(0.03474) - \sum_{n=0}^{\infty} \frac{(-1)^n}{n!} E_{n+1}\left(\frac{9}{4}\right)$$

$$= 0.06948 - 0.03476 + 0.02719 - 0.01106$$

$$+ 0.00309 - 0.00066 + 0.00012 - 0.00002 + \ldots$$

$$= 0.05338 + \varepsilon$$

Since the series is a convergent, alternating series, the value of ε is less than the first neglected term:

$$\varepsilon < \frac{E_8\left(\frac{9}{4}\right)}{7!} = 2.20 \times 10^{-6}$$

Thus, the values of W(1, 3) computed from Eqs. 6.80 and 6.74 agree to within the estimated errors for each series expansion.

CHAPTER VII

1. A well is located 200 m back from the edge of a reasonably straight seacoast. The unconfined aquifer has a permeability of 2 mm/s and an impermeable bottom boundary that is 20 m below mean sea level. The approaching, undisturbed flow on the landward side of the well has a flux velocity and depth of 0.5 m/day and 25 m, respectively. Calculate the maximum flow that can be abstracted from the well without polluting the well with sea water.

Solution:

A plan view of the flow pattern for this problem is shown in Fig. 7.8. The values of $\varepsilon = 1/40$, $K = 0.002$ m/s, $y_0 = 200$ m, $D = 20$ m and

1. Ctd. $q_0 = 0.5 \times 25 = 12.5\ m^2/day = 1.45 \times 10^{-4}\ m^2/s$ lead to the following equations when substituted into Eqs. 7.89 and 7.90:

$$0.707 = 2\xi - (1 - \xi^2)\ \ln\left|\frac{1 + \xi}{1 - \xi}\right|$$

$$\frac{Q}{0.0911} = 1 - \xi^2$$

The variable ξ is the dimensionless distance between the seacoast and the point where the interface meets the bottom aquifer boundary.

$$\xi = \frac{y}{y_0} = \frac{y}{200}$$

The solution of these equations gives

$$\xi = 0.769$$

$$y = 154\ m$$

$$Q = 0.0372\ m^3/s$$

The value of Q, of course, is an estimate of the flow that can be safely abstracted from the well. However, these calculations do not address the question of whether or not the well is capable of actually delivering this flow rate.

2. A collection gallery, parallel to a river ($\theta = 0$ in Fig. 7.19), is to be used to abstract a flow of $0.1\ m^3/s$ from an unconfined aquifer that has a transmissivity of $0.03\ m^2/s$. The gallery is 50 m from the river edge, and the drawdown at the gallery is to not exceed 1 m. What is the minimum length that is required for the gallery?

Solution:

The dimensionless flow rate is

$$\frac{Q}{TH} = \frac{0.1}{0.03(1)} = 3.33$$

Thus, the plot in Fig. 7.19 gives

$$\frac{\ell}{d} = 1.32$$

Since d = 50 m, the minimum length is

$$\ell = 1.32(50) = 66\ m$$

The minimum length can be decreased either by moving the gallery closer to the river (decreasing d) or by making the gallery deeper (increasing H).

3. A sea-water interface lies a vertical distance of 50 m below mean sea level at the edge of the sea. The aquifer is anisotropic, with principal values of the permeability tensor of 5 mm/s and 0.2 mm/s in the

3. Ctd. horizontal and vertical directions, respectively. What is the two-dimensional flow rate, q, exiting from the aquifer?

Solution:

Since x = 0 in Fig. 7.20 at the edge of the sea, Eq. 7.173 reduces to

$$\frac{K\varepsilon y}{q} = 1$$

As explained in sec. 32, this equation can be modified for anisotropic aquifers by replacing K with K_y and q with $q\sqrt{K_y/K_x}$ to obtain the result

$$q = \varepsilon y\sqrt{K_x K_y}$$

Substituting numbers gives

$$q = \frac{1}{40}\ (50)\ \sqrt{(.005)(.0002)} = 0.00125\ m^2/s$$

which means that 1.25 m^3/s exit along every 1000 m of seacoast.

CHAPTER VIII

1. The water level in a well that is 50 m from a relatively straight river edge requires 9 hrs to stabilize after a sudden drop or rise in river level. What is the value of S/T for the aquifer.

Solution:

We will assume that the aquifer can be modelled as a semi-infinite aquifer, which means that the solution is given by Eqs. 8.38 and 8.43 after setting f(t) = 1:

$$\frac{h}{h_0} = \text{erfc}\left(\frac{x}{2}\sqrt{\frac{S}{Tt}}\right)$$

The ratio of h/h_0 does not become unity in the mathematical solution until $t \to \infty$. Thus, we will set x = 50 m, t = 9 hrs = 32,400 s and examine the sensitivity of the solution by calculating values of S/T that are required to make h/h_0 equal to 0.90, 0.95 and 0.99, respectively. This gives the following result:

$\frac{h}{h_0}$	$\frac{x}{2}\sqrt{\frac{S}{T_t}}$	$\frac{S}{T}$
0.90	0.089	0.411
0.95	0.045	0.105
0.99	0.009	0.004

Since the solution for S/T is seen to be very sensitive to the ratio chosen for h/h_0, and since this ratio is probably not known very accurately, it would be much better to measure h/h_0 as a function of t after

1. Ctd. a sudden rise or drop in river level. Then the match-point method devised in problem 3 at the end of chapter V could be used. Conversely, of course, the solution for h/h_0 is relatively insensitive to errors in S/T.

2. Irrigation water is suddenly applied uniformly to a square region, 200 m along each side, that overlies a shallow, unconfined aquifer. Calculate the maximum rise in elevation of the resulting groundwater mound after 8 hrs if $S = 0.1$, $T = 0.025\ m^2/s$, $R = 10$ mm/hr and $K'/B' = 0$. Assume that the aquifer is infinite in extent.

Solution:

Since the maximum rise in elevation of the mound occurs at its geometrical center, and since $f(t) = 1$ for this problem, Eq. 8.68 reduces to

$$\frac{4Sh_0}{R} = \int_0^t \left\{2\ \mathrm{erf}\left[a\sqrt{\frac{S}{4T(t-\tau)}}\right]\right\}^2 d\tau$$

in which $h_0(t) = h(0,\ 0,\ t)$. Substituting the given numbers allows the solution to be put in the form

$$h_0(8) = 0.1 \int_0^8 \left[\mathrm{erf}\left(\frac{3.33}{\sqrt{8-\tau}}\right)\right]^2 d\tau$$

in which $h_0(t)$ is given in m and τ in hrs. Calculation of the integrand at hourly intervals gives

τ (hrs)	0	1	2	3	4	5	6	7	8
Integrand	0.82	0.86	0.89	0.93	0.96	0.99	1.00	1.00	1.00

Thus, the trapezoidal rule gives

$$h(8) = 0.1[0.82 + 2(0.86 + 0.89 + 0.93 + 0.96 + 0.99 + 1.00 + 1.00) + 1.00]\left(\frac{1}{2}\right) = 0.75\ \mathrm{m}$$

Appendix 4 Notation

Units for each variable appear in parenthesis.

$\overline{a}$ = acceleration vector (LT^{-2});

B = saturated aquifer thickness (L);

B' = aquitard thickness (L);

c = pollutant concentration (ML^{-3});

$\underline{D}$ = dispersion tensor (L^2T^{-1});

D_{ij} = dispersion tensor components (L^2T^{-1});

D_1, D_2 = principal values of the dispersion tensor (L^2T^{-1});

d = mean grain diameter (L);

$E_n(u)$ = exponential integral (dimensionless);

$\hat{e}_i$ = unit base vector (dimensionless);

$\hat{e}_n$ = unit normal vector (dimensionless);

$\hat{e}_t$ = unit tangent vector (dimensionless);

erf(z) = error function (dimensionless);

erfc(z) = complimentary error function (dimensionless);

exp(z) = exponential function (dimensionless);

g = gravitational constant (LT^{-2});

h = piezometric head (L);

i = $\sqrt{-1}$ (dimensionless);

$\hat{i}$, $\hat{j}$, $\hat{k}$ = unit base vectors in the x, y and z directions (dimensionless);

K = coefficient of permeability (LT^{-1});

K_0 = zero-order, modified Bessel function of the second kind (dimensionless);

K(k) = complete elliptic integral of the first kind with modulus k (dimensionless);

K_{ij} = permeability tensor components (LT^{-1});

K_x, K_y, K_z = principal values of the permeability tensor (LT^{-1});

K' = aquitard permeability (LT^{-1});

k_0 = intrinsic permeability (L^2);

M_i = pollutant mass in i dimensions (ML^{i-3});

$\dot{M}_i$ = pollutant mass flux in i dimensions ($MT^{-1}L^{i-3}$);

N_x, N_y = outward unit normal components (dimensionless);

n = arc length in the direction of $\hat{e}_n$(L);

P = pressure ($ML^{-1}T^{-2}$);

p = Laplace transform parameter (T^{-1});

Q = three-dimensional, volumetric flow rate (L^3T^{-1})

q = two-dimensional, volumetric flow rate (L^2T^{-1})

R = vertical recharge flux velocity (LT^{-1});

$\overline{r}$ = displacement vector (L);

r = radial coordinate (L);

S = storage coefficient (dimensionless); = three-dimensional surface (L^2);

S_s = specific storage (L^{-1});

s	= arc length along a curve (L);
T	= transmissivity (L^2T^{-1});
t	= time (T); = complex variable = $\xi + i\eta$ (L);
$\forall$	= volume (L^3);
$\overline{u}$	= flux velocity vector (LT^{-1});
$W(u, \beta)$	= leaky aquifer function (dimensionless);
$W(u)$	= well function = $E_1(u)$ (dimensionless);
w	= complex velocity potential = $\phi + i\psi$ (L^2T^{-1} or L^3T^{-1});
x, y	= Cartesian coordinates (L);
z	= Cartesian coordinate (L); = complex variable = x + iy (L);
α	= aquifer bulk coefficient of compressibility ($LT^2\ M^{-1}$);
α_1, α_2	= dispersivities (L);
β	= bulk coefficient of compressibility for water (LT^2M^{-1});
Δ	= finite-difference node spacing (L);
$\delta(x - x_0)$	= Dirac delta function (L^{-1});
δ	= shortest distance between a finite-difference node and the boundary (L);
ε	= $\rho_s/\rho - 1$ (dimensionless);
η	= Cartesian coordinate (L);
λ	= radioactive decay constant (T^{-1});
μ	= dynamic or absolute fluid viscosity ($ML^{-1}\ T^{-1}$);
ν	= kinematic fluid viscosity = μ/ρ (L^2T^{-1});
ξ	= Cartesian coordinate (L);
ρ	= fresh water mass density (ML^{-3});
ρ_s	= sea water mass density (ML^{-3});
σ	= aquifer porosity (dimensionless);
ϕ	= velocity potential (L^2T^{-1} or L^3T^{-1}); = Laplace transform of h (LT);
ψ	= stream function (L^2T^{-1} or L^3T^{-1});
Ω	= inverse velocity hydrograph function = dz/dw (TL^{-1} or TL^{-2}); and
$\overline{\nabla}$	= vector operator del (L^{-1}).

Index